DEVELOPMENTS IN THE DESIGN OF THERMAL SYSTEMS

Thermal systems are essential features in power, chemical processing, air conditioning, and other industries where heat and power are used. As the cost and complexity of designing thermal systems have increased, the need to understand and improve the design process has also grown. This book describes recent progress in the design of thermal systems.

The book begins with a brief history and outline of developments in thermal system design. Chapters then discuss computer design tools for the power and chemical industries, predicting physical properties with computational tools, the use of pinch analysis to improve thermal efficiency, applications of the exergy concept, the emerging concept of thermoeconomics, and the potential for artificial intelligence and expert systems in the design of thermal systems.

With chapters written by internationally recognized authorities, the book offers a state-of-the-art review for both researchers and practitioners in mechanical, aerospace, chemical, and power engineering.

DEVELOPMENTS IN THE DESIGN OF THERMAL SYSTEMS

Edited by

ROBERT F. BOEHM

University of Nevada at Las Vegas

CAMBRIDGE UNIVERSITY PRESS
Cambridge, New York, Melbourne, Madrid, Cape Town, Singapore, São Paulo

Cambridge University Press
The Edinburgh Building, Cambridge CB2 2RU, UK

Published in the United States of America by Cambridge University Press, New York

www.cambridge.org
Information on this title: www.cambridge.org/9780521462044

First published 1997
This digitally printed first paperback version 2005

A catalogue record for this publication is available from the British Library

Library of Congress Cataloguing in Publication data
Developments in the design of thermal systems / edited by Robert F.
Boehm.
p. cm.
Includes bibliographical references.
ISBN 0-521-46204-5 (hb)
1. Heat engineering. 2. Systems engineering. 3. Engineering
design. I. Boehm. R. F.
TJ260.D483 1997
621.402 – dc21 97-4472
CIP

ISBN-13 978-0-521-46204-4 hardback
ISBN-10 0-521-46204-5 hardback

ISBN-13 978-0-521-02005-3 paperback
ISBN-10 0-521-02005-0 paperback

Preface

This book is a milestone in the presentation of developments in techniques used to design thermal systems. On these pages is an overview of current practice in this rapidly developing field.

With roots tracing to the use of Second Law ideas for design applications decades ago, the design of thermal systems has advanced quickly in the last 20 years. The many computational tools now available make it possible to evaluate virtually all aspects of the performance of systems, from overall behavior to the details of each of the component processes. Every aspect of these types of analysis has seen significant accomplishments.

What has not been done previously is to summarize the cutting-edge trends in this field – the aim of this book. Drawing on the work of people from around the world, the book gives a good cross section of progress made to date.

Designers of thermal systems are practitioners from a variety of disciplines. Although the major contributors and users have been chemical and mechanical engineers, many others find the approaches that have been developed to be of great value. It is not unusual to find a symposium taking place on a regular basis somewhere in the world on issues related to this field.

The book starts with an outline of the major industrial thrusts that have shaped design interests in thermal systems. Summaries are then given of design trends in both the power industry (Chapter 2) and the chemical process industry (Chapter 3). Significant impacts of the rapid strides experienced in computer technology are in evidence here. Equally important is the ability to predict material properties, and Chapter 4 summarizes how this is done in modern codes. Pinch analysis, a technique that has been shown to be valuable in efficiently allocating energy in new systems, as well as in modifying existing ones, is given in Chapter 5. Applications of exergy, a concept that is an outgrowth of Second Law ideas and has become important in the design process, are summarized in Chapters 6 and 7. Topics in thermoeconomics and the application of artificial intelligence, which have been less fully developed as design tools at this time, are described in Chapters 8 and 9, respectively.

I particularly wish to thank the authors. They are busy but have taken the time to summarize carefully the key areas of this field. The definitive insights set down by these experts make this book valuable to people involved in the design of systems.

I also greatly value the assistance of Florence Padgett, editor at Cambridge University Press, in this effort. She appreciated the need for a monograph in this area. Also, she obviously has been involved with many efforts of this sort, because she understood all the excuses when deadlines slipped.

Robert F. Boehm

Biographical sketches of the authors

Adrian Bejan is the J. A. Jones Professor of Mechanical Engineering at Duke University. He received his B.S. (1972), M.S. (1972) and Ph.D. (1975) from the Massachusetts Institute of Technology. Professor Bejan is the author of eight books and more than 240 peer-reviewed articles that cover a wide variety of topics in thermodynamics, heat transfer, and fluid mechanics. He is the recipient of the Heat Transfer Memorial Award (1994), the James Harry Potter Gold Medal (1990), and the Gustus L. Larson Memorial Award (1988) of the American Society of Mechanical Engineers.

Robert F. Boehm is Professor of Mechanical Engineering at the University of Nevada, Las Vegas (UNLV), a position he has held since 1990. He was on the faculty of the University of Utah Department of Mechanical Engineering for 21 years. He holds a Ph.D. in mechanical engineering from the University of California at Berkeley. Dr. Boehm is a registered professional engineer, a fellow of the ASME, and was awarded the Distinguished Teaching Award from the University of Utah and the highest research award from UNLV, the Barrick Senior Scholar Award. He has published extensively in heat transfer, design of thermal systems, and energy conversion topics and is the author of the text *Design Analysis of Thermal Systems*. He has held professional positions with General Electric Company and Sandia National Laboratories.

Ronald P. Danner received his Ph.D. degree in chemical engineering from Lehigh University in 1965. He worked as a senior research scientist at the Eastman Kodak Company until 1967. Since that time he has been a professor of chemical engineering at the Pennsylvania State University. He served as a visiting professor at the Technical University of Denmark in 1983 and 1991. He is coauthor of five books dealing with the correlation or prediction of thermophysical properties. His research interests include phase equilibria and diffusion in polymer-solvent systems, prediction of thermodynamic properties of fluids, and adsorption of gases and liquids on solids.

Bodo Linnhoff is chairman and CEO of Linnhoff March International (United Kingdom), and honorary professor, Department of Process Integration, University of Manchester Institute of Science and Technology (UMIST). He received his Ph.D. in chemical engineering from Leeds University in 1976. From 1982 to 1994 he was professor of chemical engineering at UMIST, where he established the Process Integration Research Consortium, the Centre for Process Integration, and the Department of Process Integration. He established the basis of pinch technology as part of his Ph.D. work, and he has developed the concept further since then. Dr. Linnhoff is a chartered engineer and fellow of the Institution of Chemical Engineers.

Miguel-Angel Lozano is a professor of mechanical engineering at the University of Zaragoza, Spain, and a member of CIRCE, a center for research in power plant efficiency. The center is sponsored by the Spanish National Company of Electricity, ENDESA; the Regional Government of Aragón; and the University of Zaragoza. He is an active contributor to the field of thermoeconomics, the application of economics, and the Second Law. He is the corecipient of three ASME Edward F. Obert awards for papers on this topic, and he has published more than ninety papers in this field.

M. J. Moran is professor of engineering thermodynamics, Department of Mechanical Engineering, the Ohio State University, Columbus, Ohio. He received a Ph.D. in mechanical engineering from the University of Wisconsin (Madison) in 1967. He is a specialist in engineering thermodynamics, exergy analysis, and thermoeconomics. His texts include *Fundamentals of Engineering Thermodynamics*, 3rd ed., Wiley, 1995 (with H. N. Shapiro); *Thermal Design and Optimization*, Wiley 1996 (with A. Bejan and G. Tsatsaronis); and *Availability Analysis: A Guide to Efficient Energy Use*, ASME press, 1989. He was associate editor of the *ASME Journal of Engineering for Gas Turbines and Power* and coorganizer of several international conferences on engineering thermodynamics: Rome (1987, 1995), Florence (1990, 1992), Beijing (1989), Stockholm (1996). He was also a U.S. Department of Energy fellow from 1993 to 1995.

Rudolphe (Rudy) L. Motard is professor and former chairman of chemical engineering at Washington University in St. Louis, which he joined in 1978 after 21 years at the University of Houston. He holds a D.Sc. in chemical engineering from Carnegie-Mellon University and an undergraduate degree from Queen's University in Canada. He has coauthored thirty-seven publications, including two books, mostly on computer applications in chemical engineering. His pioneering work in chemical process simulation earned him the AIChE, CAST Division award in 1991. His current research interests include process synthesis and simulation, engineering databases, and process monitoring. He is a consultant to the AIChE, Process Data Exchange Institute, and a fellow of AIChE. He was chair of CAST in 1995.

Manoj Nagvekar received his Ph.D. degree in chemical engineering from the Pennsylvania State University in 1990. Since then, he has been with the M. W. Kellogg Company in Houston, Texas, and is presently a senior process engineer in the Technical Data Group of the Process Engineering Department. He has been involved with several aspects of thermodynamic data work including analysis, modeling, and correlation of data; software development; and modification and application of commercial simulation programs for process design.

Beniamino Paoletti is project manager of Decision Support Systems at Alitalia, Rome, Italy. He holds an M.S. in philosophy from the University of Roma 1, where he did postgraduate work in the fields of mathematical logic and applications of AI techniques to the formalization of natural languages. He joined the Department of Operational Research of Alitalia in 1988, where he coordinated advanced applications in the fields of combinatory optimization and graph theory, with specific regard to scheduling and operation control. Since 1990, Dr. Paoletti has been an adjunct researcher at the University of Roma 1, where he cooperates in a long-rage research project on expert systems applications to the design of thermal systems.

Robert M. Privette earned his B.S. and M.S. degrees in mechanical engineering, having received the latter from Purdue University in 1986. He has more than 10 years of experience in hydraulic and thermal-hydraulic systems. He has worked in the U.S. defense and power generation industries and is currently employed by Babcock & Wilcox (B&W) in their Research & Development Division in Alliance, Ohio. With B&W he has done experimental research related to nuclear production reactors and has managed the Power Systems Evaluation Section, which used process simulation software to evaluate and design various power generation systems. He currently manages a demonstration program for ten-kilowatt solid oxide fuel cell systems and evaluates new technologies of interest to B&W and it's parent company, McDermott International.

Enrico Sciubba is a professor in the Mechanical and Aeronautical Engineering Department of the University of Roma 1, La Sapienza, where he teaches thermal sciences. His specific research fields relate to CFD applications to turbomachinery flows and modeling and simulation of thermodynamic cycles and processes. He has authored or coauthored more than seventy internationally published papers on these topics, including about ten concerning expert systems applications to the design and optimization of thermal processes. Dr. Sciubba holds an M.E. from the University of Roma (1972) and a Ph.D. from Rutgers University (1981). He taught thermal sciences at Rutgers and C.U.A. (Washington D.C.) and is involved in several international research projects in applied fluid dynamics and thermodynamics.

Calvin Spencer received his Ph.D. degree in chemical engineering from the Pennsylvania State University in 1975. Since graduation he has worked at the M. W. Kellogg Company in Houston, Texas, where he is presently chief technology engineer of the Technical Data Group, which is responsible for the integrity of all thermophysical property and phase equilibrium data used in process design and simulation. He has spearheaded the development of comprehensive data packages for a broad range of technologies, including a number of first-of-a-kind and specialty chemical processes, and the integration of these models with the major commercial simulation programs. He is the coauthor of several papers on the correlation and estimation of physical properties and phase equilibrium.

Antonio Valero is the chairman of the Department of Mechanical Engineering at the University of Zaragoza, Spain, and the director and founder of CIRCE, a center for research in power plant efficiency, presently composed of a team of sixty researchers. CIRCE is a joint institution sponsored by the Spanish National Company of Electricity, ENDESA; the Regional Government of Aragón; and the University of Zaragoza. He is one of the main contributors to the development of the topic of thermoeconomics, the merging of concepts from economics and the Second Law. He is a corecipient of three ASME Edward F. Obert awards related to this topic, and he has published more than ninety papers in this field. Since 1986 he has been a member of the worldwide Second Law Analysis Conferences.

1

Introduction and trends in the thermal design field

R. BOEHM
University of Nevada at Las Vegas

1.1 Brief history of thermal system design

1.1.1 Early developments

Thermal processes have been in existence since the earliest phases of the formation of the earth. However, as humans gained more insights into the elusive concept of heat, and its connection to the ability to do work, the idea of making use of a thermal system was born. No precise definition of this term exists, but it generally is taken to be the linking of heat and work processes to produce desired results.

Although some thermal system concepts have been in use since the earliest of recorded history – Hero's engine is one such example – little orderly development was evident in these early days. Knowledge of the applications of gunpowder in China was brought to the western world in the 1300s and enabled development of a number of thermal systems.

Among people who wished to exploit the technology, work was a better understood concept. By about the 1400s, several water-driven machines were in use for applications such as ventilating mines and irrigation. Pumps were a fairly well-developed technology by the 1600s. Vacuum pumps were also being developed in this period.

Some of the first formal studies of heat involved attempts to measure temperature. Galilei's work in the late 1500s is attributed to be among the earliest, leading to the development of the mercury-in-glass thermometer by Fahrenheit in the early 1700s. This development then led to calorimetry efforts, which became well understood in the 1800s.

1.1.2 The impetus due to engine development

Steam engines were the first major thermal systems, applying heat to produce work. Although some work began much earlier, steam engine development started in earnest in the 1700s. Newcomen's engine in the early part of that century established England as the major industrial force in the world. The use of a separate condenser, double-acting pistons, compounding, and engine speed regulators followed by the end of the century. Significant improvements in operational aspects made the use of this type of technology widespread.

About the same time, heat transfer was first being understood. In the late 1700s, combustion processes became better understood, and the phlogiston concept was shown to be incorrect. Latent heat, specific heat, ideal gas, and conductive and convective heat transfer were also studied in the 1700s.

Thermal systems have played an important role in production processing since the start of the industrial revolution. Some of the initial applications for these systems took the form of simple engines or relatively straightforward manufacturing processes.

Early development of devices followed an evolutionary approach. Developer B would use the same ideas as developer A but would incorporate some generally simple modifications. Developer C would then modify design B in just a few ways. This approach yielded two desirable characteristics: the new device would probably perform nearly as expected because it was a simple modification of one whose operation was known, thus saving wasted investments; and the new device could honestly be claimed to be better than the existing one, thus ensuring a market niche. Another benefit was that small changes usually did not cause major problems if the devices did not work as well as expected. The development of engines, such as the early steam engines and internal combustion engines, is clearly an example of this approach. Also, an element of this approach plays a major role in modern engine design.

1.1.3 200 fruitful years

Basic science and its application in technology and engineering has advanced quickly over the last 200 years. The scientific understanding of both thermodynamics and heat transfer developed, and this tended to aid the concept evolution with approaches that had more basis than simple trial and error. Much of the focus was on how to make the overall processes better by attempting to define ultimate performance limitations.

Carnot and his contemporaries firmed up many of the ideas that are the basis of the *Second Law of Thermodynamics* during the mid-1800s. Gouy (1889) and Stodola (1898) drew upon these ideas. They are often credited with the initial contribution related to the topic known variously as exergy, availability, maximum work, and several other names. The idea of exergy allows a uniform basis for comparison of energy forms (for example, heat, work, chemical processes) and their quality (for example, the source or sink temperature involved in heat transfer processes). More precise insights to the definitions and implications of exergy are given later in this book in the separate chapters by Bejan and by Moran. Little interest beyond theoretical developments was generated in the early 1900s.

In the 1930s and 1940s engineers and scientists began to apply analytical tools with much more intensity, and, as a result, the numbers of these tools available and the general understanding of their application began to grow significantly. It was in this period that Keenan revisited and developed further the ideas of exergy. This work was examined in a series of papers and in his textbook on thermodynamics (Keenan 1941). Even at this time, relatively little interest was generated in this concept. It took the energy crisis of the 1970s to stimulate additional applications of this idea.

As scientific calculators and computers became a reality, the use of the analytical tools in much more concentrated and powerful forms became a possibility. Both the analysis techniques and the tools to carry them out on a large scale were to have a profound effect on engineering design, even though engineering design was still in the developing stages. Analysis tools were becoming available to allow the designer to examine many what-if questions. Where before a single solution would suffice, it now became possible to find a whole family of solutions. The best of these solutions could then be picked for the actual design.

1.1.4 Developments in the power and chemical industries

Two major industries in which these developments had an impact were the thermal power plant industry (TPPI) and the chemical process industry (CPI). The manufacturers of refrigerators and heat pumps share some aspects in common with the first of these industries, and comments made about the TPPI generally apply to these manufacturers also. Although the approaches to design processes between these two categories were similar, distinct differences also existed. Similarities included the need to understand basic processes. For example, the need to be able to design, or specify the performance of, a heat exchanger was a common aspect. Another need was to develop basic cost estimating information.

A fundamental and significant difference between the TPPI and the CPI applications was that CPI plants could differ in many ways, depending upon the products sought and the feedstocks used. Virtually an infinite number of variations were possible. In the TPPI were differences between plant designs, but the variations were less numerous. Sure, the design of a coal-fired steam power plant was significantly different from that of a boiling-water nuclear power plant, but they shared many fundamental aspects. Both used the fundamental Rankine power cycle, for example.

In the TPPI field, companies usually specialized in certain market sectors. As one example, Westinghouse developed a large organization in Pennsylvania that designed pressurized-water nuclear-reactor systems. Westinghouse developed simulation codes that could be used for design, but these codes tended to be specialized to the pressurized-water reactor product line. In general, the TPPI companies developed codes that were detailed and that were often treated as proprietary. In current use within the TPPI are codes for detailed combustion process design, performance analysis of transient behavior of power plants, rotating machinery performance, optimization approaches for heater placement, and a variety of others. These codes often followed a specialized and evolutionary approach related to the mode the product development process took.

In contrast to the depth that was found in the TPPI design, CPI plant requirements challenged the designers in breadth aspects. CPI applications found a large range of possible plant designs. It is true that some plant designs needed only small modifications, perhaps a scale up, on a previous design. On the other hand, a company such as DuPont might need to develop a plant to manufacture a new product never before produced in any significant amounts. Although this plant would contain many conventional unit operations found in other plants, the combination of these unit operations might differ considerably from those used before. Unit operations are a grouping together of processes of like form to facilitate analysis of their performance. Thermal processes play an important role in these types of systems, but not the dominant role usually found in mass transfer considerations.

The concept of unit operations used in the CPI was not referred to in the same terms in the TPPI. These types of processes are in the TPPI, however. Various types of heat exchangers including mixers (the open feedwater heaters) and cooling towers, pumps, expanders (turbines), and evaporators (boilers) are some of the components that make up this list. However, the importance of unit operations in chemical plant design is clear, and a significant portion of the chemical engineer's training focuses on this aspect.

1.1.5 Flow sheeting

The philosophy of unit operations and their application to plant analysis led to the concept of flow sheeting. In this approach, more generalized than approaches found historically in the TPPI, various generic unit operations (blocks) are connected into an overall plant design. Heat, work, and mass-flow streams are then connected to the various blocks as appropriate. The flow sheeting approach had been used in the design of CPI plants for a long time, and the format formed the basis of a system analysis using computers. Similar approaches were in fact used in other fields, such as electrical engineering and the TPPI, but these were referred to by other names. Excellent reviews of the early approaches to flow sheeting were given by Westerberg et al. (1979) and Rosen (1980). The chapter by Motard in this monograph gives current insights into this important field.

A series of software products have been developed based on the flowsheeting concept. An early version is a program called FLOWTRAN (see, for example Rosen and Pauls 1977). Similar codes were developed in a variety of applications, including versions for programmable calculators (see, for example, Blackwell 1984 and for more specific applications such as, for example, solar systems, Dutre 1991). Much more powerful forms of flow sheeting include the code ASPEN (1982), which was developed with U.S. Department of Energy funding and has been commercially developed further. This code is extremely powerful in that it can be used for analyzing a wide variety of systems and materials. Several aspects of the application of this modeling technique will be described in later chapters that underline this statement. Applications in both the CPI and the TPPI are included.

1.1.6 Property evaluation

In order to perform system analysis, material properties must be predicted. In the early days of the power plant calculations, all that was needed was a Mollier diagram or steam tables. For refrigeration systems, the situation was only a bit more complicated, because although several refrigerants were in common use, the number was limited to just a few. Refrigerant manufacturers usually furnished a P-h diagram or refrigerant tables to aid the analyst. Although the efforts started earlier, significant efforts were expended in the 1970s to develop computer-based property evaluation routines. Many companies in the TPPI and refrigeration equipment industries developed their own property routines. Now, many basic thermodynamic texts have a personal computer disk included that can be used for prediction of thermodynamic properties of water and common refrigerants. Many of these predictions are accomplished with the use of specialized empirical curve fits, and most of these were facilitated with thermodynamic concepts. Perhaps one of the foremost examples of these is the Benedict, Webb, Rubin equation (Benedict et al. 1940; 1942). This equation has found much application over the years.

A generalized program such as ASPEN PLUS, a later version of ASPEN, must have the ability to predict properties for a large number of compounds. This number can range into the thousands. Obviously, the development of specialized curve fits for every compound is not an efficient approach to this need. The development of appropriate property prediction methods has become a field of great importance to the whole modeling and computer-design effort. The lack of this aspect can be the single biggest barrier to a modeling effort (Glasscock and Hale 1994). A chapter by Danner, Spencer, and Nagvekar is included later in this monograph addressing some property prediction characteristics of these modern design codes.

1.1.7 Current developments

The oil embargoes of the 1970s focused attention on the efficient use of energy, emphasizing this aspect in plant design that had not been emphasized previously. Up until that point, energy had generally been viewed as a small and almost unimportant aspect of plant operation. Many factors had contributed to this earlier view, including the feeling in some quarters that certain new approaches to the generation of electricity might make this form of power "too cheap to meter." Almost overnight after the first oil shock, energy conservation became important. Attention refocused on concepts that had been developed earlier, including exergy. Exergy was particularly valuable because it allowed a uniform basis for comparison between disparate processes. Moran (1982), Szargut et al. (1988), and others fleshed out approaches that had been identified many years earlier to allow the determination of the ultimate energy value of chemical processing, heat transfer, power generation, and a variety of other physical processes using a single basis of comparison.

About that time, courses in the design of thermal systems were being developed at universities. Although thermal design had existed as outgrowths of fluids, heat transfer, thermodynamics, and other courses much earlier, now specific attention was focused on this topic at many schools. One of the earliest textbooks developed for these courses was the book by Stoecker (1971). As represented by this book, thermal system design involved primarily thermal modeling, optimization, and economics. The first two aspects were dealt with heavily from an analytical (as contrasted with numerical) basis. Several related texts have followed, including ones by Hodge (1985), Boehm (1987), Guyer (1989), Janna (1993), and Bejan et al. (1996). In these treatises, a variety of approaches have been taken ranging from stressing design-related heat transfer, fluid, and thermodynamic approaches to ones that include many of the topics addressed in the present monograph. Of course, these texts hold some topics in common with books in traditional chemical engineering design, such as Peters and Timmerhaus (1991) and Ulrich (1984).

Although early work with exergetic concepts emphasized analysis of systems, the ability to examine processes has always been present. In recent years, the focus has been more on the basic elements of this aspect. Bejan has been a leader in this work, and he has described many aspects of this field (Bejan 1996). By using fundamental applications of exergy concepts to heat transfer, fluid flow, thermodynamics, and other processes, he has formulated tools to allow developers to focus on the details of process design and operation that can affect device performance. With individual components operating at peak performance, obviously the overall system performance should be more nearly optimal. He has contributed a chapter to this monograph related to these issues.

Another valuable analysis approach, particularly for complex systems, involves the concept termed by its developer, Linnhoff (1993), as *pinch technology*. Application of the concept gave, among other items, insights about the economical use of waste heat from one part of a plant for supplying heat demands elsewhere in the same plant. This tool has become valuable in analyzing the design of new plants as well as in formulating energy conservation approaches for existing plants. Although some have claimed that pinch technology does not offer anything different from other forms of Second Law analyses, the formal techniques of pinch technology have been developed in a format that may be better used by practitioners. Pinch technology is described in Chapter 5.

Newer tools to aid in the design of thermal systems are being developed each year. Concerns about energy crises have subsided since the mid-1980s, taking some of the momentum from this work. Many governments and other organizations that sponsored research in this area have significantly decreased funding for related topics. But work continues. Efforts in this area are needed to make processes and plants more economically competitive for an ever-shrinking world marketplace.

One of the newer tools that has been gaining popularity over the past several years in thermoeconomics. This topic is an extension of Second Law analysis where costs become the ultimate comparison basis. Some pioneering work in this field by Valero and Lozano is also presented in this text as Chapter 8.

1.2 Future trends in design systems

1.2.1 Computer-aided design

What does the future hold? The whole area of computer-aided design is moving rapidly, and computing machines are becoming more powerful day by day. These two factors make this question difficult to answer. However, some current trends are clear.

Graphical interfaces will continue to be enhanced, allowing codes to become even more user-friendly. There has been a strong move from the representation of a system and its performance as simple lists of numbers and letters, the characteristic of the early codes, to graphical representations that can be better understood by the novice. Much more efficient use of these types of routines will be the outcome, including easier interpretation of results.

Codes will become more intelligent. The possible impacts of this aspect are far-reaching. Like improved graphical interfaces, this change should help the user to understand the various implications of large problems. Complex chemical plants, for example, may have many input and many output streams of both materials and energy. As a result, how complicated plants should be optimized may not be clear. Companies are developing codes that allow the user better insights about this aspect. Also, because the various equation systems are nonlinear, early versions of optimizing codes required guesses at the final results that were close to the final answers. If one missed by too large a factor in the guesses, some programs were not able to generate a solution. Efforts to make codes more robust to handle this problem are ongoing.

More effort will be made to develop the ability to analyze transient behavior in steady-state codes. Of course, this aspect is necessary for control issues, and codes that deal with this type of performance have been developed. However, the codes designed for transient analysis often have not been useful for optimizing designs on a steady-state basis. Codes that combine these aspects in an efficient manner will become more common.

The use of plug-in modules, written by the user, will also become more common. Although many codes have had this option for some time, the use of these will become much more convenient. Commercial developers are making available plug-in shells of various processes that can be easily adapted or modified, if the user desires.

All these aspects make the application of the types of codes currently available easier to use. What will the next, really new, generation of codes involve?

1.2.2 Artificial intelligence

Probably the use of artificial intelligence (AI) will become a big factor in new codes. Much more effort will be directed to developing methods whereby machines can take on more of the design work. Initial phases, within the expert system (ES) realm of AI are already being applied. ES techniques, described in Chapter 9 by Paoletti and Sciubba, are valuable when a less experienced person needs to design a fairly common process or system. Some less experienced designers may not even know the right questions to pose about a design. These advanced codes can furnish the right questions. The codes may also be able to offer possible educated answers to these questions. ES-enhanced codes may also free up the experienced designer to focus on more novel or difficult aspects of layout and specification.

Truly new approaches will follow within the AI realm. Ultimately it is possible that the user will simply specify the plant outputs desired and any limitations on the inputs. The code will then be able to perform the complete design based on some definition of the optimal choice. Included will be the ability of the code to place blocks in various locations within the system to achieve the desired result. Once this front-end strategy has been developed, a trial flow sheet will result. The trial flow sheets can than be evaluated using current approaches. Optimizing evaluations, again defined within the general context of the code, can be performed. This truly automated design situation is quite a way off for all but the simplest of systems.

1.3 Basic elements of thermal system design

What are general characteristics of thermal systems? Many aspects have to be brought together to perform an analysis. Some of these are shown in Fig. 1.1. Some generalized characteristics of thermal system design (TSD) include the following:

- TSD is the application of thermodynamics, heat transfer, mass transfer, fluid mechanics, and cost information to the manufacture of a product or products. These products can be physical quantities or energy quantities.
- TSD involves the use of a number of individual processes to form a system of interest.
- Any individual process element may require a great deal of design insight and analysis before a complete specification of that element is made.

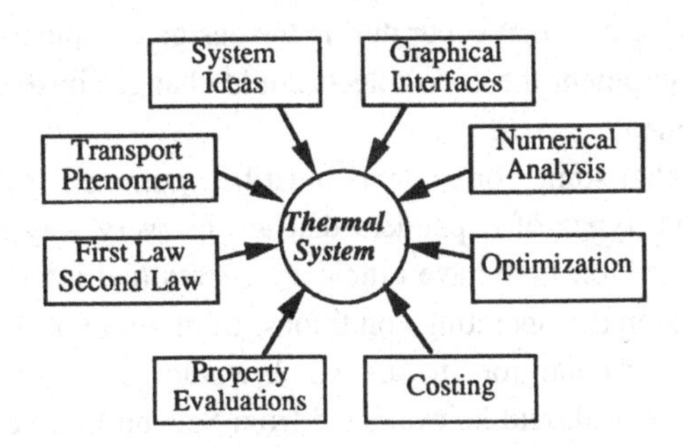

Fig. 1.1. Capabilities in several areas have to be brought together to develop the analysis necessary to design a thermal system.

- Processes may be configured together in many ways that can accomplish the same desired product.
- The designer views some ways that the system can be specified as preferable to others. Hence some kind of optimization is usually required.
- A final system design then results from the analysis. This design can include detailed aspects down to specifying details on all components and laying out the piping diagrams.

In its general form, TSD involves a large number of components (referred to below as blocks) operating as a system. The specification of the various components and how they are tied together can be an involved process. How a design of a given plant should be evaluated is often unclear. A plant can have so many inputs and outputs that how to define the objective function(s) for the given design may not be clear. Although some constraints on the design are easily identified, others may be difficult to specify.

Several basic aspects have to be incorporated in codes that perform these kinds of tasks. In general, the system must be broken down into a series of components or processes. Of course the fundamental physics of the processes have to be specified for categories of like processes in general ways. Then specific details of a particular component, operating at particular conditions, can be determined from the physics.

Many variations of components may perform similar functions, and choices between these components are often needed. One example is choosing the proper type of heat exchanger, from the thousands of models that are commercially available, to apply in a particular circumstance. Factors used in selecting one type of heat exchanger compared with another must be known, and these factors must be applied during the design process. Limitations on any particular device must be listed as constraints on that block. These, in turn, could constrain the complete design. Design codes should be able to handle automatically any variations in device type that occur during the design computations. Variations in the type of equipment the code selects could change constraints associated with that component.

It is possible that some components must be optimized at the block level. For example, two types of expanders that are in every way appropriate for a particular application may have efficiency curves that intersect at different points depending on the operating conditions. Both might have to be analyzed for any given set of conditions to see which would perform with the highest efficiency. Efficiency should be evaluated from Second Law (exergy) ideas to make the evaluation consistent when this device is compared with other devices in the overall design.

Not only do the physics of a given block need to be handled appropriately, but the costs of the device must also be known as a function of the parameters of importance. This parameter list could be large, including size, materials of manufacture, operating range, estimated year of purchase, projected inflation rate, and shipping and installation costs, as well as a number of other factors. There could be cost factors that reflect aspects of connecting the device of interest to other devices, such as piping runs between components. These other devices cannot be fully evaluated until the design is more complete.

Components should then be configured together into a system. Further, the outputs from preceding devices have to be correctly identified as inputs to succeeding devices. Of course the overall system design must result in operation that yields the correct specifications of the overall plant outputs while using the specified plant inputs. To achieve these specifications, some iterations are required.

All the design requirements might be met using the sequential-modular approach, which is the way that classical thermodynamics problems have been worked. In this approach, each block is handled in an order that allows it to have flows and other conditions previously calculated or otherwise known for other blocks. This calculation order is not usually clear a priori in complex systems, and a solution technique of this type may not even be possible.

Because complexities almost certainly exist, a simultaneous approach is used. The simultaneous approach is much more general, and it is the solution approach of choice. In this approach, the governing equations for all components are formulated and solved simultaneously. Known parameters can be readily applied, as can component connectivity information. This system of equations is usually nonlinear. Finding a solution may require having some close approximations of that solution. Several solutions will often exist. Some will be preferable to others.

To determine the system performance through calculations, thermodynamic, thermophysical, and chemical properties are required. Property values are needed to a sufficient degree of accuracy over a wide range of operating conditions. Extensive ensembles of codes may be necessary to enable evaluation of all design options. Because a general simulator will need to predict properties for literally thousands of compounds, the amount of code can be large. Even designs that involve only one or two fluids (say a steam power plant) may require a sizable computation scheme, if the fluids are used in the saturation region.

When the system of equations is determined, including all the auxilliary information related to properties and costs, the numbers of independent variables and equations could be large. The system will usually be sparse (most equations will have only a few variables), and the equations will usually be nonlinear. Possibly, variables may outnumber equations. From this system, tens of thousands of possible solution sets could exist, with some being arbitrarily different from others.

Appropriate optimizing routines have to be used to determine which of the solutions best meets the desired plant output. This aspect raises several problems. One is how to define the optimal solution. Although the ultimate objective function is usually going to be the net revenue stream, determining this quantity may not be easy. If the system is large and complex, defining the optimal solution also can be difficult. Many inputs and outputs other than the primary quantities given in the initial problem definition may exist. Some monetary values obviously have to be assigned to them.

Another problem in finding the optimum is related to the computations involved. Are they too numerous for the computing machine to be used for this effort? Not everyone has access to a supercomputer. There has been a special thrust to attempt to develop routines that perform appropriate computations on a PC. Special techniques have to be used in these types of solutions.

Not the least of the issues in the solution approach is the robustness of the method. Nonlinear optimization approaches on large systems often require that one guess the answers. If these guesses are too far from the actual answers for the system being considered, many numerical approaches cannot converge on the desired solution. As a result, the code may have to have some initial computational approach built in that will reasonably estimate these initial guesses.

The solution approach also has to handle highly constrained systems, because TSD almost always deals with systems of this type. Some may be inequality constraints, and some may be equality constraints. These can complicate the general solution techniques because inappropriate handling of constraints may introduce severe "bumpiness" in the outputs as a function of input variations.

1.4 Simulators

Individuals and groups have been at work for many years incorporating the concepts outlined above into workable codes. The result is a rich array of tools available to the designer. I will not attempt to summarize these codes at this point. This topic is covered in several chapters that follow.

The usefulness of these codes extends far beyond the design and layout of processing or power systems. Consider the sequence of milestones in the design and operation of a plant as shown in Fig. 1.2. Each of these has opportunities for saving time and money through the use of simulators. Although a cost is associated with the use of simulators, proper use of these can offer benefits that far outweigh this cost.

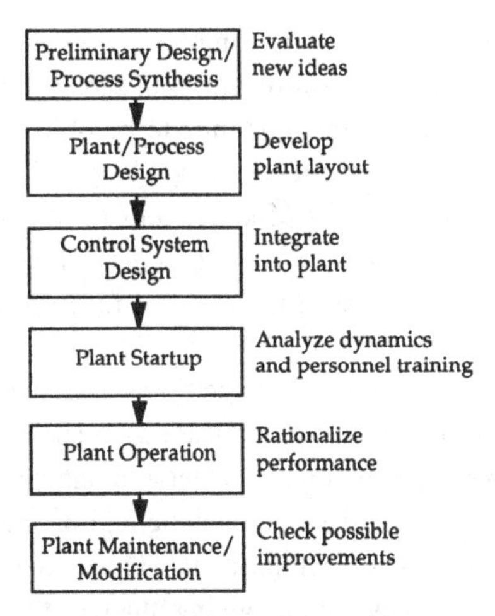

Fig. 1.2. The major elements of bringing a plant from the conceptual stage to mature operation are shown. Simulators can aid with the tasks shown at right.

During the preliminary design stage, fundamental options may need to be evaluated. One possibility is to handle all of this phase by something only slightly more than back-of-the-envelope evaluations. Another is to perform simplified analyses on a simulator. These analyses could give good insights about the impact of the new concepts and could include the development of simple economic evaluations of the options.

The second block in Fig. 1.2 is widely recognized as an element where simulators can be of value.

Several codes allow the dynamic performance of the plant to be simulated, generally done by time-stepping steady-state evaluations, allowing the results of one time sequence to input new variables to the next. With simulations of this type, the control systems of the plant can be specified. A wide range of scenarios can be considered to make sure that the specified control systems are sufficiently robust.

Plant startup is also, of course, a transient situation. The use of the same simulators as for the control system design can be valuable in this phase of the development of a facility. A related issue of immense importance is that these tools can be used for training operational personnel in the startup procedures. Graphical outputs from these codes can be made to represent the plant monitor outputs, giving the operations crew considerable background in how to rectify abnormal situations.

Once the plant is in operation, simulation tools can be used for rationalizing performance. Efficiency values or product yield as a function of input parameters can be checked against the simulator predictions. Assumptions in the design can be easily checked. Malfunctions of subsystems can be identified.

As the plant matures, simulators continue to be of value. Routine maintenance, such as checking the performance degradation of heat exchangers due to fouling, can be easily evaluated. Modifications of the basic plant to install new pollution control equipment, to boost product output, or to change some other fundamental process can be easily checked for impacts on other aspects of the plant.

The biggest problem facing groups in getting started with simulators is determining which ones to use for the applications desired. Too often the case is that software companies claim their product will handle seemingly all problems. Reading the sales literature on computational fluid mechanics codes, for example, should convince anyone of this claim. The wrong codes can consume a great deal of time and money without giving much useful information. Some codes are much more user-friendly than others, saving time both in developing the in-house ability to use the codes as well as in managing the day-to-day operations.

Wise choice of the computational products and appropriate training of the applications people should yield significant benefits. Simulators are a key to higher productivity.

References

ASPEN (1982). Computer-Aided Industrial Process Design, the ASPEN Project, Final Report. U.S. Department of Energy Report MIT-2295T0-18.

Bejan, A. (1996). *Entropy Generation Minimization*. Boca Raton, Florida: CRC Press, Inc.

Bejan, A., Tsatsaronis, G. & Moran, M. (1996). *Thermal Design & Optimization*. New York: John Wiley & Sons, Inc.

Benedict, M., Webb, G. & Rubin, L. (1940). An empirical equation for thermodynamic properties of light hydrocarbons. I. Methane, ethane, propane and n-butane. *Journal of Chemical Physics*, **8**, 334–45.

Benedict, M., Webb, G. & Rubin, L. (1942). An empirical equation for thermodynamic properties of light hydrocarbons and their mixtures. II. Mixtures of methane, ethane, propane and n-butane. *Journal of Chemical Physics*, **10**, 747–58.

Blackwell, W. (1984). *Chemical Process Design on a Programmable Calculator*. New York: McGraw-Hill Book Company.

Boehm, R. (1987). *Design Analysis of Thermal Systems*. New York: John Wiley & Sons, Inc.

Dutre, W. (1991). *Simulation of Thermal Systems*. Dordrecht: Kluwer Academic Publishers.

Glasscock, D. & Hale, J. (1994). Process simulation: the art and science of modeling. *Chemical Engineering*, November, 82–9.

Gouy, G. (1889). About available energy (in French). *Journal de Physique*, **8**, 501–18.

Guyer, E. (ed.) (1989). *Handbook of Applied Thermal Design*. New York: Mc Graw-Hill Book Company.
Hodge, B. (1985). *Analysis and Design of Energy Systems*. Englewood Cliffs, New Jersey: Prentice Hall.
Janna, W. (1993). *Design of Fluid Thermal Systems*, Boston PWS-KENT Publishing Company.
Keenan, J. H. (1941). *Thermodynamics*, New York: John Wiley & Sons, Inc.
Linnhoff, B. (1993). Pinch Analysis—a state-of-the-art overview. *Transactions of the Institute of Chemical Engineers*, **71**, Part A, 503–22.
Moran, M. (1982). *Availability Analysis: A Guide to Efficient Energy Use*. Englewood Cliffs, N.J.: Prentice Hall.
Peters, M. & Timmerhaus, K. (1991). *Plant Design and Economics for Chemical Engineers, Fourth Edition*. New York: McGraw-Hill Book Company.
Rosen, E. (1980). Steady state chemical process simulator: a state-of-the-art review. In *Computer Applications in Chemical Engineering*, ed. R. Squires & G. Reklaitis, ACS Symposium Series 124. Washington, D.C.: American Chemical Society.
Rosen, E. & Pauls, A. (1977). Computer aided chemical process design: the FLOWTRAN system. *Computers and Chemical Engineering*, **1**, 11–21.
Stodola, A. (1898). The cycle processes of the gas engine (in German). *Z. VDI*, **32**, 1086–91.
Stoecker, W. (1971). *Design of Thermal Systems*. New York: McGraw-Hill.
Szargut, J., Morris, D. & Steward, F. (1988). *Exergy Analysis of Thermal, Chemical, and Metallurgical Processes*. New York: Hemisphere Publishing Corporation.
Ulrich, G. E. (1984). *A Guide to Chemical Engineering Process Design and Economics*. New York: John Wiley & Sons, Inc.
Westerberg, A., Hutchinson, H., Motard, R. & Winter, P. (1979). *Process Flowsheeting*. Cambridge: Cambridge University Press.

2

Computer-aided process design trends in the power industry

ROBERT M. PRIVETTE
Babcock & Wilcox

2.1 Introduction

Competition within the power industry has encouraged industry developers, suppliers, and operators to reexamine options for providing cost-effective electricity. One approach to improving generation efficiency involves system optimization and evaluation using process analysis tools. These tools are used to evaluate factors such as plant performance, environmental emissions, and controllability. A systems approach, based on computer-aided process analysis, provides an ideal tool for use in quantifying costs, benefits, and risk in power generation development projects.

A power plant model can be as simple or as complex as is necessary to address the needs of the analysis. System models can be limited to predicting the release of heat during fuel combustion or can include detailed representations of fuel handling, preparation, consumption, and emission processes as illustrated in Fig. 2.1. Power consumption resulting from auxiliary systems can be identified, and performance sensitivity to component behavior can be quantified. Specific subsystems and components can be modeled to answer specific risk-related questions.

Such modeling is increasingly being used to answer project development questions, direct repowering and refurbishing projects, and optimize plant operations. Instrumental in this trend is the use of computer-aided process design tools that provide an efficient and capable means for the system designer and system developer to accomplish these goals.

Process simulators were originally used in the chemical and petroleum industries on large stand-alone mainframe computers. Now, with recent advances in computer technology, significantly more powerful and more user-friendly programs are available for even the smallest portable laptop computers. This development has resulted in much broader access and application of these tools to answer questions and assess performance in circumstances that would not have been feasible twenty years ago (Glasscock and Hale 1994). This breadth of use has developed to include processes and systems associated with the power generation industry. Cycle analysis tools have significantly affected the industry by increasing the capability of engineers to rapidly assess the merits of process design, operation, and equipment changes.

Computer-aided process design refers to the application of computer-based cycle analysis software to the design and analysis of a power generation system or subsystem. The term *cycle analysis tool* in this context refers to a computer software program that allows the user to analyze the performance of a desired power system (or cycle) comprising a variety of interconnected components. Components may include boilers, pumps, heat exchangers, fans, turbines, and any other equipment that affects the flow of energy or material through a process.

In addition to process components, the interconnecting streams carrying energy and material are also represented. For instance, the schematic in Fig. 2.1 includes an incoming coal stream, combustion air, and makeup water as well as exhaust stack gas. The flow of heat and work would also be represented by including the exchange of heat between gas, air, and water streams and the extraction of shaft work at the steam turbine. Combustion of the coal can be represented with the accompanying heat release resulting in an increase in the furnace exhaust gas energy.

The amount of modeling detail can vary significantly depending on the requirements of the analysis. A model designed to accurately predict the off-design performance of a component or system may require correlations for the component performance over a range of pressure, temperature, and flow rates. These modeling details are generally not necessary in a model developed to simply predict performance at one design condition.

Although cycle analysis tools are commonly used for such tasks as predicting power plant performance, comparing alternative plant designs and technologies, predicting performance based on alternative fuels, optimizing cycle operation, and assessing component performance degradation, they can also be used to analyze various what-if scenarios. For example, "What if a fouled heat exchanger, with degraded heat transfer characteristics, is replaced by a new heat exchanger? How much operational time is necessary before the cost of hardware replacement is made up by the savings resulting from increased process efficiency?" As another example, a power project developer may ask the question "What if an integrated gasification combined cycle (IGCC) plant is built and used to provide electricity in place of a traditional pulverized coal plant? How do the plant efficiencies and emissions compare?"

Questions such as these involve performance predictions for processes comprising multiple interconnected and interdependent components. Such

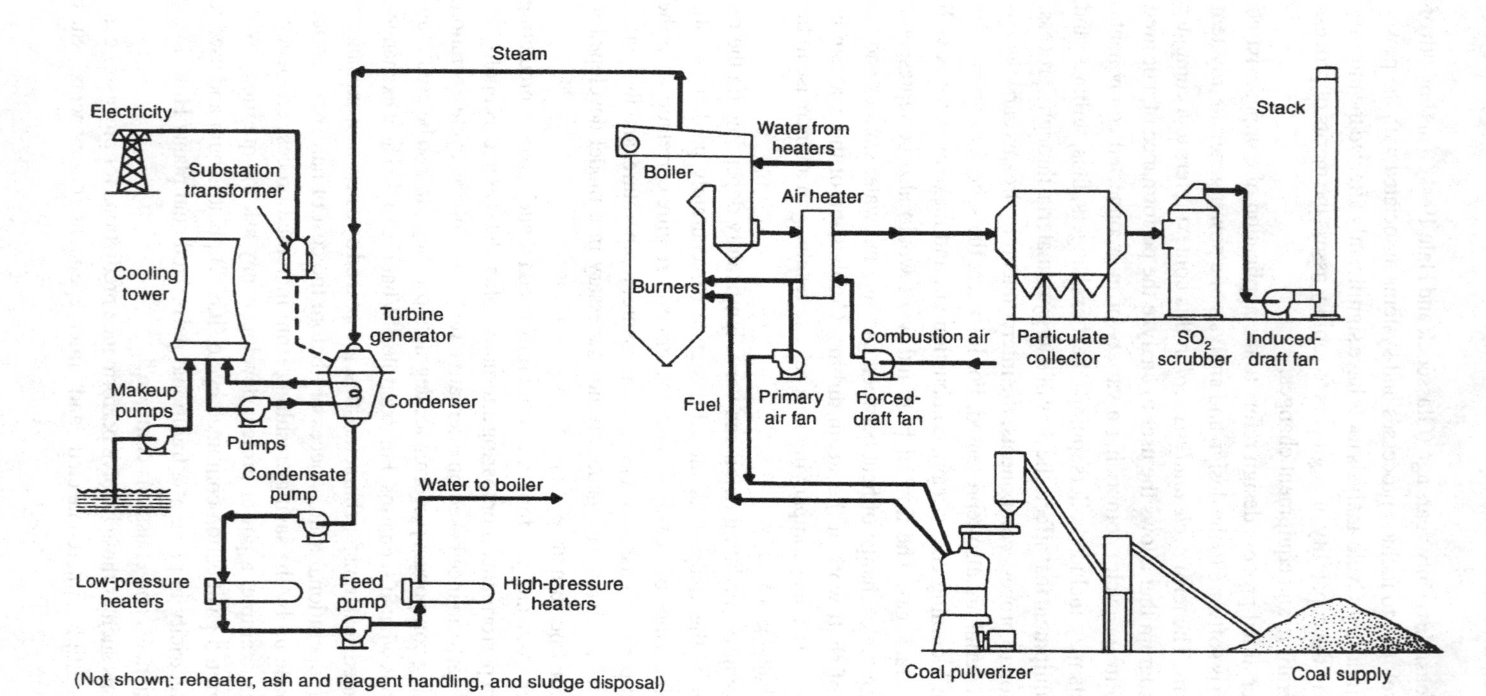

Fig. 2.1. Coal-fired utility power plant schematic.

predictions are not typically available through analysis tools intended for single components such as heat exchanger design tools or pump sizing programs. Furthermore, such performance predictions are often strongly dependent on actual hardware performance data. Empirical correlations representing measured performance data can often be included in a system model to improve the validity of the predictions. In this way, the use of process analysis tools complements the use of more traditional component analysis tools which are frequently based largely on empirical correlations (Linnhoff and Alanis 1989). Later in this chapter, we will describe the complementary roles of system analysis tools, and component performance tools, noting the reason why both are required for accurate and reliable system design activities.

Cycle analysis tools are most often based on fundamental conservation equations for heat and material balance and can be used to model a wide variety of systems involving closed and open thermodynamic cycles and chemical processes. These tools belong to an ever-increasing list of computer-aided engineering tools. Computer-aided design (CAD) programs and computer-aided manufacturing (CAM) programs are perhaps the best-known computer-aided engineering tools. Computer-aided process engineering (CAPE) tools, however, have made an equally significant impact on the process industry of which power generation can be included as a subset.

Of course the analysis of a power plant's thermodynamic performance does not require the use of a computer tool. In fact the analysis of a standard Rankine cycle is generally assigned as homework in any college thermodynamics course. The importance of the computer implementation of existing analysis procedures lies simply in the degree of detail and breadth of analysis that can be brought to bear on a problem in a relatively short time. Sophisticated methods for predicting stream property data, for example, eliminate the need to spend time manually interpolating between tabular values. Ideally more time can be spent considering the implications of performance predictions and less time worrying about the convergence of a recycle calculation.

Cycle analysis tools are available to address both steady-state and dynamic (transient) performance analysis of power systems. Increasingly, issues of plant cycling, controllability, and safety are being addressed through dynamic system simulation. Such dynamic analysis tools, once regarded as fragile and unreliable, have been developed to the point where they are now generally robust, flexible, and practical for use in analyzing systems in which time dependency is an issue (Wozny and Jeromin 1994) (de Mello and Ahner 1994).

As with any type of computer-aided engineering, the use of a computer program does not reduce the importance of user experience and insight. In fact, in many ways, the importance of these factors is heightened as a result of the ability to analyze complex, highly-detailed analyses of systems. Care must be exercised to avoid the tendency to accept computer-generated solutions without the appropriate scrutiny and reasonability checks. This is serious issue and requires diligence to ensure representative results. Such scrutiny must include checks of data entry, basic energy, and material balances, as well as modeling assumptions. This last issue, the use of appropriate modeling assumptions, is perhaps the easiest to overlook and can be the most costly. Inappropriate modeling assumptions can easily result in excessive complexity and unrealistic predictions. In the end, a thorough technical understanding of the fundamental mechanisms involved, a practical understanding of the operational limitations, and God-given insight have no replacements. Such an understanding is almost always strengthened by an in-depth understanding of the physical equipment being modeled. Expert systems and sophisticated modeling can rarely take the place of human experience.

This brief introduction has presented descriptions of some of the benefits and pitfalls that accompany cycle analysis. The balance of the chapter will focus on how these tools are used in the power generation industry and includes a list of commonly used programs.

2.2 Power plant applications

Over the last several years, numerous papers have been published that describe the use of cycle analysis for tasks such as

- comparing and contrasting alternative power plant designs and technologies,
- predicting thermodynamic performance with various alternative fuels (Kettenacker and Hill 1992),
- predicting future performance and supplementing costly boiler tests (Elliott 1994),
- optimizing cycle operation (Tsatsaronis et al. 1992) (Parker and Hall 1994) (Lie 1994),
- providing guidance in prioritizing and executing equipment testing (Kisacky et al. 1992),
- identifying plant equipment with a particularly strong effect on plant efficiency (Kettenacker 1988),
- assessing the impact of component performance degradation (Jain et al. 1989), and
- identifying the impact of proposed corrective actions (Naess, Mjaavatten and Li 1994).

Power industry applications include a wide range of components and systems from flue gas desulfurization (FGD) equipment to pulverized coal power plants, gas turbine combined cycle (GTCC) systems, and IGCC systems (Tsatsaronis and Lin 1993). Individual component design can be enhanced through development based on the component's effect on overall system performance. System modeling can be an integral tool in this process, allowing the engineer to make design decisions and trade-offs in overall design, performance, and system operation (Moore 1995).

As an example, consider an efficiency improvement option for a fossil-fired power plant. Typically, such plants are designed so that the exhaust gas temperature never drops below the acid dew point. This strategy avoids issues associated with corrosion and costly equipment replacement costs. The reduced risk, however, comes at a penalty in performance. Clearly, any heat contained in the exhaust gas represents an irrecoverable loss.

A condensing heat exchanger can allow the designer to avoid limitations resulting from acid-based material wastage in the exhaust stream, thereby allowing the exhaust gas to be cooled below the acid dew point and maximizing the heat recovery potential. Risk associated with material corrosion is minimized by providing a protective coating on all components exposed to the cool exhaust gas. With a condensing heat exchanger, additional heat can be recovered from the exhaust gas and integrated into the process.

In assessing this efficiency enhancement option, recovery and integration of the newly available heat must be analyzed to quantify the overall benefit. Decisions to implement improvement options such as this require an understanding of the financial benefits, which may include decreased fuel costs, and reduced gas scrubbing equipment duties. Alternative uses for the waste heat could include feedwater or makeup water heating, air preheating, or heating of auxiliary water for cogeneration applications. These alternatives can be modeled and evaluated for feasibility and performance enhancement and can provide input to economic models used to evaluate the cost trade-offs of such a system improvement option.

In addition to providing flexible and powerful assessment capability, the use of such tools provides additional benefit as part of a consistent system evaluation methodology. Many times system enhancement opportunities can be difficult to compare directly as a result of vastly different generation technologies, different resource assumptions, or a combination of both. Appropriate modeling of the resources and infrastructure available at the prospective plant site and the equipment necessary to modify or upgrade them for use in the system (for instance, boosting gas pressure or pulverizing coal) can help ensure that system

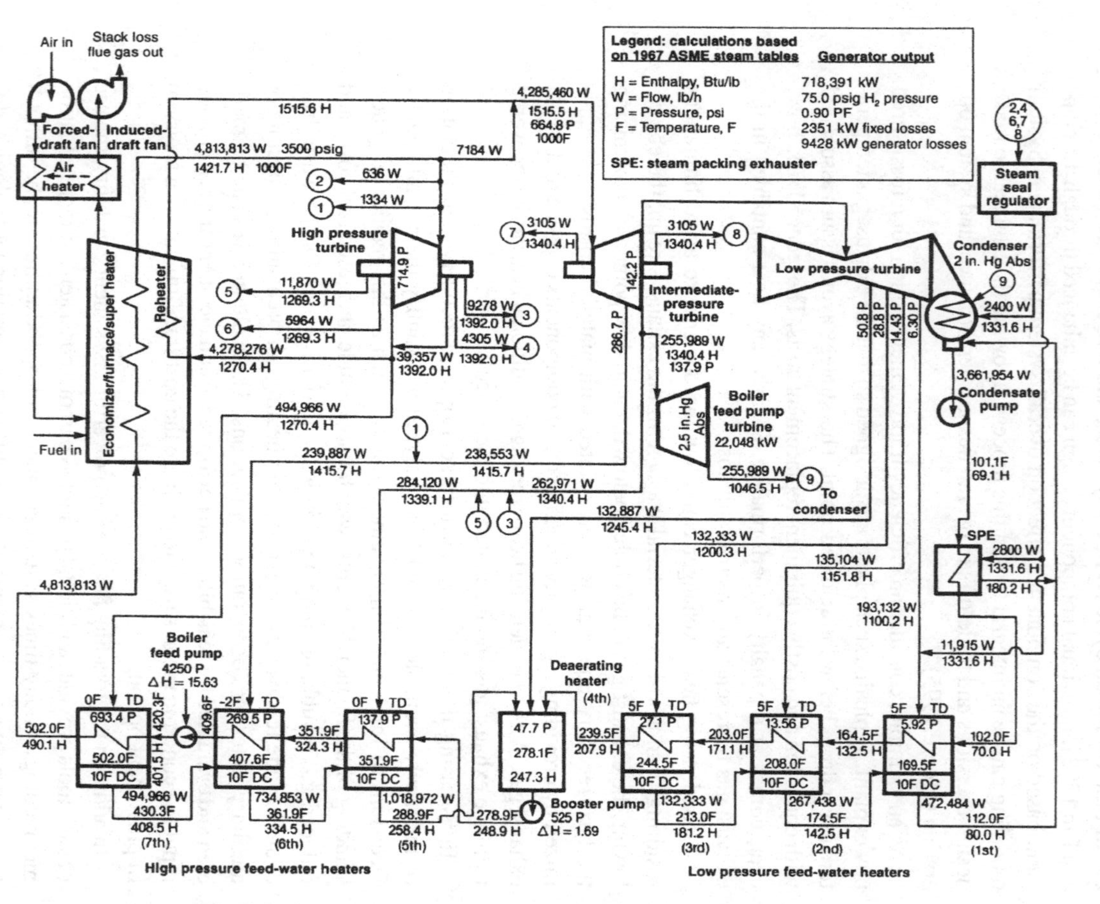

Fig. 2.2. Supercritical pressure, 3500 psig turbine cycle heat balance.

performance results are comparable. Other benefits can include

- consistency among system analysis methods;
- access to system performance predictions by diverse segments including component designers, marketing, finance, and others;
- potential for shared modeling software (corporate licensing);
- shared modeling experience;
- shared experience with analyses of particular cycles; and
- increased business opportunities.

Boiler and steam cycle modeling are common applications for cycle analysis tools. Fig. 2.2 shows the result of a typical system calculation for a supercritical pressure steam cycle. The diagram shows the pressure, temperature, enthalpy, and flow rate for the steam cycle. The air–gas side of the system is shown in limited detail. System designers may investigate alternative arrangements for feedwater heaters, vary the number of reheat stages, or look for other ways of increasing system efficiency (Elliott 1989).

Such analyses must include hardware performance details in order to accurately predict overall performance. For instance, traditional air heater designs typically have air–gas leakage that, if neglected, will result in overly optimistic predictions of system performance. When practical, performance information should be based on operating or vendor data. Although vendor data will always be the preferred source for reliable performance information, at times such information is not available. In such instances, the experience and insight of the system designer can make the difference between unrealistically optimistic or pessimistic performance predictions and realistic predictions. This aspect is perhaps the most critical of system analyses. Although the computer tools provide a means for automating calculations that may have been impractical because of time requirements in the past, the basis and assumptions behind the calculations and system modeling are left to the judgment and experience of the engineer. Inexperienced users should be aware of such pitfalls and should rely on equipment vendor performance information, test data, or accepted industry rules of thumb in developing new system models (Wilbur 1985) (Bartlett 1958) (Spencer, Cotton and Cannon 1963) (Korakianitis and Wilson 1994).

The system represented in Fig. 2.2 contains several different material streams. Consider the fuel (coal, oil, etc.), combustion air, exhaust gas, water, steam, condenser water, and other minor material streams. For each stream involved in the calculation, conditions such as flow rates and temperatures are typically required to complete the heat balance. More important, methods of estimating material property for several of the key streams such as the water, air, and fuel can significantly affect the resulting performance predictions. Many modeling

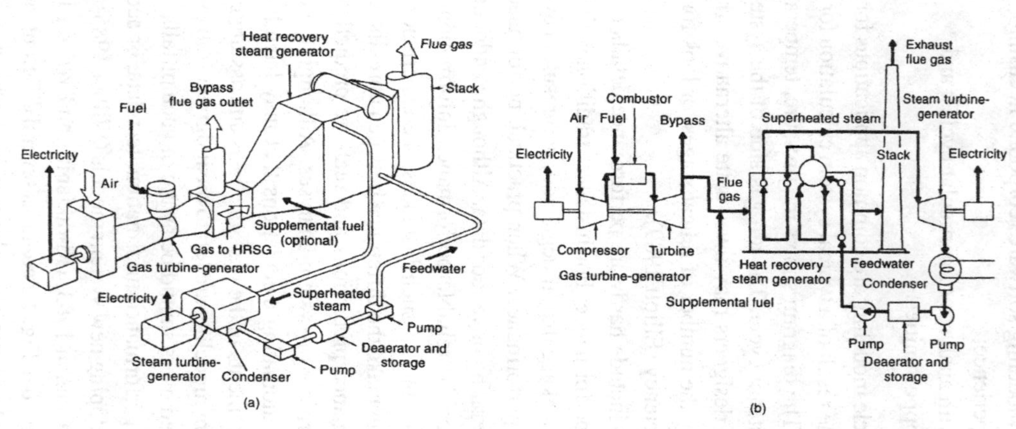

Fig. 2.3. Simplified combined cycle schematics.

tools provide extensive correlations for calculating material properties. Some limit the number of correlations used in a given analysis. This limitation can force the user to use a nonoptimal correlation. For instance, in a system with both air and steam, the use of a single correlation such as the 1967 ASME steam correlation would not represent the properties of air sufficiently. Physical property correlations should be considered when selecting a process analysis tool. If the tool is prespecified, the design and use of the system model should recognize and consider the impact of property estimation on the results (Boston, Mathias and Watanasiri 1984).

Another modeling feature that should be considered is solids handling. If solids such as coal are important to the application, and if their form is the basis for certain process operations such as sizing, the operations models must be able to handle solids. Process analyses that consider the effects of the size of solids on the process performance may require additional assumptions about the distribution of solids, or a unit's separation performance, to complete the analysis. Other considerations, such as the maximum number of components that can be included in a single model, the variety of methods for calculating process fluid properties, and the robustness of calculational algorithms, are also important when applying process analysis tools to system design and evaluation.

The simulation of processes involving fuels such as coal or petroleum oils many times requires the user to specify a representative composition or property model. This requirement is due to the inexact nature of the stream composition. Coal, for instance, can vary widely in composition, heating value, and other parameters that are important for certain modeling applications. Pseudocomponent descriptions, in which the essential property information is supplied by the user, provide a common technique for analyzing systems involving materials that are otherwise unknown to the simulator.

Another application commonly represented using process analysis tools is the repowering of aging boiler equipment using updated boiler technology, combustion turbines, or reciprocating engines. In addition, recent interest in GTCC systems, employing a combustion turbine in conjunction with a heat recovery steam generator, has resulted in significant interest in modeling these systems. In fact, several modeling tools are either dedicated or derived from GTCC system analysis, including GT Pro, and Gatecycle.

Figure 2.3 shows a simple component layout for a GTCC system and also shows a process representation of the same system. The formulation of the process model can vary widely depending on the intended use of the modeling results; however, identification of the key components involved in the system that affect the performance aspects being investigated is the first step in developing any process model. In Fig. 2.3 material streams input to the gas turbine system include fuel (natural gas, syngas, etc.) and air. Combustion of the fuel and air produces hot gas, which is expanded through the turbine and routed to a heat recovery steam generator (HRSG). The heat remaining in the hot exhaust gas is extracted in the HRSG before the gas exits the system through a stack. A water–steam loop within the HRSG absorbs the heat from the hot gas as the gas passes over heat-exchanging tubes. The resulting steam is then expanded through a steam turbine.

The system shown in Fig. 2.3 includes a steam cycle with no reheat; however, a common alternative is to extract some of the partially expanded steam from a high-pressure section of the steam turbine and direct it back into the HRSG for reheat. Single, double, and even triple reheat steam cycles are common. Reheat has long been recognized as a cost-effective technique for increasing the efficiency of the steam cycle. The ability to simulate reheat cycles is common but not universal among process simulators.

Figure 2.3 shows supplemental fuel entering the HRSG. Auxiliary burners, fired using supplemental fuel, are used to provide added steam generating flexibility and increase capacity. Auxiliary burners in the HRSG partially decouple the gas turbine and steam turbine and allow generation of more steam than would be possible given the exhaust conditions of the gas turbine alone. The flexibility of a simulator's ability to easily model such features should be considered when selecting a product.

Gas turbine systems are high on the list of power generation technologies modeled using cycle analysis tools. These include both simple cycle (gas turbine alone) and combined cycle (gas turbine and steam turbine) configurations. Some analysis tools presently include databases for performance characteristics for 50 or more current gas turbine models and various reciprocating engine models (GT Pro, Gatecycle). Predicting the performance of gas turbines and other systems using alternative fuels such as oil and syngas can be important for evaluating and designing systems for specific user sites.

Recent interest in minimizing emissions from power plants has resulted in a heightened interest in using process analysis tools to integrate components such as flue gas generation equipment (boilers or gas turbines, for example), with cleanup systems such as wet and dry scrubbers or selective catalytic reduction (SCR) units (Hilaly and Sikdar 1996). These are examples of products that are many times designed independently based primarily on mutual boundary conditions. A system analysis capability to assess the total system can lead to design enhancements presently unavailable to the independent design process. The requirement to model gas cleanup equipment is important because in many cases it requires analysis of solids or detailed chemistry, which is not universal among cycle analysis tools.

2.3 End user requirements

The predictive capabilities of process analysis tools can benefit a variety of users for addressing issues from conceptual design to plant shutdown. Architect/engineer firms may use such tools to gain insight into system feasibility, optimization, and operability (Narula 1995). Equipment suppliers can predict and evaluate system benefits available through new components, operating procedures, and process configurations (Paul 1994). Operators use process analysis to compare control alternatives, support system diagnosis and troubleshooting, and monitor plant performance relative to predicted performance (Elliott 1994).

The applications for computer-aided process analysis fall generally into one of three categories:

- System design: Scoping and detailed analysis of various system configurations.
- Component design: Effect of specific components on overall plant performance.
- Marketing and sales: What-if analysis of various technologies, operating scenarios, and presentations.

Although tools are available that can accomplish all three of these objectives, the trade-offs involved with such codes in terms of ease of use, and detailed modeling of custom components such as heat recovery steam generators, can make a single code inappropriate to meet all the needs. Table 2.1 compares typical considerations and applications for the three user categories. This table

Table 2.1. *Categories of cycle analysis use*

User	Important features	Typical development time
System design	Up-to-date equipment data Modeling flexibility Accuracy	2 days to 6+ weeks
Component design	Ability to customize calculations Ease of use Accuracy	Less than 2 days
Marketing and sales	Ease of use Presentation of quality results and reports Minimal computer requirements Economics	2 hrs

highlights the different uses of cycle analysis tools and suggests different software specification requirements depending on the user's end application of the tool.

The cost of cycle analysis tools can vary significantly, ranging from a one-time purchase price of $500 to an annual lease cost of $20,000 or more per year, with dynamic simulators typically costing more than steady-state simulators. The benefits available from these programs, however, can quickly pay for their costs many times over. Consider using a $10,000 computer program to analyze and optimize a small seventy megawatt plant configuration such that the net efficiency is improved from thirty two to thirty four percent. If fuel costs are $3 per million British thermal units, then the fuel savings would pay for the cost of the program nearly twenty times over within the first year. Thus, cycle analysis tools can be extremely cost-effective, and it is important that power industry engineers understand and take maximum benefit from the recent advances in cycle analysis and thermodynamic modeling technology. Such tools should be considered as part of an organization's overall power system design, analysis, and diagnosis methodology.

The key software capabilities generally sought include

- heat balance and process analysis calculations for multicomponent systems based on a range of chemical and thermodynamic technologies;
- accurate material and operation modeling with robust computational algorithms;
- industry acceptance;
- maintenance, support, and training;
- ease of use, generally including a graphical user interface (GUI);
- the capability to run on a PC or workstation.

Most process analysis tools provide a GUI in which icons representing various unit operations such as heat exchangers, pumps, condensers, and boilers are available to the designer to arrange into a network representing the system of interest. Following analysis of the network, graphical output of network diagram (process flow diagram) and system properties as well as detailed stream analyses for all process flows are generally available. Although most process analysis tools provide graphical representations of the modeled system, the quality of output can vary.

Many times the plant specification that a system designer works to is general, perhaps specifying the plant capacity and technology but leaving significant detail to the designer to work out. For this reason, flexibility in modeling a variety of hardware configurations is important in a cycle analysis tool. For instance, the ability to model subcritical and supercritical cycles, multiple reheat cycles, multiple turbine extractions and feedwater heat streams, dry and wet condenser cooling, various air preheating techniques, variable steam pressure operations and resulting turbine performance, and off-design point conditions all represent the type of modeling flexibility that may be important. Some tools provide nearly unlimited customization capabilities by allowing user-written FORTRAN subroutines to be linked to the program. Others allow the user to develop custom operations based on the existing capabilities of the program. These types of flexibility are important and allow the experienced user to adjust to changing needs and new technologies.

An example of such customizations includes the desire to include existing component performance programs within a system analysis. FORTRAN-based, component design tools are common in the power industry, and such tools are typically based on performance models that have been enhanced and finely tuned based on detailed operating experience. Although many times the incorporation of such detailed component performance models is not feasible or practical because of computing time limitations, simplified models can be included to incorporate the basic performance of such items.

Some applications require the ability to incorporate plant data into the performance models of equipment. In addition, the process of reconciling data to determine system operation is offered by some programs and can be helpful during system checkout activities. Although some codes accommodate user data, this capability is not universal and could represent an important consideration if plant data or monitoring and analyzing plant performance is important.

Engineers involved with detailed component design and performance need the capability to modify detailed aspects of the equipment design. In most cases, this work is done using computer programs written in-house incorporating proprietary correlations and performance factors. Such codes have generally been developed over a number of years and incorporate a great deal of experience. Cycle analysis tools are generally not well suited to replace these detailed component design tools. However, the value that a cycle analysis tool can offer is the ability to integrate the component into the overall system performance (Ong'iro et al. 1995).

Traditionally, in-house component design programs allowed little or no modeling of the total system. The addition of a cycle analysis tool to the component design process can provide benefits such as an increased understanding of the components role in system performance, an appreciation for the need to balance operation of the whole system, and potential alternative uses or design configurations for the component hardware to optimize performance.

Sales engineers typically need to quickly identify, on a comparison basis, the advantages and shortcomings of various power generation options, for example, assessing the performance benefit available from a larger fan or comparing the performance of reciprocating engines with gas turbines. This capability can be useful in discussions with plant operators and can help engineers better understand developers' needs. The software requirements to meet these objectives can be quite different from those for a design engineer.

Ease of use and the ability to produce meaningful results in a short time is vital for such applications. Minimal training should be required to use the software. Complexity typical of a LOTUS spreadsheet is an appropriate target.

The technologies required by sales engineers are similar to those of the system designers; however, the level of detail required in modeling the system is generally not as great. Also, the accuracy of the tool is generally not as important as ease of use and the ability to get rough or ballpark answers. One of the primary benefits such tools offer to the sales engineer is a resource to help examine various customer operating scenarios and to propose alternative operating conditions.

Transient analysis is substantially more complicated than steady-state analysis because of modeling complexity, equipment heat capacities, and control systems (Massey and Paul 1995). In addition, in many cases, the types of tools available for transient analysis are different from those for steady-state calculations. Control system design and analysis tools such as MATLAB focus on the dynamics of practically any system modeled mathematically (Ordys et al. 1994). As with the original steady-state process simulators, the chemical and petroleum industries are at the forefront in applying such tools to control system design, startup/shutdown operation, and control optimization (Naess et al. 1993) (Murray and Taylor 1994).

An example of a more advanced use of process analysis is the on-line optimization of plant operations based on the selection of process control parameters using a value descriptor (Krist et al. 1994). Detailed simulations of an existing plant process are used, together with plant data describing operating conditions, to tune a model to accurately represent the current plant conditions. The tuned simulation is then used to predict control set points that optimize performance relative to the defined value description. This extension of the simulation from agreement with current conditions (a result of tuning) to predict more optimal conditions is based on the assumption that the simulation will predict plant behavior with high accuracy at conditions close to those at which the simulation was tuned. The control set points either may be displayed for operator information or may be directly downloaded to plant control systems.

Because process analysis tools offer such a broad range of benefits, the diversity of people interested in understanding the results of such an analysis is also broad. Effective communication of system analysis results is a point that is commonly overlooked and that is arguably the most critical step in realizing the full benefit of such tools. In many cases decisions based on system analysis results involve business, financial, legal, strategic, and other considerations, and the ability to communicate modeling results and conclusions is key to sound decision making.

The technical focus of most tools has resulted in little development of capabilities for producing flexible model representations and analysis reports. Although some tools allow results to be exchanged through import and export procedures, it remains largely up to the talents and creativeness of the user to produce figures, reports, and summaries that can be used for presentation. The use of stand-alone drawing tools and spreadsheets is common when analysis results are required. This is an area in which improvement in the capabilities of future software tools could enhance the growth of process analysis in the power generation and other industries.

2.4 Software tools

Information on specific cycle analysis software programs is available from a variety of sources including vendor and professional contacts, trade journals, and publications such as *Computer Selects* reviews and *Chemical Engineering Progress' Annual Software Directory* (Simpson, 1995). Industry conferences can also be valuable for exchanging information and modeling experiences.

A relatively new source of information is the World Wide Web. Sites such as Novem (http://dutw239.wbmt.tudelft.nl/PItools/index.html) provide valuable information on cycle analysis tools and opportunities to exchange experiences with other users.

A list of some common, commercially available cycle analysis tools is shown below. The tools shown vary significantly in their application, complexity, and cost. A description of each program is provided to help distinguish defining features and capabilities. Pricing is described using the following identifiers ($ < $1,000; $$ < $10,000; $$$ > $10,000).

EndResult ($): Sega, Inc., (800) 531-1950 EndResult is a package of spreadsheet addins, analysis tools, and example spreadsheet problems. It is available in formats for Lotus and Excel. Target applications include boilers, gas and steam turbines, and condensers. The package allows component heat performance predictions as well as evaluations of entire plant energy cycle performance. Specific tools include calculation of boiler efficiency and air heater performance, mixed gas thermophysical properties, mixed fuel ultimate analysis, thermocouple and

RTD tables, and boiler drum and condenser level calibrations. EndResult can operate under DOS 2.1 or higher with a minimum of 640 kilobytes of RAM.

Steambal/energy analyst ($): Thermal Analysis Systems Co., (708) 439-5429 Steambal and the Energy Analyst are BASIC programs that allow the analysis of industrial steam power cycles by performing mass and energy balances to find heat required and power produced. Steambal consists of two types of module subroutine, one that describes equipment and another used to describe stream properties. The user creates a cycle model by connecting a series of equipment module subroutines, which automatically call the property subroutines. The Energy Analyst allows calculation of insulation economics, combustion analysis, pipe network flow rates, steam heater and steam tube performance, gas turbine performance, and others.

Heatnet: National Engineering Laboratory, UK, (44) 0 1355-272143 Heatnet is designed for ease of use, targeting both the specialist and nonspecialist engineer. The program features pinch analysis, assessment of minimum total capital cost, and optimization. Versions are available for both PCs and Unix workstations. PC versions require a 386 processor and coprocessor and at least four megabytes of memory.

Boiler ($): Joseph D. Barnes, (814) 724-4615 Boiler is one of a package of power system tools that can be run on a 286 PC using DOS versions 2.0 or better. It can be used to compute heat balances for fuel-fired steam boilers and can be used to predict boiler heat losses, efficiency, and combustion products.

Design II: WinSim Inc., (713) 414-6711 Design II was developed for those in the petrochemical, chemical, and refinery industries. It allows heat and material balances for a variety of processes and includes built-in unit operations such as chemical and equilibrium reactors, heat exchangers, fired heaters, pumps, compressors, and turbine expanders. An extensive set of thermodynamic property correlations is available focused on the hydrocarbon and chemical applications.

F-Cycle, C-Cycle, I-Cycle ($$$): Encotech, Inc., (518) 374-0924 These heat balance programs provide the capability to analyze conventional fossil-type units, gas turbine combined cycle units, combined fossil and gas turbine units, and industrial steam cycles employing multiple boilers or turbines. F-Cycle can analyze fossil units with a fired boiler and zero, one, or two reheats. It includes steam turbine performance models, boiler and air heater system models, cooling system and pump calculations, and generator performance models. C-Cycle can be used to analyze gas turbine combined cycle units. It calculates complete heat balances for the gas turbine, HRSG, steam turbine, cooling system, and feed-water systems. A library of gas turbine engines is included. The programs will run on a 286 PC or higher. A math coprocessor is not required but is recommended.

GTPro and GTMaster ($$$): Thermoflow Inc., (617) 237-5573 GTPro provides the ability to perform heat balances and analyze gas turbine combined-cycle systems based on preliminary hardware descriptions and thermodynamic inputs. The program includes a database of more than fifty gas turbine engines. Steam systems can range from simple heat recovery boilers to multipressure steam-injected systems and combined cycles with up to four process steam extractions. GTMaster allows calculation of system performance under varying off-design conditions. An available economic analysis module allows calculation of cash flow for the life of a project. Computer requirements include DOS 3.0 or better, with 640 kilobytes of RAM, a math coprocessor, and a VGA monitor. Thermoflow Inc. also has programs for reciprocating engines (RECIPRO) and advanced gas turbines (GASCAN+).

SteamPro and SteamMaster ($$$): Thermoflow Inc., (617) 237-5573 SteamPro and SteamMaster use an approach similar to that of GTPro and GTMaster for analyzing and designing conventional steam power plants with or without cogeneration. Subcritical or supercritical steam cycles with up to two reheats can be modeled. Up to thirteen steam streams can be extracted from the turbine, and the program allows up to ten stages of feed-water heating. The program includes a preliminary boiler design model. SteamMaster computes cycle performance under varying off-design conditions. Computer requirements include DOS 3.0 or better, with 640 kilobytes of RAM, a math coprocessor, and a VGA monitor.

Ex-Site ($$): Energetic Systems, Inc., (510) 233-0963 Ex-Site enables the user to model fossil-fired and nuclear steam power plants. The program can be used to predict steam extraction conditions as well as HP, IP, and LP steam turbine efficiencies. The program includes flow stratification effects and can model degraded turbine nozzles using a flow area correction factor. A DOS operating system is required.

MAX ($$$): Aspen Technology, Inc., (617) 557-0100 MAX is a graphical process simulation system designed for chemicals and petrochemicals. It can be used to support calculations of stream flashes, mass and energy balances, column rating, and property calculations for a wide range of chemical systems. Unit operations include heat exchangers, pumps, single and multistage compressors and expanders, equilibrium, yield-based and stoichiometric reactors, and equipment heating and cooling curves. MAX supports customization through in-line FORTRAN for user models and stream reports and is upwardly compatible with the company's ASPEN PLUS process simulator.

HYSIM ($$$): Hyprotech, Ltd., (800) 661-8696 HYSIM is an interactive, graphical flow sheet for simulating plant design and process evaluations in the gas processing, refining, petrochemical, and chemical industries. The program includes operations for heat exchangers, compressors, two- and three-phase separators, valves, and various reactors. It also has the capability to perform pinch analysis. Customization is available through the use of an internal programming language or through linking with external FORTRAN or C modules. DOS, VMS, or Macintosh platforms are supported, with a minimum four megabytes of RAM required.

PEPSE ($$$): NUS Corp., (208) 529-1000 PEPSE is a graphical heat balance program for predicting thermal power plant and heat recovery steam cycle performance. Component models allow detailed analysis of all major and minor streams including sealing stream leakages and moisture loss. PEPSE is particularly suited for maintenance of existing steam systems, allowing the user to input plant test data. Calculation models include steam and combustion turbines, feedwater heaters, condensers, heat exchangers, and fluid bed combustors.

GateCycle ($$$): Enter Software Inc., (415) 322-6610 GateCycle is a graphically based heat balance program designed to predict both design-point and off-design performance of various gas turbine-derived systems including combined cycles, cogeneration systems, and combined heat-and-power plants. The program is based on a library of 120 hardware performance models including models for a fossil boiler and steam drum that complement the original gas turbine combined cycle plant models. GateCycle includes a library of more than 100 specific gas turbine models representing various manufacturers' equipment. GateCycle was developed in conjunction with the Electric Power Research Institute (EPRI) and runs under the DOS operating system, requiring 550 kilobytes of memory and 10 megabytes of disk space.

ASPEN PLUS ($$$): Aspen Technology, Inc., (617) 577-0100 ASPEN PLUS is a general-purpose, graphically based process simulation system. Although intended primarily for the chemical and petrochemical industries, it can also be used to analyze power generation systems. Among the features of ASPEN

PLUS are a large set of physical property models, with the capability to handle solids directly. SteamSys, an integrated add-in module to ASPEN PLUS, provides detailed steam turbine performance modeling. In addition to performance modeling, financial analysis including estimates of capital, operating costs, and profitability are included. ASPEN PLUS supports in-line FORTRAN for customization of user models and user stream reports. It also interfaces to AutoCAD and HTFS and HTRI programs. Versions of ASPEN PLUS are available for most computing platforms.

SOAPP (State-of-the-Art Power Plant): EPRI, (510) 943-4212 SOAPP has been developed in conjunction with the Electric Power Research Institute (EPRI) and is intended to provide a comprehensive information base for analyzing, designing, and costing key power generation technologies. Originally envisioned to include modules for gasification combined cycles, gas turbine combined cycles, circulating fluidized beds, and fossil steam systems, the complete set of capabilities is still being developed. SOAPP consists of technology modules, which provide stand-alone technology information, and the SOAPP Workstation, which accesses and integrates the various modules. The program is extensive in scope, providing significantly more than simple system simulation capability; however, currently available power generation technologies are limited.

ChemCAD III ($$$): Chemstations, Inc., 800-CHEMCAD ChemCAD III is a general-purpose process simulation program. A library of equipment includes heat exchangers, pumps, compressors, heaters, and expander/turbines, and users can add additional custom operations. ChemCAD features solids handling and interfaces with AutoCAD. It runs under DOS and requires 580 kilobytes of memory.

PD-PLUS ($$): Deerhaven Technical Software, (617) 229-2541 PD-PLUS was developed for steady-state flow sheet calculations related to the chemical, petrochemical, and refinery industries. A 32-bit version runs on PCs with a 386 processor and at least 1.3 megabytes of extended memory. A sixteen-bit version is available for older machines. PD-PLUS uses a keyboard input for model generation. Unit operations are available for heat exchangers, compressors, and pumps as well as for other industry specific operations. The source code is available in FORTRAN for special applications.

EnviroPro designer ($$): Intelligen, Inc., (201) 622-1212 EnviroPro is a process simulator designed for the environmental industry. Models include wastewater treatment and other pollution control processes. SuperPro extends EnviroPro's capabilities through additional manufacturing process unit operation models. Both feature a graphical user interface and run on PCs under Windows version 3.1. At least four megabytes of memory and eight megabytes of disk space are required.

2.5 Transient simulators

SPEEDUP($$$): Aspen Technology, Inc., (617) 577-0100 SPEEDUP is a general-purpose dynamic simulation package used in the chemical, petrochemical, refining, gas-processing, power, and food industries. It shares a common physical property estimation system with ASPEN PLUS. The program is equation-based and uses keyboard input to build system models. Extensive physical property calculation methods are available, and built-in models are available for common unit operations and control elements. SPEEDUP is available for PCs and workstations.

D-SPICE: Fantoft UK Limited, (44) 01705-231881 D-SPICE is a graphically based dynamic simulation tool for use in analyzing transient process plant operation. Best suited for gas process applications, it provides links for thermodynamic property calculation and can be customized with user-written FORTRAN modules. DSPICE runs on Unix workstations.

SACDA MASSBAL ($$$): SACDA Inc. (519) 640-6557 MASSBAL is a steady-state and dynamic simulation system designed for use in evaluating new or existing industrial processes. It features a graphically based process flow-sheeting program called CADSIM, which can be used to build system models. Process optimization, equipment sizing, and data reconciliation capabilities are provided. Training courses are available.

PC TRAX ($$$): Trax Corp. PC TRAX is a transient system simulator designed for the fossil power generation industry. Unit operations available include pulverized coal boilers, gas turbines, and balance of plant components.

Modular Modeling System ($$$): Framatone Technologies. Modular Modeling System (MMS) provides a library of more than 180 modules for transient system simulation. Unit operations are available for fossil boilers, and balance of plant components. MMS runs on both personal computers and workstations, and requires 4 megabytes of memory, and 8 megabytes of disk space.

HYSYS ($$): Hyprotech, Ltd., (800) 661-8696 HYSYS provides integrated steady-state and dynamic modeling through a graphical user interface. It is best suited for the petroleum industry but can be used for simulating some power generation systems. HYSYS features a multivariable optimization for process design, integrated dynamic modeling for control strategy development, and interfacing with digital control system software. HYSYS runs on a PC under MS Windows.

2.6 Other software tools

TAG Supply: EPRI, (415) 855-2722 TAG Supply incorporates the financial analysis methodologies of the *EPRI Technical Assessment Guide*. The methodology allows financial analysis of various power generation technologies including fossil boilers, integrated gasification combined cycles, gas turbine combined cycles, and others. TAG Supply features technology-specific cost figures that can be used to compare different technologies on a common basis. TAG Supply is currently available only to EPRI members.

2.7 Conclusions

A systems approach, based on computer-aided process design and analysis, provides an ideal means for evaluating power generation options. Within the range of power generation application, three categories were identified that represented common needs. The categories – system design, component design, and marketing and sales – each had different software requirements that would be best met by examining system analysis tools based on the individual requirements of the types of calculations to be performed.

Power industry applications include a wide range of components and systems from FGD equipment to pulverized coal power plants, GTCC systems, and IGCC systems. Modeling flexibility is generally important so that a wide range of technologies can be considered. In addition, robust calculation algorithms, flexibility in methods for calculating process fluid physical properties, and flexibility to customize unit operations are important features. Ease of use and vendor support are also important considerations.

Detailed design methods that have been developed on the basis of years of experience are unlikely to be replaced by commercial analysis codes; however, these design codes may be supplemented by cycle analysis tools to provide an integrated systems approach to component design. In addition, a systems approach to product design offers the additional benefit of allowing the product designer to better understand the trade-offs and available options.

Cycle analysis tools can be a great benefit to those who provide products and services to the power industry. The recent availability of substantial modeling power on laptop computers and workstations, combined with the increasing need within the power industry to achieve high efficiency and low cost, provides an ideal situation for application of cycle analysis tools to a variety of operations within the power industry.

References

Bartlett, R. (1958). *Steam Turbine Performance and Economics*. New York, McGraw-Hill. University Microfilms International, Out-of-Print Books, Ann Arbor, MI.

Boston, J., Mathias, P. & Watanasiri, S. (1984). Effective utilization of equations of state for thermodynamic properties in process simulation, *AIChE Journal*, **30**, 182–186.

deMello, F. P. & Ahner, D. J. (1994). Dynamic models for combined cycle plants in power system studies. *IEEE Transactions on Power Systems*, **9**, pp. 1698–1708.

Elliott, T. C. (1989). *Standard Handbook of Powerplant Engineering*. New York, McGraw-Hill.

Elliott, T. C. (1994). Use of power simulators goes beyond training. *Power*, March, pp. 32–44.

Glasscock, D. A. & Hale, J. C. (1994). Process simulation: the art and science of modeling. *Chemical Engineering*, November, Vol. 101, No. 11, pp. 82–89.

Hilaly, A. K. & Sikdar, S. K. (1996). Process simulation tools for pollution prevention. *Chemical Engineering*, February, Vol. 103, No. 2, pp. 98–104.

Jain, Padgaonkar, Kessler, Gloski, & Kozlik (1989). Plant thermal analysis and data trending using THERMAC. 1989 EPRI Heat Rate Improvement Conference. Electric Power Research Institute (EPRI), Palo Alto, CA. EPRI-GS-6989, pp. 4B.1–4B.15.

Kettenacker, D. (1988). Use of an energy balance computer program in the plant life cycle. 1988 EPRI Heat Rate Improvement Conference. Electric Power Research Institute (EPRI), Palo Alto, CA. EPRI-GS-6635, pp. 6B.41–6B.53.

Kettenacker, D. & Hill, D. (1992). Verifying and predicting boiler performance for various low sulphur coal blends using an energy balance computer program. EPRI 1992 Heat Rate Improvement Conference. Electric Power Research Institute (EPRI), Palo Alto, CA. EPRI-TR-102098, pp. 35.1–35.12.

Kisacky, R., Travers, T., Tripp, L., Wong, F. & Diaz-Tous, I. (1992). The computerized performance test and diagnostic program. EPRI 1992 Heat Rate Improvement Conference. Electric Power Research Institute, (EPRI) Palo Alto, CA. EPRI-TR-102098, pp. 47.1–47.13.

Korakianitis, T. & Wilson, D. G. (1994). Models for predicting the performance of Brayton-cycle engines. *Journal of Engineering for Gas Turbines and Power*, Vol. 116, No. 2, April, pp. 381–388.

Krist, J. H., Lapere, M. R., Wassink, S., Groot, Neyts, R. & Koolen, J. L. A. (1994). Generic system for on-line optimization and the implementation in a benzene plant. *Computers in Chemical Engineering*, **18**, N Suppl., pp. S517–S524.

Lie, A. B. K. (1994). Exergy analysis: calculating exergies with ASPEN PLUS. Presented at 1994 ASPENWORLD Conference, Boston, MA.

Linnhoff, B. & Alanis, F. J. (1989). A system's approach based on pinch technology to commerical power station design. Presented at the ASME Winter Annual Meeting.

Massey, N. D. & Paul, B. O. (1995). Solving dynamic problems on a steady–state simulator. *Chemical Processing*, Vol. 58, No. 3, March, pp. 74–77.

Moore, T. (1995). Desktop design for advanced power plants, *EPRI Jounal*, March/April, pp. 24–30.

Murray, S. & Taylor, B. (1994). Integrating dynamic simulation into process engineering applications using PROMACE. Presented at 1994 ASPENWORLD Conference, Boston, MA.

Naess, L., Mjaavatten, A. & Li, J. O. (1993). Using dynamic process simulation from conception to normal operation of process plants. *Computers in Chemical Engineering*, Vol. 17, Nos. 5–6, May–June, pp. 585–600.

Narula, R. G. (1995). Salient design considerations for an ideal combined cycle power plant. *Heat Recovery Systems & CHP*, **15**, 97–104.

Ong'iro, A. O., Ugursal, V. I., Al Taweel, A. M. & Blamire, D. K. 1995. Simulation of combined cycle power plants using the ASPEN PLUS shell. *Heat Recovery Systems & CHP*, **15**, 105–113.

Ordys, A. W., Pike, A. W., Johnson, M. A., Katebi, R. M. & Grimble, M. J. (1994). *Modeling and Simulation of Power Generation Plants*. New York, Springer–Verlag.

Parker, S. & Hall, S. (1994). Pinch technology: process analysis and design. Presented at 1994 ASPENWORLD Conference, Boston, MA.

Paul, B. O. (1994). Software designs complex processes. *Chemical Processing*, January.

Simpson, K. L. (1995). *1996 Chemical Engineering Progress Software Directory*. December, p. 74.

Spencer, R. C., Cotton, K. C. & Cannon, C. N. (1963). A method for predicting the performance of steam turbine–generators. *ASME: Journal of Engineering for Power*, October.

Tsatsaronis, G., Lin, D., Pisa, S. & Tawfik A. (1992). Improvement of heat rate and cost of electricity in IGCC power plants. EPRI 1992 Heat Rate Improvement Conference. Electric Power Research Institute (EPRI), Palo Alto, CA. EPRI-TR-102098, pp. 44.1–44.22.

Tsatsaronis, G. & Lin, D. (1993). Cost optimization of an advanced concept for generating electric power. American Chemical Society. *Proceedings of the 28th Intersociety Energy Conversion Engineering Conference*, Vol. 2, pp. 299–2.104.

Wilbur, L. C. (1985). *Handbook of Energy Systems Engineering*. New York: John Wiley & Sons, Inc.

Wozny, G. & Jeromin, L. (1994). Dynamic process simulation in industry. *International Chemical Engineering*, **34**, 159–177.

3

Automated design of chemical process plants

RUDOLPHE L. MOTARD

Washington University

3.1 Principal activities in design

With its roots in research and development, the design of chemical process plants is prompted by new chemical science, expanding or evolving markets for chemically derived products or intermediates, or a need to retrofit existing facilities. Retrofitting may be influenced by environmental regulation or simply a need to make plants more energy-efficient.

The steps involved in the life-cycle ownership and stewardship of a manufacturing facility are shown in Table 3.1.

A more formal representation of the activities briefly summarized in Table 3.1 has recently been prepared at the University of Missouri-Rolla in conjunction with the American Institute of Chemical Engineers, Process Data Exchange Institute (PDXI) (Book et al., 1994). An activity model based on the US Air Force (1981) ICAM Architecture (IDEF0) was completed for the front-end design activities, commonly referred to as the process engineering phase of process design. The objective of PDXI has been the design of a data exchange protocol for the sharing of data among the various organizations and groups involved in the engineering of chemical plants.

3.1.1 An activity model for process engineering

The top level view of the activity model is shown in Fig. 3.1. The entities that enter from the left of the diagram indicate the materials and intellectual property that will be employed in the design and operation of the process. Those that exit from the right are the products, including the useful materials for consumption and the material residues of the process operations and, most important, the plant life-cycle documentation that will need to be maintained at least through the decommissioning of the plant. The entities that enter from the top are the constraints supporting the activity of process design, such as the fundamental

Table 3.1. *Life-cycle design of a chemical manufacturing facility*

Process synthesis
Research and development
Process alternatives
Socioeconomic comparisons
Process design
Models and performance data
Material and energy balances
Designing and rating process
Equipment
Design analysis and cost optimization
Process engineering analysis
Hazards analysis
Control and safety systems
Startup and acceptance procedures
Documentation
Maintenance and operations
Decommissioning

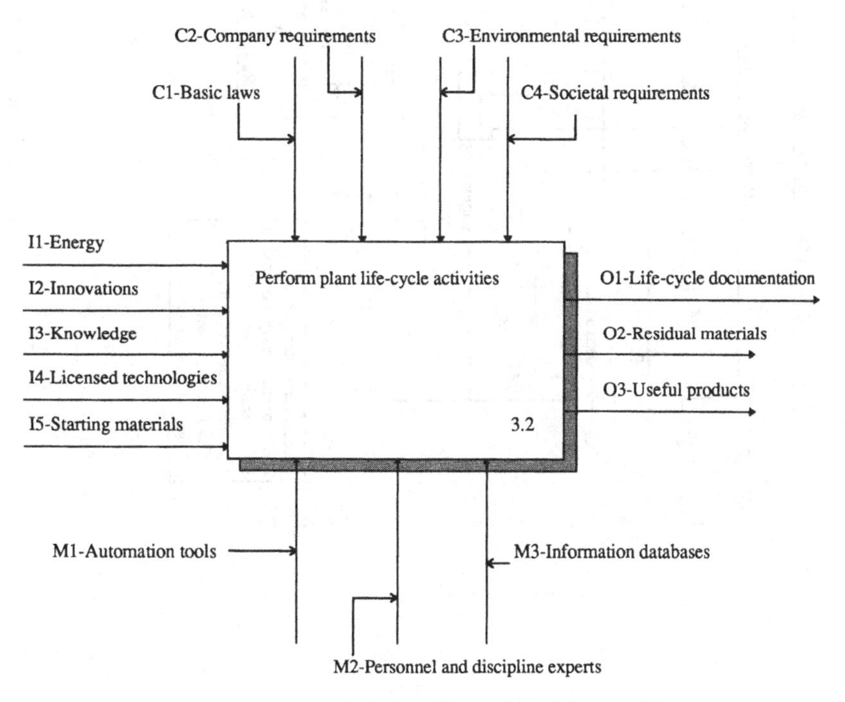

Fig. 3.1. Top level of plant life-cycle activity model.

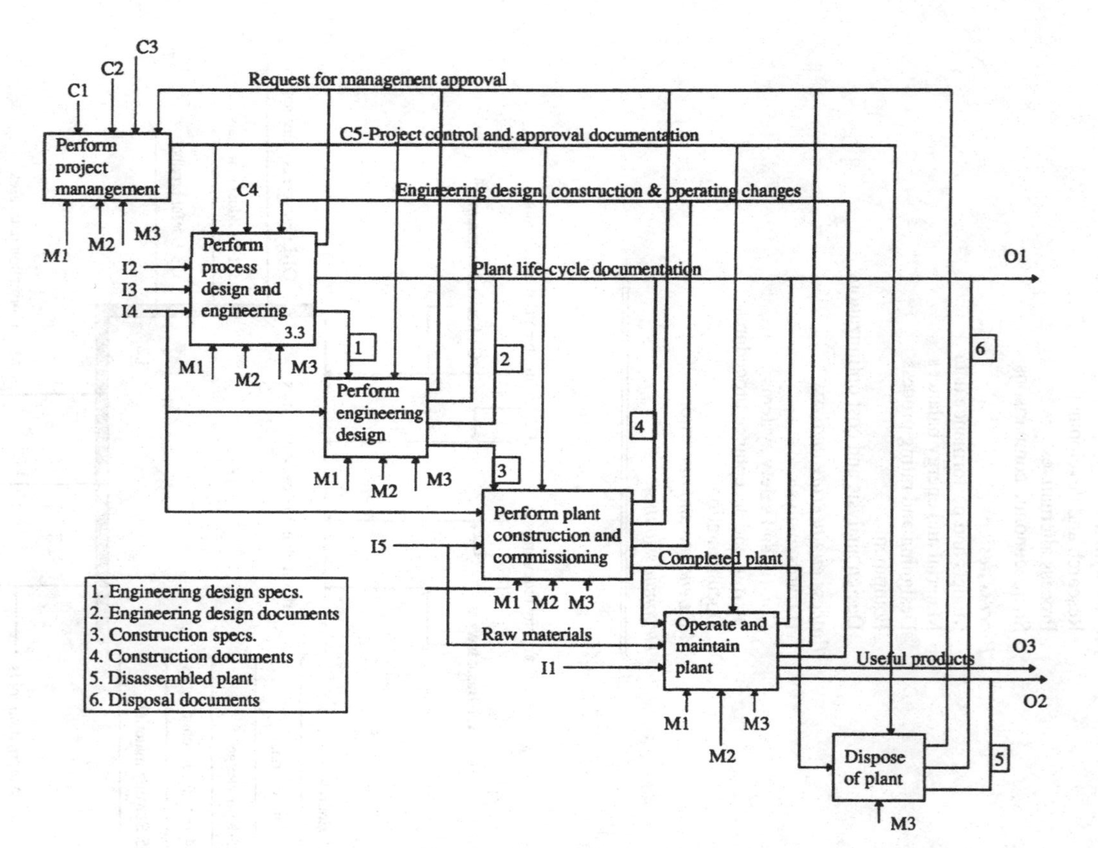

Fig. 3.2. Perform plant life-cycle activities.

Table 3.2. *Steps in PDXI activity model*

Perform project management
Define project procedures and guidelines
Manage project
Obtain company management approval
Perform process design and engineering
Synthesize process
Design process
Analyze process engineering
Perform engineering design
Define modules
Define module systems
Procure equipment
Generate construction specifications
Perform plant construction and commissioning
Devise manufacturing plan
Design special tools, jigs, and fixtures
Select manufacturing sites
Manufacture and construct plant
Perform final assembly
Operate and maintain plant
Dispose of plant

laws of nature and the requirements placed on the designer by the company, the regulatory agencies, and society in general. Finally, those that enter from the bottom are the resources to be used in design, such as automation tools, databases, and expert personnel. The major documents to be managed are the contractual documents, drawings, and design reports; the regulatory compliance information; and any operating and performance data that support the operation of the plant during its lifetime or outlive it.

Within the box shown in Fig. 3.1 are the six major life-cycle activities described in Fig. 3.2 using the IDEF0 methodology. Again the inputs and outputs are arranged clockwise from the left, inputs, constraints, outputs or products, and resources. The principal activities have been reorganized or renamed to reaggregate some of the substeps shown in Table 3.1, but the major steps are essentially the same, the terms being more action-oriented. In Table 3.2 the boxes shown in Fig. 3.2 are expanded to show the relationship with Table 3.1.

The distinction must be made between the chemical engineer's notion of process engineering and engineering design, or project engineering. Most of the activities in process engineering encompass the materials in the process and the operating conditions of temperature, pressure, and composition. Flow rates and volumes are also considered in this phase along with the overall material

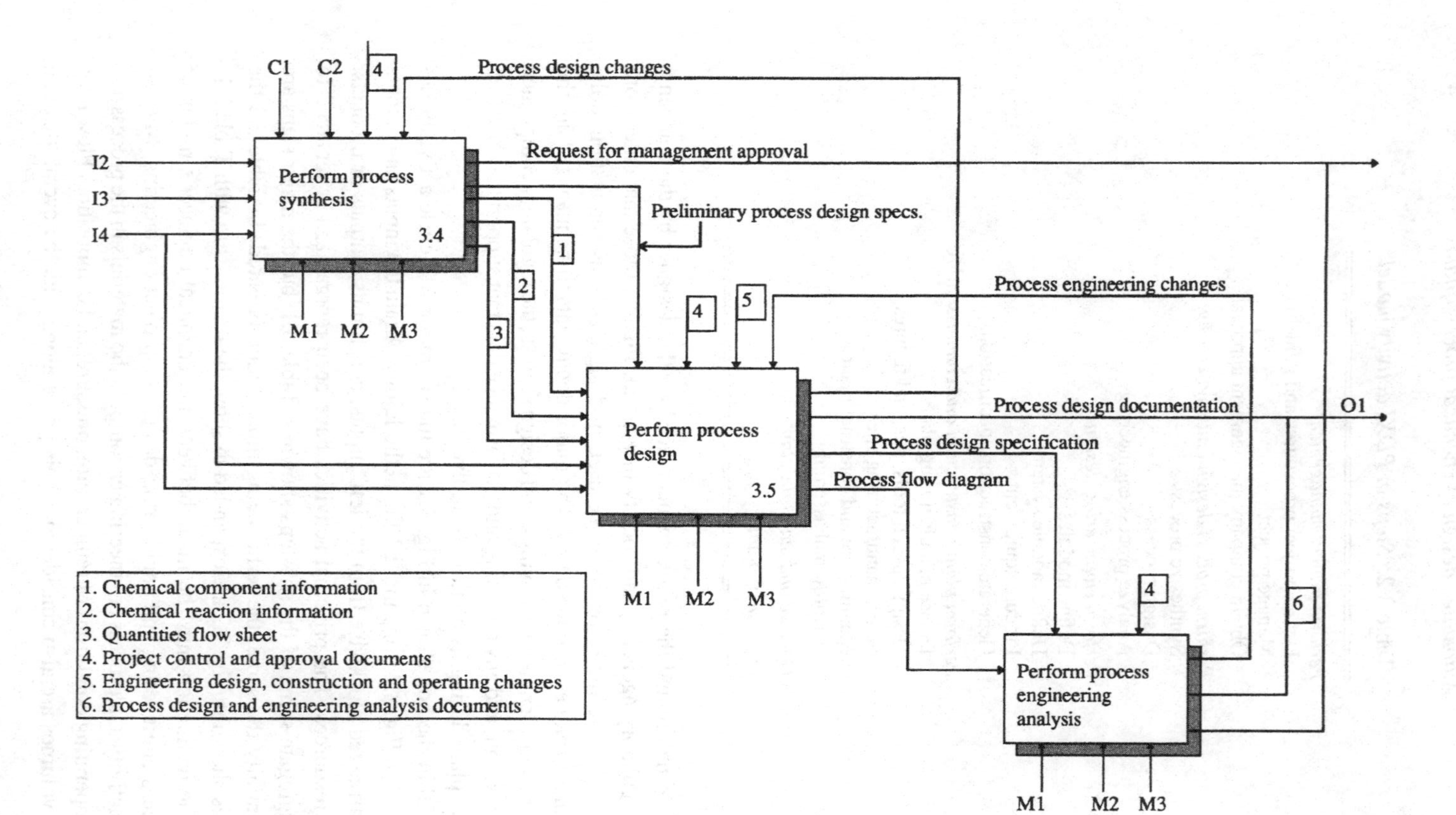
C1
C2
4
Process design changes
I2
I3
I4
Perform process synthesis
3.4
Request for management approval
Preliminary process design specs.
1
2
3
M1
M2
M3
4
5
Process engineering changes
Perform process design
3.5
Process design documentation
O1
Process design specification
Process flow diagram
M1
M2
M3
4
6
Perform process engineering analysis
M1
M2
M3
1. Chemical component information
2. Chemical reaction information
3. Quantities flow sheet
4. Project control and approval documents
5. Engineering design, construction and operating changes
6. Process design and engineering analysis documents

Fig. 3.3. Perform process design and engineering.

and energy balances. All of the latter are of primary interest to a chemical engineer or process engineer. The rest of the design activities bring in other disciplines associated with the components that hold the materials in process, such as pipes, valves, vessels, pumps, compressors, heat exchangers, towers, and reactors along with the structures that support the hardware and the utility systems that supply the plant.

3.1.2 Process synthesis

Figure 3.3 is the activity model for process design and engineering, and Fig. 3.4 is the model for process synthesis, the front end of a design project. The foundation of process development is most certainly the scientific base of chemistry and physics and their application to physical and chemical transformations. Experience and existing technology are also important facets in the creation of new manufacturing schemes. Given a problem statement or needs analysis, process synthesis proceeds in an organized way through three phases: representation, strategy, and evaluation. We will return to these phases later.

In what follows we will concern ourselves mainly with continuous processes, which are by far the largest component of modern chemical industry. What is an industrial chemical process? It is a conglomerate of vessels, pipes, valves, pumps, compressors, heat exchangers, furnaces, chemical reactors, and many accessories for control and surveillance that effect changes in the physical and chemical characteristics of the materials being processed. The major unit operations are species separation; stream mixing and division; raising or lowering of pressure with pumps, compressors, and valves; raising or lowering of temperature through heat exchange, cooling, or heating; and changing chemical structure through chemical reactors. We will not be discussing the synthesis of reaction paths and reaction systems because it is the most difficult and least developed of the organized techniques of process synthesis. We assume that the process chemistry has been developed in the laboratory before the process engineer is asked to invent a process.

The major features of a process then will be the energy management hardware, the separation sequences, and the integration of these two major aspects of chemical processing within the context of a total flow sheet. The product of this phase is a process flow sheet, still lacking many of the mechanical details of the plant.

3.1.3 Process design

Figure 3.5 is the activity model for process design introduced in Fig. 3.3. After process synthesis, this phase is the heart of process engineering. The first two

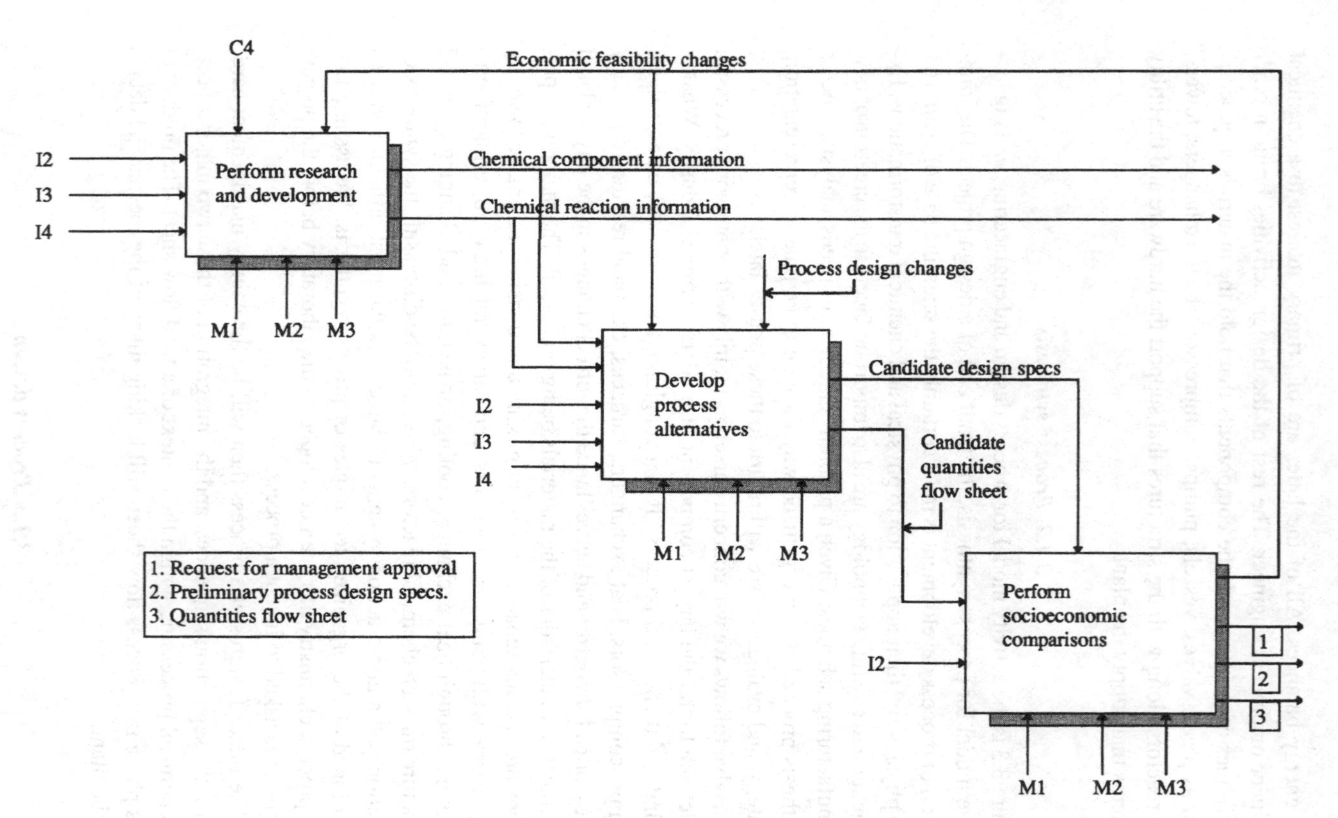
C4
Economic feasibility changes
I2
I3
I4
Perform research and development
Chemical component information
Chemical reaction information
M1
M2
M3
Process design changes
Develop process alternatives
Candidate design specs
Candidate quantities flow sheet
Perform socioeconomic comparisons
1
2
3
1. Request for management approval
2. Preliminary process design specs.
3. Quantities flow sheet

Fig. 3.4. Perform process synthesis.

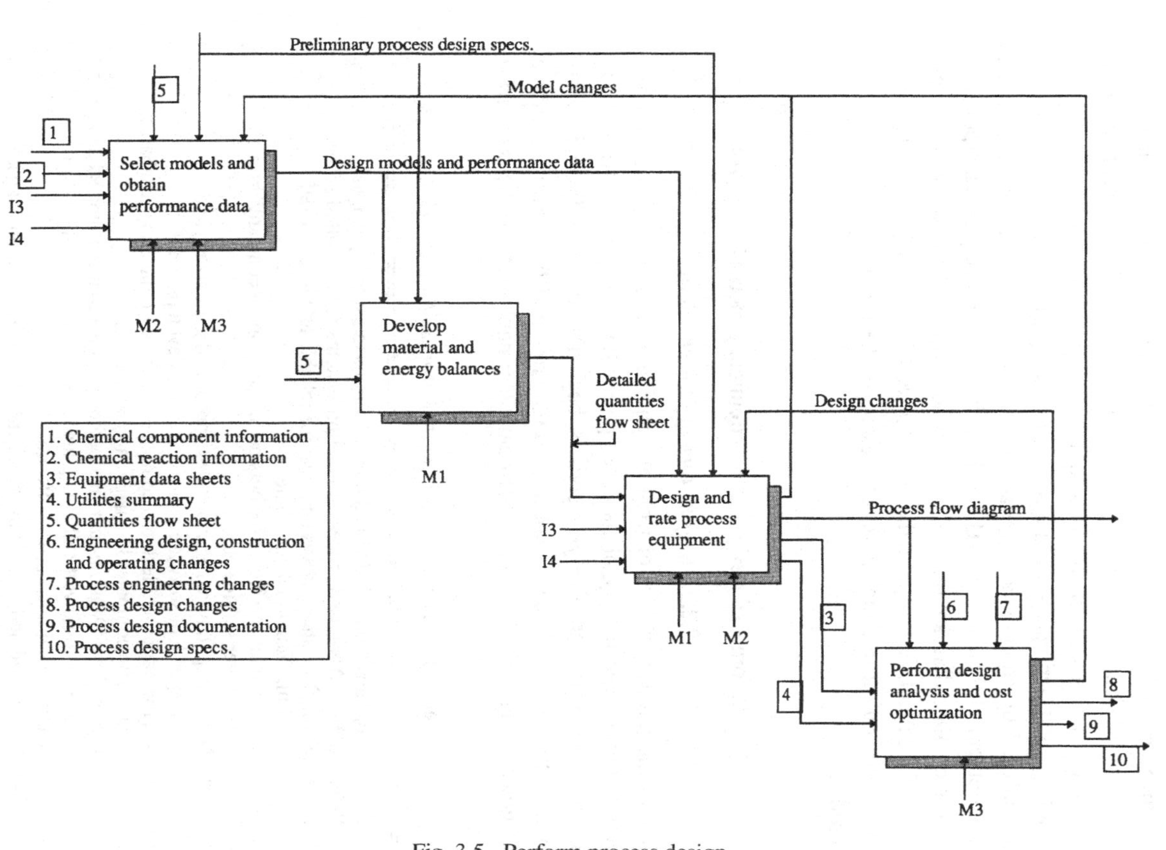
Preliminary process design specs.
Model changes
1
2
I3
I4
5
Select models and obtain performance data
M2
M3
Design models and performance data
Develop material and energy balances
5
M1
Detailed quantities flow sheet
Design changes
Design and rate process equipment
I3
I4
M1
M2
Process flow diagram
3
4
6
7
Perform design analysis and cost optimization
M3
8
9
10
1. Chemical component information
2. Chemical reaction information
3. Equipment data sheets
4. Utilities summary
5. Quantities flow sheet
6. Engineering design, construction and operating changes
7. Process engineering changes
8. Process design changes
9. Process design documentation
10. Process design specs.

Fig. 3.5. Perform process design.

steps in process design revolve around automated flow-sheeting systems, which have been under development for twenty five years. From the detailed quantities flow sheet, including all temperature, pressure, and composition variables, one can then proceed to rate-, cost-, and size-specific equipment. The latter may in fact be handed off to companies specializing in the design technology for certain classes of equipment, as would certainly be true for heat exchanger detailing and fractionation columns. Pumps and compressors, however, would be analyzed by equipment manufacturers for the privilege of bidding on the procurement of their product. The end result of process design is a piping and instrumentation diagram that adds a lot of accessories to the major equipment items of the plant. At this point the project is ready for profitability analysis and cost optimization in anticipation of presenting the proposal for management approval.

3.2 Automated design in the chemical process industries

3.2.1 Brief history of flow sheeting

One of the major activities of front-end engineering is process simulation or flow sheeting. The history of its development for computer analysis breaks down conveniently into five-year periods, beginning in 1955. Table 3.3 summarizes these stages. One cannot discuss the growth of flow sheeting without acknowledging the contribution that computers have made to the supporting

Table 3.3. *History of automated process flow sheeting*

1955–59	Programs are for individual unit operations in assembly language.
1960–64	FORTRAN-based process simulation systems are of limited scope.
1965–69	Some flow-sheeting systems become reliable and widely used within individual companies.
1970–74	Commercial services grow outside the developing company through computer service bureaus.
1975–79	New computational architectures appear and markets expand. Independent software houses take over development.
1980–84	Microcomputing, desk-top computer graphics, engineering databases, and new computer languages appear and grow.
1985–89	Personal workstations and distributed systems grow. Automated design tools become vertically integrated. Expert system facilities are introduced into design software.
1990–94	Object-oriented paradigms invade the design desk. Graphical user interfaces proliferate. Object-oriented database support is more powerful. Influence of AI techniques grows.

infrastructure of process design. Notable has been the growth of databases and models for computing the thermophysical and transport properties of substances and their mixtures.

3.2.2 General capabilities of automated flow-sheeting systems

As indicated in Table 3.3, automated tools for general process engineering application became available in 1970. A number of large petroleum refining and petrochemical companies had developed in-house systems that matured and then disappeared when the cost of maintenance became prohibitive in the 1980s. Practically all process simulation today is done on vendor-developed software, although the genesis of such software owes much to the experience and earlier investments of time and talent by private operating companies.

The unit operations concept in chemical engineering has been a boon to the profession since its invention in the 1920s. It has also enabled a quick start on the design of software of great flexibility in the simulation of complex flow sheets. By 1970 the first so-called sequential modular flow-sheeting systems were on the market (FLOWTRAN 1970) (DESIGN 1970) built around the unit operations concept. Tracing the development from 1970 would be tedious because estimates show that at various times at least 100 process simulation packages have been developed at universities, research institutes, operating companies, and software houses. Nevertheless, at the Foundations of Computer-Aided Process Design (FOCAPD) conferences the theme in 1980 (Evans) was primarily the status and prospects for the then maturing sequential modular systems. At about the same time, Westerberg et al. (1979) published a monograph outlining the essential elements of process flow sheeting. Three years later the emphasis at the second FOCAPD had moved to engineering databases (Winter and Angus 1983) and equation-based flow-sheeting systems (Perkins, 1983). The latter are only now reaching the robustness necessary for widespread commercial use.

By 1989 the third FOCAPD (1990) took up new challenges in artificial intelligence applied to process synthesis, new mathematical tools, and new computer environments. The proliferation of computer-generated graphics attested to the explosive growth of personal computing. The FOCAPD series is a rich source of bibliographies on CAD in process design.

The latest (1994) *Chemical Engineering Progress Software Directory* lists many large and small purveyors of flow-sheet simulation software. Each has a particular specialization, whether it be refineries, natural gas plants, mineral and metallurgical operations, pulp and paper plants, bioprocessing, or another specialization. Some are primarily mainframe packages, and others are compact enough to fit PCs. The most critical aspects of flow-sheet simulation are the robustness of the computations in the context of difficult phase behavior, complex reaction-phase equilibrium problems (for example, reactive distillation) and the convergence of flow-sheet level recycle calculations. Substantial progress has been made in all areas. Growing precision in the description of the physical, equilibrium, and thermal properties of mixtures has contributed to increased confidence in the validity of computer-generated flow-sheet analysis.

More recently, the development of graphic front ends and links to spreadsheeting, databases, and drafting programs have helped to integrate flow-sheet simulation into the engineer's daily routine. Rapid progress in the theoretical basis for solving large systems of nonlinear equations and differential-algebraic models is allowing easy transitions to global optimization and dynamic modeling of complex plants.

3.2.3 Flow sheeting example

In sequential flow-sheet simulation, the very nonlinear nature of chemical process calculations made it necessary to solve the connected unit operations one at a time. Moving output stream descriptions from one unit to another, the output streams would become input stream descriptions. If materials were recycled the information structure became cyclic and the sequential modular approach could become time-consuming unless attenuated by a convergence acceleration algorithm packaged with the process model. An example will serve to illustrate the flavor and potential of flow sheeting.

The American Institute of Chemical Engineers publishes a contest problem in process design each year. The 1985 problem involved the production of styrene monomer from toluene and methanol. This approach was fairly innovative because it avoided two major problems in conventional styrene manufacture. Eliminating benzene as a feedstock avoided the risk of human exposure to one of the most dangerous carcinogenic chemicals. Because styrene in the proposed process is produced in one step rather than two, the very high energy cost associated with the dehydrogenation of ethylbenzene, namely the high volume of dilution steam, was also avoided. The reaction scheme is as follows:

Toluene + methanol $\leftrightarrow$ styrene + water + hydrogen (primary reaction)

Toluene + methanol $\leftrightarrow$ ethylbenzene + water (secondary reaction)

The reaction has already been studied in the laboratory, and a typical reactor condition is

Inlet temperature	525°C
Inlet pressure	400 kPa absolute
Conversion	0.82 mols of toluene reacted per mol feed
Yield	0.72 mols of styrene per mol of toluene reacted

For preliminary analysis, the pressure drop estimates in Table 3.4 may be assumed. The anticipated production of styrene is 300,000 metric tons per year, with no more than 300 parts per million (PPM) of ethylbenzene. The onstream time is estimated at 95% (8,320 hours per year). Ethylbenzene will not be recycled, although this possibility should be investigated in the laboratory. A metric ton is 1,000 kilograms, and the molecular weight of styrene is 104. Putting all quantities on an hourly basis will require the production of $(300{,}000 \times 1{,}000)/(8320 \times 104) = 346.71$ kilogram mols per hour of styrene.

The yield of styrene is 0.72 mols per mol of toluene reacted; hence, the toluene feed is at least $346.71/0.72 = 481.54$ kilogram mols per hour. The methanol feed rate equals that for toluene at the reactor inlet. Unreacted toluene and methanol will be recycled to the reactor to conserve the intake of fresh raw materials. Additional toluene and methanol feed needs to be provided, however, for the losses that will occur in byproduct streams. The synthesis of the flow diagram at this point is straightforward, as shown in Fig. 3.6.

The separation system lumped into one box on Fig. 3.6 is one of the more complex aspects of this simulation. Because the reactor effluent contains water and hydrogen, we can reasonably anticipate that when the effluent is cooled we will still have a gaseous phase containing most of the hydrogen; an aqueous

Table 3.4. *Equipment pressure drop, kPa*

Fired heater	66
Reactor	70
Shell or tube side of heat exchanger	13
Vacuum condensers	5
Other major equipment	13
Distillation, pressure column	1.0 per stage
Distillation, vacuum column	0.6 per stage

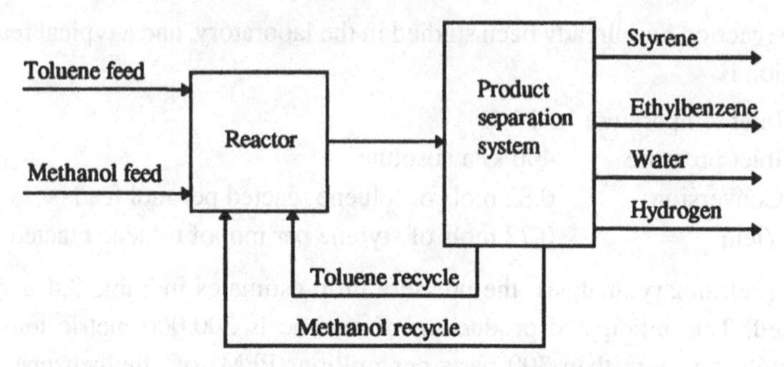

Fig. 3.6. Flow diagram for styrene process.

phase containing most of the water, some methanol, and a trace of organics; and an organic phase with everything else. This area is where the modern flow-sheeting systems have made considerable progress, namely in providing thermodynamic models that can represent such complex phase equilibria. The ASPEN PLUS system that we will illustrate is one of several commercial systems that can handle these problems. It is selected here because it is available and familiar to me. I am not implying endorsement or that it is better than others. All have similar features.

We may now put more details on the separation process. Leaving the reactor, the effluent will be cooled to obtain three phases. The hydrogen-rich phase will be sent to fuel. The aqueous phase will be distilled to recover most of the methanol for recycling. The organic phase, containing methanol, water, toluene, ethylbenzene, and styrene, will need to be distilled twice to make two products, styrene and ethylbenzene, plus a mixture of all the rest to be recycled to the reactor. Really, the only question with regard to synthesizing the fractionation of the organic phase is whether we split the mixture first between toluene and ethylbenzene or between ethylbenzene and styrene.

Applying the rules developed by Lu and Motard (1982), we will confine ourselves to comparing these two splits solely on the basis of the minimum number of stages required for each option. For the toluene–ethylbenzene split the approximate volatility ratio, α, between toluene as a distillate and ethylbenzene as a residue is 2.28. α, in its simplest interpretation, is approximated by the ratio of the vapor pressures of the light and heavy key components. The same ratio for the ethylbenzene–styrene split is $\alpha = 1.31$. The lower the α, the more difficult is the separation, requiring a taller column with more trays. Compounding the problem is that we wish to produce almost pure styrene. The recovery ratios (rec_i) for the two options are

- toluene in distillate, $rec_d = 99\%$, that is, toluene flow in distillate relative to flow in feed,

- ethylbenzene in residue, $rec_r = 99\%$,
- ethylbenzene in distillate, 99.92%,
- styrene in residue, 99%.

The minimum stage calculation follows Fenske's equation,

$$N_{\min} = \frac{\log\left[\dfrac{rec_d \times rec_r}{(1 - rec_d) \times (1 - rec_r)}\right]}{\log \alpha}$$

The results for the two options are

- toluene–ethylbenzene, $N_{\min} = 11$,
- ethylbenzene–styrene, $N_{\min} = 43$.

There is no question that the toluene–ethylbenzene split should be favored to reduce the scale of the styrene recovery column.

Recovering the styrene product first looks attractive nevertheless to avoid prolonged processing of the styrene, which will tend to polymerize at temperatures higher than 145°C when it exceeds fifty mol percent of the mixture. Because 145°C is the normal boiling point of styrene, that is, at one atmosphere pressure, the column that makes almost pure styrene (<300 PPM of ethylbenzene) will surely operate under some level of vacuum. This reason is simply another why we should reduce the size of the styrene column as much as possible. Thus, toluene and lighter components will be removed first and sent to the reactor, and then the other two products will be recovered. A possible simplified flow diagram for the expanded process is shown in Fig. 3.7.

Now, it is necessary to represent the flow diagram in terms of the modular flow-sheeting system that we will use and the symbols that are commonly used in process engineering. Such a diagram is Fig. 3.8, generated by the graphic user interface of ASPEN PLUS, called Model Manager, where we have also

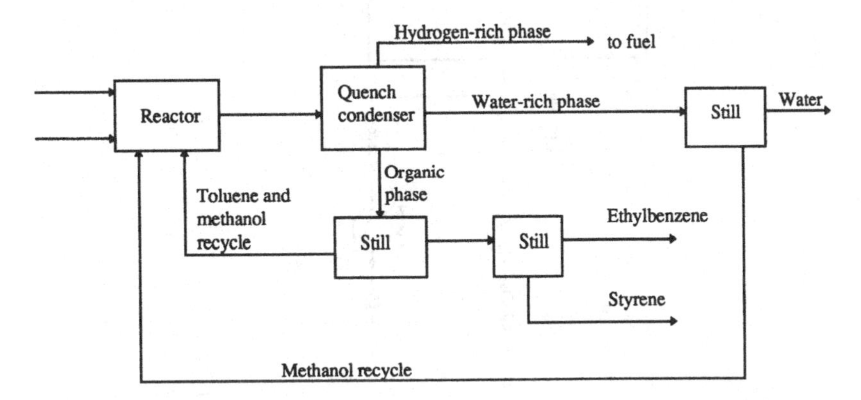

Fig. 3.7. Expanded flow diagram for styrene.

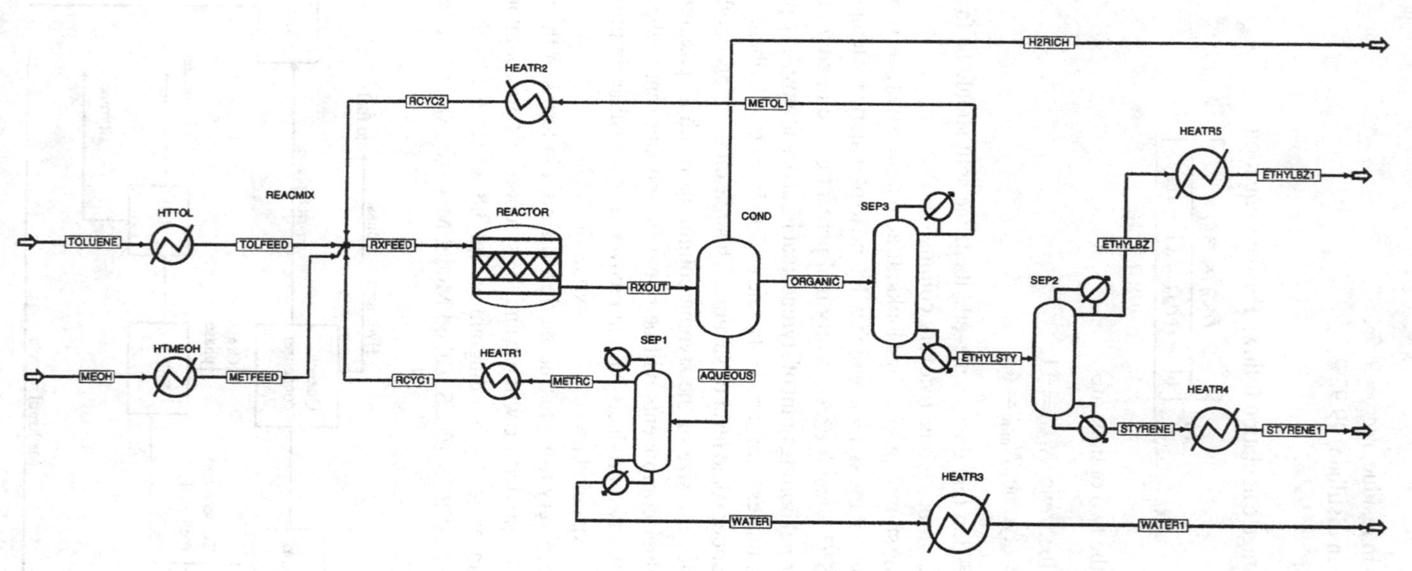

Fig. 3.8. Flow sheet for styrene process.

added heater units. In fact, the heater units can also adjust the pressure of a stream so no pumps are needed, for the moment. The ASPEN PLUS modules are coded in the diagram in capital letters, and the streams have been given symbolic names. Thus, HEATER, MIXER, RSTOIC (stoichiometric reactor), and FLASH3 (three-phase separation) are all unit operation modules in the system. TOLUENE, RXFEED, RCYC1, etc., are user-provided names for the streams. The batch input file to the simulator and the abbreviated and edited results are shown in Appendixes 3.1 and 3.2.

Normally a flow diagram such as Fig. 3.8 would include a stream table, but for considerations of scale we have separated the two, and the stream table appears as Fig. 3.9. This table gives the composition, condition, and enthalpy flow in each major stream. From these data we will be able to synthesize the energy recovery system or heat exchanger network. From the stream table we detect that the purity specification on the styrene product has been met. But there are some questions:

- Is a loss of 6.98 kilogram mols per hour of methanol in the hydrogen purge acceptable?
- How about a loss of 0.37 kilogram mols per hour in the waste water?
- The toluene–ethylbenzene column, SEP3, has a reboiler with seventy-two percent styrene at 147.5°C. Should it be operated under vacuum, because it is currently atmospheric?

The answers to these questions will not be elaborated on here, only commented on. The loss of methanol in the hydrogen by-product is a serious economic give away if the hydrogen is to be burned. If the hydrogen is to be sent to another plant as supplemental synthesis gas for methanol production, then further cleanup is not needed.

One cannot dump water with 0.08 percent methanol into the environment. However, it is likely that biological treatment processes could quickly eliminate the problem. Indeed one might produce protein for animal feed from the microorganisms that metabolize methanol.

The question of the pressure on SEP3 needs further study because the conditions for the thermal polymerization of styrene are necessarily fuzzy.

3.3 Process synthesis in energy integration

We have alluded to some of the heuristic techniques for synthesizing process structure in our discussion on the sequence of fractionators. Expert systems for

Styrene from Methanol and Toloene											
Stream ID		AQUEOUS	ETHYLBZ	ETHYLBZ1	ETHYLSTY	H2RICH	MEOH	METFEED	METOL	METRC	ORGANIC
Phase		LIQUID	LIQUID	LIQUID	LIQUID	VAPOR	LIQUID	VAPOR	MIXED	MIXED	LIQUID
Component mol flow	KMOL/HR										
HYDROGEN		0.001	—	—	—	351.182	0.000	0.000	0.830	0.001	0.830
METHANOL		73.906	0.000	0.000	—	6.981	495.105	495.105	26.395	73.537	26.395
WATER		486.763	0.000	0.000	—	5.848	0.000	0.000	20.344	4.868	20.344
TOLUENE		0.020	1.059	1.059	1.059	1.119	0.000	0.000	104.871	0.020	105.930
ETHYLBNZ		0.005	135.902	135.902	136.011	0.561	0.000	0.000	1.374	0.005	137.385
STYRENE		0.014	3.502	3.502	350.212	0.972	0.000	0.000	0.109	0.014	350.321
Total Mol Flow	kgmol/hr	560.708	140.463	140.463	487.282	366.663	495.105	495.105	153.923	78.444	641.206
Temperature	Deg C	38.000	114.733	38.000	147.540	38.000	20.000	524.000	42.506	101.633	38.000
Pressure	kPa	374.000	55.000	55.000	115.000	374.000	570.000	570.000	110.000	360.000	374.000
Enthalpy	mmkcal/hr	−37.346	0.296	−0.194	11.163	−0.587	−28.627	−20.080	−2.467	−4.370	5.977

Styrene from Methanol and Toloene											
Stream ID		RCYC1	RCYC2	RXFEED	RXOUT	STYRENE	STYRENE1	TOLFEED	TOLUENE	WATER	WATER1
Phase		VAPOR	VAPOR	VAPOR	VAPOR	LIQUID	LIQUID	VAPOR	LIQUID	LIQUID	LIQUID
Component mol flow	KMOL/HR										
HYDROGEN		0.001	0.830	0.832	352.012	—	—	0.000	0.000	—	—
METHANOL		73.537	26.395	595.034	107.283	0.000	0.000	0.000	0.000	0.370	0.370
WATER		4.868	20.344	25.212	512.962	0.000	0.000	0.000	0.000	481.895	481.895
TOLUENE		0.020	104.871	594.818	107.067	—	—	489.927	489.927	—	—
ETHYLBNZ		0.005	1.374	1.379	137.949	0.109	0.109	0.000	0.000	—	—
STYRENE		0.014	0.109	0.123	351.303	346.710	346.710	0.000	0.000	—	—
Total Mol Flow	kgmol/hr	78.444	153.923	1217.397	1568.578	346.819	346.819	489.927	489.927	482.265	482.265
Temperature	Deg C	524.000	524.000	523.282	524.000	126.578	38.000	524.000	20.000	140.245	38.000
Pressure	kPa	570.000	570.000	400.000	400.000	60.000	60.000	570.000	570.000	365.000	365.000
Enthalpy	mmkcal/hr	−3.241	1.371	−5.714	−1.781	10.237	8.873	16.235	1.449	−31.921	−32.997

Fig. 3.9. Stream table for styrene process.

separation problems are still primitive because so many of the decisions are guided by problem-specific issues, as we saw in our flow-sheeting example.

Heat exchanger network synthesis has been automated more successfully because of the homogeneity of the problem. A recent review paper by Prof. Bodo Linnhoff (1993) gives a state-of-the-art summary of the substantial progress made in developing heat-integrated processes using pinch analysis. The same ideas will be exploited here to construct a heat exchanger network for our example. The computer package employed to apply pinch analysis is ADVENT, also a product of Aspen Technology.

Earlier we indicated that the three steps in systematic process synthesis are representation, strategy, and evaluation. In heat exchanger network synthesis we begin the representation with temperature versus enthalpy diagrams. The hot streams or heat sources to be cooled are composited into a single profile, and the same is done with the cold streams and heat sinks to be heated. Pinch technology, wherein a global minimum approach temperature is set by the designer, tells us what the minimum cold and hot utility requirements are. This process is called target setting. No network can be assembled with lower utility consumption without violating temperature constraints. Another target is based on the heuristic that the same area for exchange is cheaper in larger exchangers than in smaller ones. So we strive for the minimum number of heat exchange units. This number can be determined by assuming that all matches between streams will tick off, or take one of the matched streams to its target temperature. Thus, the matching of streams pair by pair to tick off one in each pair is the strategy. A network that meets the two targets is thought to be a high-quality design (evaluation).

Preliminary attempts to integrate the styrene process case study flow sheet led to some rethinking about the details of the process. The styrene recovery from its mixture with ethylbenzene is a demanding fractionation step when done in one column. The high specification on the purity of styrene, namely less than 300 PPM of ethylbenzene, leads to an expensive column (sixty six theoretical stages) and high energy consumption dictated by a high reflux ratio requirement (almost twenty). So it was decided to split the purification of styrene into two columns and to integrate these columns in a multieffect configuration in the style developed by Andrecovich and Westerberg (1985). Both columns produce a bottom stream of specification-grade styrene; however, the distillate from the first column (SEP2A) becomes the feed to the second (SEP2B). The recovery fraction (bottoms/feed) of the styrene in the first column is set low. The net result is that the heat duty in the condenser of the first column can be supplied by the reboiler of the second. The energy required to operate the two columns is roughly one-half of that required for the single column. This result is the one envisioned by Andrecovich and Westerberg.

Table 3.5. *Summary of heating and cooling duties*

Full Name	Abbreviation	Type	Supply Temp.°C	Target Temp.°C	Heat Duty MMkcal/h
SEP1 Condenser	SEP1CND	HOT	102.50	101.50	2.726
SEP1 Reboiler	SEP1REB	COLD	139.75	140.75	3.784
SEP2A Condenser	SEP2ACD	HOT	131.00	130.00	14.697
SEP2A Reboiler	SEP2ARB	COLD	137.50	138.50	14.338
SEP2B Condenser	SEP2BCD	HOT	108.50	107.50	14.200
SEP2B Reboiler	SEP2BRB	COLD	119.50	120.50	13.970
SEP3 Condenser	SEP3CND	HOT	56.50	55.50	4.477
SEP3 Reboiler	SEP3REB	COLD	147.05	148.05	7.308
Reactor Primary Condenser	CND	HOT	524.00	38.00	29.880
Heater 1	HEATR1	COLD	101.60	524.00	1.010
Heater 2	HEATR2	COLD	56.50	524.00	3.660
Heater 3	HEATR3	HOT	140.25	38.00	0.888
Heater 4	HEATR4	HOT	131.00	38.00	1.603
Heater 5	HEATR5	HOT	108.00	38.00	0.462
Toluene Preheater	HTTOL	COLD	20.00	524.00	14.760
Methanol Preheater	HTMEOH	COLD	20.00	524.00	8.231

Figure 3.10 is the revised configuration for SEP2 (Fig. 3.8) in the energy-integrated process. Figure 3.11 is the composite, H-versus-T, diagram for our example flow sheet, showing the heating and cooling composite profiles. The estimates of the heating and cooling requirements to bring all streams to their target temperatures are shown in Table 3.5 as obtained from the ASPEN simulation. Also shown, in Table 3.6, is the cascade diagram or problem table results that indicate how the heat flows through the various temperature intervals (from hotter to colder) in the network. The latter information becomes the basis for the grand composite diagram that allows one to decide at what temperature levels the utility heating and cooling duties are to be inserted into the network. The grand composite is shown in Fig. 3.12.

Finally, the ladder network with all stream matches and heat loads is shown in Fig. 3.13.

3.4 Electronic data interchange

The exchange of design data between different phases of process synthesis and design is an active area of research. This area is also of intense interest to various standards organizations responsible for defining processes that move data from one application to another without redundancy and with total confidence in the accuracy of the transfer. For instance, detailed mechanical design of equipment

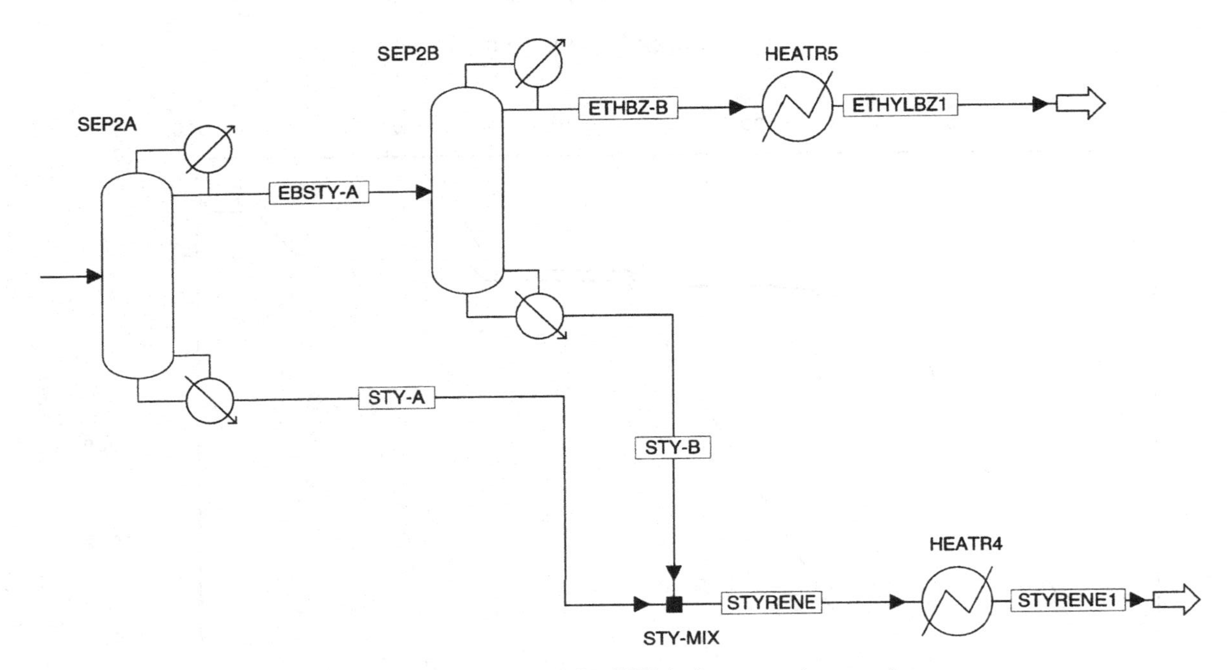

Fig. 3.10. Revised configuration for SEP2 in the energy-integrated process.

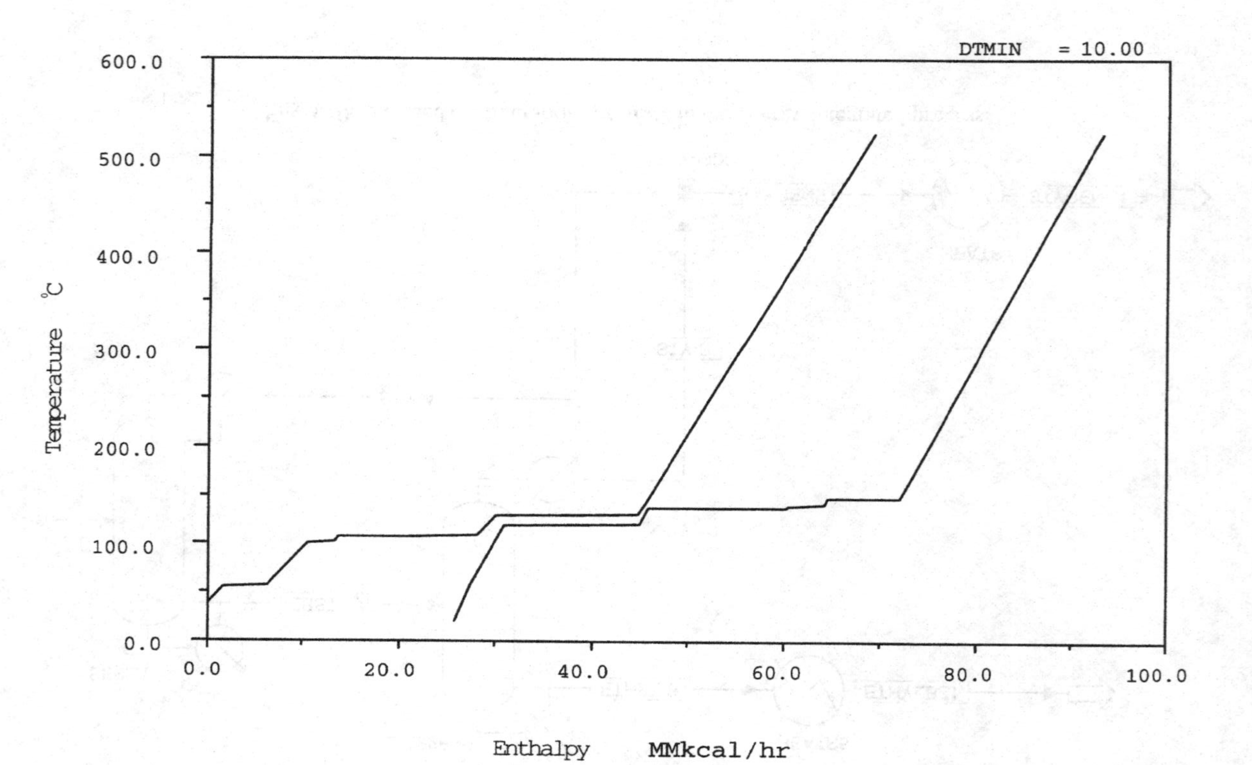

Fig. 3.11. Composite, H-versus-T, diagram.

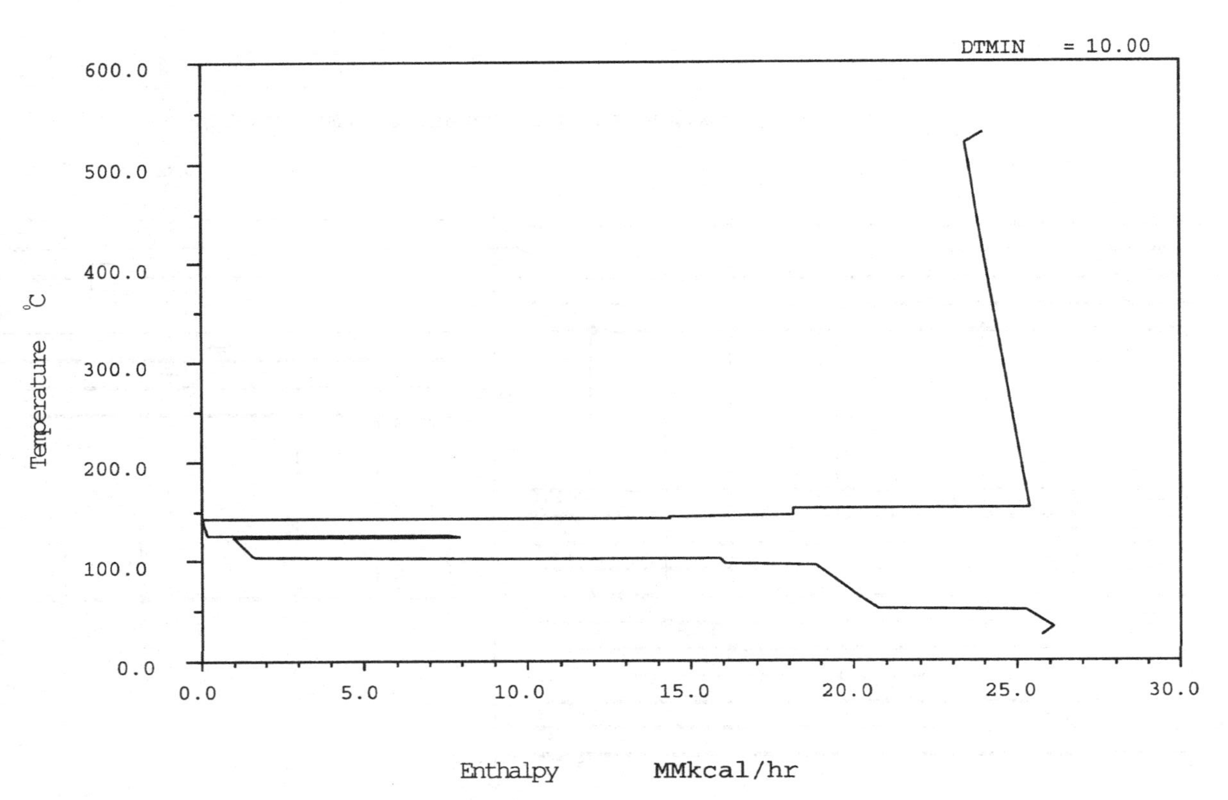

Fig. 3.12. The grand composite.

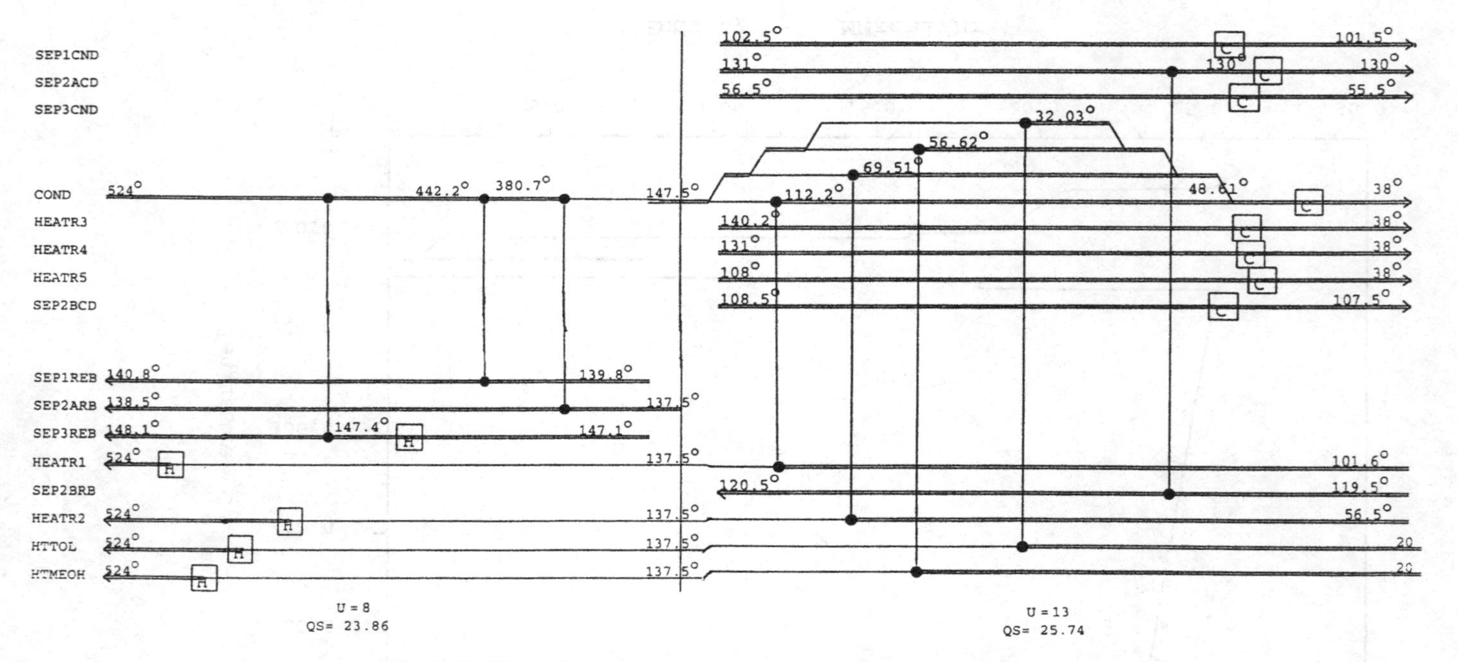

Fig. 3.13. The ladder network with all stream matches and heat loads.

Table 3.6. *Utility targets and energy cascade*

Interval Temperature °C	Net Energy Consumption MMkcal/h	Accumulated Energy	Comments
534.00	0.5584	23.8633	Minimum hot utility
524.00	–2.0656	23.3049	↑
158.05	7.3024	25.3705	Hot end:
157.05	–0.0356	18.0681	only overall
150.75	3.7784	18.1037	heating
149.75	–0.0071	14.3253	allowed
148.50	14.3324	14.3324	↓
147.50	–0.0409	0.	Pinch temperature
			No energy flux past here
140.25	–0.1325	0.0409	↑
131.00	–7.3643	0.1734	
130.50	–0.3793	7.5377	
130.50	6.9692	7.9170	Cold end:
129.50	–0.5650	0.9478	only overall
111.60	–0.1053	1.5128	cooling
108.50	–7.1170	1.6181	allowed
108.50	–7.1203	8.7351	
107.50	–0.2028	15.8554	
102.50	–2.7670	16.0582	
101.50	–1.4195	18.8252	
66.50	–0.4839	20.2447	
56.50	–4.5254	20.7286	
55.50	–0.8467	25.2540	
38.00	0.3649	26.1007	↓
30.00		25.7358	Minimum cold utility

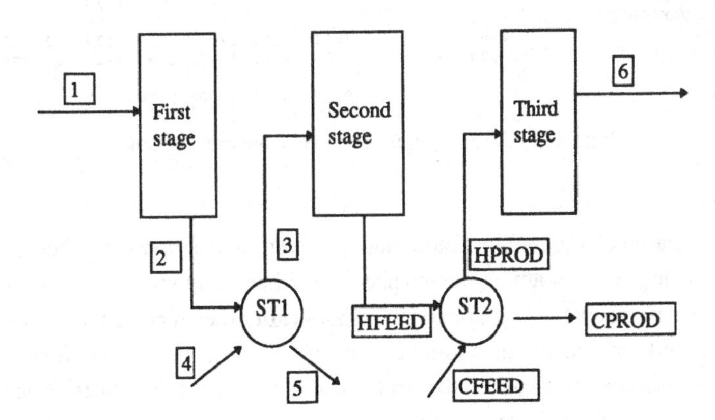

Fig. 3.14. Three-stage compressor with intercoolers.

FLOW-SHEET SECTION

FLOW-SHEET CONNECTIVITY BY STREAMS

STREAM	SOURCE	DEST	STREAM	SOURCE	DEST
1	----	STAGE1	4	----	ST1
CFEED	----	ST2	2	STAGE1	ST1
HFEED	STAGE2	ST2	6	STAGE3	----
3	ST1	STAGE2	5	ST1	----
HPROD	ST2	STAGE3	CPROD	ST2	----

FLOW-SHEET CONNECTIVITY BY BLOCKS

BLOCK	INLETS	OUTLETS
STAGE1	1	2
STAGE2	3	HFEED
STAGE3	HPROD	6
ST1	2 4	3 5
ST2	HFEED CFEED	HPROD CPROD

STREAM SECTION

STREAM ID	1	2	3	4	5
FROM :	----	STAGE1	ST1	----	ST1
TO :	STAGE1	ST1	STAGE2	ST1	----
PHASE:	VAPOR	VAPOR	VAPOR	LIQUID	LIQUID
COMPONENTS: KMOL/SEC					
WATER	0.0	0.0	0.0	1.2680	1.2680
ETHYLENE	0.1890	0.1890	0.1890	0.0	0.0
TOTAL FLOW:					
KMOL/SEC	0.1890	0.1890	0.1890	1.2680	1.2680
STATE VARIABLES:					
TEMP K	288.7055	437.5944	310.9277	294.2611	307.9351
PRES N/SQM	5.0663+05	1.2412+06	1.2412+06	1.0133+05	1.0133+05
ENTHALPY:					
WATT	9.8484+06	1.1240+07	1.0029+07	-3.6260+08	-3.6139+08

STREAM ID	6	CFEED	CPROD	HFEED	HPROD
FROM :	STAGE3	----	ST2	STAGE2	ST2
TO :	----	ST2	----	ST2	STAGE3
PHASE:	VAPOR	LIQUID	LIQUID	VAPOR	VAPOR
COMPONENTS: KMOL/SEC					
WATER	0.0	1.6247	1.6247	0.0	0.0
ETHYLENE	0.1890	0.0	0.0	0.1890	0.1890
TOTAL FLOW:					
KMOL/SEC	0.1890	1.6247	1.6247	0.1890	0.1890
STATE VARIABLES:					
TEMP K	466.4833	294.2611	307.1925	460.9277	310.9277
PRES N/SQM	7.5994+06	1.0133+05	1.0133+05	2.8979+06	2.8979+06
ENTHALPY:					
WATT	1.1558+07	-4.6460+08	-4.6313+08	1.1496+07	1.0029+07

Fig. 3.15. ASPEN input for three-stage compressor.

such as heat exchangers and fractionating towers is frequently farmed out to hardware design companies. Design needs for other components such as pumps and compressors may be completed by vendors after they receive the traditional transfer medium, namely an equipment specification sheet. Looking forward to all-electronic communication, the design and manufacturing industries need to establish protocols that are unambiguous for exchanging design specifications and product designs.

```
ISO-10303-21;
HEADER;
FILE_DESCRIPTION((),'Three-stage compressor example - Aspen simulation component
 and stream data');
FILE_NAME('pdxi.stp','1995-01-12 13:08:25',(),(),'PDXI Prototype API
 Working Version',' ',' ');
FILE_SCHEMA(('PDXI_api_prototype_schema'));
ENDSEC;
DATA;
#0=COMPONENT('WATER',(),(),(#2,#5,#8,#11,#14),(#45,#78,#111,#142,#175,#208,#239,#2
72,
#305,#336,#369,#402,#433,#466,#499,#530,#563,#596,#627,#660,#693,#724,#757,#790,
#821,#854,#887,#918,#951,#984));
#1=DEFINED_POINT_PROPERTY(*,$,.CRITICAL_TEMPERATURE_PP.);
#2=APPROVED_PROPERTY_VALUE(#1,#3);
#3=ATTRIBUTE_MEASURE(647.13,$,$);
#4=DEFINED_POINT_PROPERTY(*,$,.CRITICAL_PRESSURE_PP.);
#5=APPROVED_PROPERTY_VALUE(#4,#6);
#6=ATTRIBUTE_MEASURE(22055000,$,$);
#7=DEFINED_POINT_PROPERTY(*,$,.CRITICAL_COMPRESSIBILITY_FACTOR_PP.);
#8=APPROVED_PROPERTY_VALUE(#7,#9);
#9=ATTRIBUTE_MEASURE(0.229,$,$);
#10=DEFINED_POINT_PROPERTY(*,$,.CRITICAL_VOLUME_PP.);
#11=APPROVED_PROPERTY_VALUE(#10,#12);
#12=ATTRIBUTE_MEASURE(0.0559478,$,$);
#13=DEFINED_POINT_PROPERTY(*,$,.AVERAGE_MOLECULAR_WEIGHT_PP.);
#14=APPROVED_PROPERTY_VALUE(#13,#15);
#15=ATTRIBUTE_MEASURE(18.01528,$,$);
#16=COMPONENT('ETHYLENE',(),(),(#17,#19,#21,#23,#25),(#47,#80,#113,#144,#177,
#210,#241,#274,#307,#338,#371,#404,#435,#468,#501,#532,#565,#598,#629,#662,#695,
#726,#759,#792,#823,#856,#889,#920,#953,#986));
#17=APPROVED_PROPERTY_VALUE(#1,#18);
#18=ATTRIBUTE_MEASURE(282.34,$,$);
#19=APPROVED_PROPERTY_VALUE(#4,#20);
#20=ATTRIBUTE_MEASURE(5040000,$,$);
• • • •
#25=APPROVED_PROPERTY_VALUE(#13,#26);
#26=ATTRIBUTE_MEASURE(28.05376,$,$);
#27=PROCESS_SIMULATION('pdxi_prototype_sim',.F.,$,$,(),(#28,#125,#222,#319,#416,
#513,#610,#707,#804,#901));
#28=MATERIAL_STREAM('1',#27,$,$,$,#29,.MAT_FORM_UNSPECIFIED.,(#60,#93),());
#29=MATERIAL_FLOW(#30,$,$,$,#31,$,$,#32,(),(),(),());
#30=MOLE_FLOW_MEASURE(188.99682,$,$);
#31=MATERIAL('1_BULK',$,#44,(#29),(#32));
#32=THERMODYNAMIC_STATE(#31,$,#34,#33,(#35),(#41),(#38),(),());
#33=PRESSURE_MEASURE(506625,$,$);
#34=TEMPERATURE_MEASURE(288.705554,$,$);
#35=THERMODYNAMIC_PROPERTIES(#32,#37,$,$,$,$,$,$,$,#36,$,$,$,$);
#36=MOLE_HEAT_CAPACITY_MEASURE(42032.2001,$,$);
#37=MOLE_ENTHALPY_MEASURE(52108.7233107943,$,$);
#38=PVT_PROPERTIES(#32,$,$,$,#40,#39,$,$,$,$,$);
#39=ATTRIBUTE_MEASURE(-1e+035,$,$);
#40=MASS_DENSITY_MEASURE(5.92100323,$,$);
#41=TRANSPORT_PROPERTIES(#32,#42,$,#43);
#42=THERMAL_CONDUCTIVITY_MEASURE(-1e+035,$,$);
#43=VISCOSITY_MEASURE(-1e+035,$,$);
#44=MIXTURE('1_mix',(),(),(#49,#51,#53,#55,#57),(),(#45,#47));
#45=MIXTURE_COMPOSITION_ELEMENT(#46,$,$,#0,#44);
#46=RATIO_MEASURE(0,$,$);
#47=MIXTURE_COMPOSITION_ELEMENT(#48,$,$,#16,#44);
```

Fig. 3.16. Partial STEP/SDAI file for three-stage compressor.

The American Institute of Chemical Engineers (AIChE) has joined in this international effort through the foundation of its own PDXI. The activities of this project begun in 1991 were recently summarized in *Chemical Engineering Progress* (1994). The process of implementing the desired exchange capabilities is well under way for the CPIs through the specification of a neutral repository that captures data according to the international standard called STEP (Standard for the Exchange of Product Model Data, 1992). The demonstration project at PDXI uses STEP to produce a neutral file that can then be read by any software containing a STEP-compliant interface. The file can also be read by domain experts, that is, engineers and designers. The file specification is called Standard Data Access Interface Specification – SDAI (STEP 1992).

An example of a STEP-neutral file is abbreviated here for a simple flow sheet consisting of a three-stage ethylene compressor with intercooling between the first two stages, Fig. 3.14. The ASPEN PLUS results, highly abbreviated, are shown in Fig. 3.15. These results were transferred to a STEP file as shown in capsule form in Fig. 3.16. These figures are derived from a prototype process data exchange software package distributed by AIChE.

The STEP file could then be handed over to a heat exchanger manufacturer to complete the functional design of the intercoolers and to a compressor manufacturer to do likewise for the compressor. Of course, the specification of a compressor or heat exchanger involves more than the process simulation data, so that the STEP file would need to be augmented with other engineering specifications by the project engineers before it would be ready to go to vendors for bids. Much of this part can be automated as well. If all else fails, editors particularly suited to reviewing and upgrading the STEP file are available.

References

ADVENT Process Synthesis and Optimization, Aspen Technology Inc., Ten Canal Street, Cambridge, MA 02141.

American Institute of Chemical Engineers, Student Contest Problem, 345 East 47 Street, New York, NY 10017, (1985).

Andrecovich, M. J. & Westerburg, A. W. (1985) A simple synthesis method based on utility bounding for heat-integrated distillation sequences. *AIChE Journal* **31**, 363–75.

ASPEN PLUS Process Simulation, Aspen Technology, Inc., Ten Canal Street, Cambridge, MA 02141.

Book, N., Sitton, O., Motard, R. L., Blaha, M. R., Maia-Goldstein, B., Hedrick, J. & Fielding, J. (1994). *The Road to a Common Byte*, Chemical Engineering New York: McGraw-Hill.

Chemical Engineering Progress, Annual CEP Software Directory, American Institute of Chemical Engineers, 345 East 47th Street, New York, NY 10017 (1994).

FLOWTRAN, Monsanto Company, St. Louis Missouri (1970). Note: this package was removed from the industrial market in 1973 although it served the academic community for many years beyond that.

DESIGN, Chemshare Corporation, P.O. Box 1885, Houston, Texas 77251 (1970).

Evans, L. B. (1980). Advances in Process Flowsheeting Systems. In *Foundations of Computer-Aided Chemical Process Designed.* ed. R. S. H. Mah & W. D. Seider, Vol. 1, pp. 425–69. New York: American Institute of Chemical Engineers.

Foundations of Computer-Aided Process Design, *Proceedings of the Third International Conference FOCAPD*, ed. J. J. Siirola, I. E. Grossmann & G. Stephanopoulos, New York, Elsevier (1990).

Linnhoff, B. (1993) Pinch analysis: a state-of-the-art overview. *Transactions of IChemE*, **71**, Part A, 503–22.

Lu, M. D. & Motard, R. L. (1982). A strategy for the synthesis of separation sequences. In *Understanding Process Integration*, symp. ser. no. 74, pp. 141–51, Rugby, U.K.: Institution of Chemical Engineers.

Perkins, J. D. (1983). Equation-based flowsheeting. In *Proceedings Second International Conference on Foundations of Computer-Aided Process Design*. ed. A. W. Westerberg & H. H. Chien, pp. 309–67, Ann Arbor, MI: CACHE Publications.

STEP, Product Data Representation and Exchange, Part 22, Standard Data Access Interface Specification, ISO 10303, James E. Fowler, Editor, NIST, Gaithersburg, MD 20899, (1992).

United States Air Force, "ICAM Architecture Part II, Vol. IV – Function Modeling Manual (IDEFO)," AFAWL-TR-81-4023, USAF Wright Aeronautical Laboratories, Wright-Patterson AFB, Ohio (1981).

Westerberg, A. W., Hutchison, H. P., Motard, R. L. & Winter, P. (1979). *Process Flowsheeting*. Cambridge, UK: Cambridge University Press.

Winter, P. & Angus, C. J. (1983). The database frontier in process design. In *Proceedings Second International Conference on Foundations of Computer-Aided Process Design*. ed. A. W. Westerberg & H. H. Chien, pp. 75–127, Ann Arbor, MI: CACHE Publications.

Appendix 3.1 Aspen Plus Simulator Input File

```
;
TITLE "STYRENE FROM METHANOL AND TOLUENE"
;
; definition of units: metric, kilopascals and celsius
;
IN-UNITS MET PRESSURE=KPA TEMPERATURE=C DELTA-T=C PDROP=KPA
;
; all streams are conventional
;
DEF-STREAMS CONVEN ALL
;
; batch run time limit at 60 sec
;
RUN-CONTROL MAX-TIME=60.0
;
DESCRIPTION "
        ASPEN PLUS SIMULATION OF 1985 AICHE STUDENT CONTEST PROBLEM"
;
;
; AIChE DIPPR library to be searched for binary interaction coefficients
;
DATABANKS DIPPRPCD / BINARY

PROP-SOURCES DIPPRPCD / BINARY
;
; chemical species to be processed and their aliases
;
```

Appendix 3.1 (continued)

```
COMPONENTS
    H2 H2 HYDROGEN /
    METH CH4O METHANOL /
    H2O H2O WATER /
    TOL C7H8 TOLUENE /
    EB C8H10-4 ETHYLBNZ /
    STY C8H8 STYRENE
;
; flow sheet topology
;
FLOWSHEET
    BLOCK HTMEOH IN=MEOH OUT=METFEED
    BLOCK HTTOL IN=TOLUENE OUT=TOLFEED
    BLOCK REACMIX IN=METFEED TOLFEED RCYC2 RCYC1 OUT=RXFEED
    BLOCK REACTOR IN=RXFEED OUT=RXOUT
    BLOCK COND IN=RXOUT OUT=H2RICH ORGANIC AQUEOUS
    BLOCK SEP2 IN=ETHYLSTY OUT=ETHYLBZ STYRENE
    BLOCK HEATR2 IN=METOL OUT=RCYC2
    BLOCK SEP1 IN=AQUEOUS OUT=METRC WATER
    BLOCK HEATR1 IN=METRC OUT=RCYC1
    BLOCK SEP3 IN=ORGANIC OUT=METOL ETHYLSTY
    BLOCK HEATR3 IN=WATER OUT=WATER1
    BLOCK HEATR4 IN=STYRENE OUT=STYRENE1
    BLOCK HEATR5 IN=ETHYLBZ OUT=ETHYLBZ1
;
; property sets to be used: SYSOP4 unless specified as SYSOP18 in
; individual process units
;
PROPERTIES SYSOP4 / SYSOP18
;
; feed stream flows and initial conditions
;
STREAM MEOH
    SUBSTREAM MIXED TEMP=20.0 PRES=570.0
    MOLE-FLOW METH 484.0

STREAM TOLUENE
    SUBSTREAM MIXED TEMP=20.0 PRES=570.0
    MOLE-FLOW TOL 481.540
;
; parameters for process blocks
;
BLOCK REACMIX MIXER
    PARAM PRES=400.0

BLOCK SEP2 DSTWU
    PROPERTIES SYSOP18
    PARAM LIGHTKEY=EB RECOVL=0.9992 HEAVYKEY=STY RECOVH=0.01 &
        PTOP=55. PBOT=60. RR=-1.60

BLOCK SEP3 DSTWU
    PROPERTIES SYSOP18
    PARAM LIGHTKEY=TOL RECOVL=0.99 HEAVYKEY=EB RECOVH=0.01 &
        PTOP=110. PBOT=115. RR=-1.4 RDV=0.01
```

Appendix 3.1 (continued)

```
BLOCK HEATR1 HEATER
    PARAM TEMP=525.0 PRES=570.0

BLOCK HEATR2 HEATER
    PARAM TEMP=525.0 PRES=570.0

BLOCK HTMEOH HEATER
    PARAM TEMP=525.0 PRES=570.0

BLOCK HTTOL HEATER
    PARAM TEMP=525.0 PRES=570.0

BLOCK COND FLASH3
    PROPERTIES SYSOP18
    PARAM TEMP=38.0 PRES=374.0 L2-COMP=H2O

BLOCK SEP1 DSTWU
    PROPERTIES SYSOP18
    PARAM LIGHTKEY=METH RECOVL=.9950 HEAVYKEY=H2O RECOVH=.010 &
        PTOP=360.0 PBOT=365.0 RDV=.050 RR=-1.30

BLOCK REACTOR RSTOIC
    PARAM TEMP=525.0 PRES=400.0 NPHASE=1
    STOIC 1 MIXED METH -1.0 / TOL -1.0 / H2 1.0 / H2O &
        1.0 / STY 1.0
    STOIC 2 MIXED METH -1.0 / TOL -1.0 / H2O 1.0 / EB &
        1.0
    CONV 1 MIXED TOL .59040
    CONV 2 MIXED TOL .22960

BLOCK HEATR3 HEATER
    PARAM TEMP=38. PRES=0

BLOCK HEATR4 HEATER
    PARAM TEMP=38. PRES=0

BLOCK HEATR5 HEATER
    PARAM TEMP=38. PRES=0
;
; set feed rate of methanol equal to that of toluene at the
; reactor input
;
DESIGN-SPEC METRATE
    DEFINE MEORAT MOLE-FLOW STREAM=RXFEED SUBSTREAM=MIXED &
        COMPONENT=METH
    DEFINE TOLRAT MOLE-FLOW STERAM=RXFEED SUBSTREAM=MIXED &
        COMPONENT=TOL
    SPEC "MEORAT" TO "TOLRAT"
    TOL-SPEC "0.001*TOLRAT"
    VARY MOLE-FLOW STREAM=MEOH SUBSTREAM=MIXED COMPONENT=METH
    LIMITS "480" "500"
;
; set toluene flow rate to produce desired styrene rate
;
```

```
DESIGN-SPEC TOLRATE
    DEFINE STYRAT MOLE-FLOW STREAM=STYRENE SUBSTREAM=MIXED &
        COMPONENT=STY
    SPEC "STYRAT" TO "346.71"
    TOL-SPEC "1E-3"
    VARY MOLE-FLOW STREAM=TOLUENE SUBSTREAM=MIXED COMPONENT=TOL
    LIMITS "481" "490"

CONV-OPTIONS
    PARAM TOL=.0010
    WEGSTEIN WAIT=3 QMIN=-10000.0 ACCELERATE=3

STREAM-REPOR INCL-STREAMS=METFEED TOLFEED RCYC1 RCYC2 RXFEED &
    RXOUT H2RICH AQUEOUS ORGANIC ETHYLSTY METOL STYRENE ETHYLBZ &
    METRC WATER WATER1 STYRENE1 ETHYLBZ1
;
;
;
;
;
```

Appendix 3.2 Simulation results

FLOW-SHEET CONNECTIVITY BY STREAMS

STREAM	SOURCE	DEST	STREAM	SOURCE	DEST
MEOH	----	HTMEOH	TOLUENE	----	HTTOL
METFEED	HTMEOH	REACMIX	TOLFEED	HTTOL	REACMIX
RXFEED	REACMIX	REACTOR	RXOUT	REACTOR	COND
H2RICH	COND	----	ORGANIC	COND	SEP3
AQUEOUS	COND	SEP1	ETHYLBZ	SEP2	HEATR5
STYRENE	SEP2	HEATR4	RCYC2	HEATR2	REACMIX
METRC	SEP1	HEATR1	WATER	SEP1	HEATR3
RCYC1	HEATR1	REACMIX	METOL	SEP3	HEATR2
ETHYLSTY	SEP3	SEP2	WATER1	HEATR3	----
STYRENE1	HEATR4	----	ETHYLBZ1	HEATR5	----

FLOW-SHEET CONNECTIVITY BY BLOCKS

BLOCK	INLETS	OUTLETS
HTMEOH	MEOH	METFEED
HTTOL	TOLUENE	TOLFEED
REACMIX	METFEED TOLFEED RCYC2 RCYC1 RXFEED	
REACTOR	RXFEED	RXOUT
COND	RXOUT	H2RICH ORGANIC AQUEOUS
SEP2	ETHYLSTY	ETHYLBZ STYRENE
HEATR2	METOL	RCYC2
SEP1	AQUEOUS	METRC WATER
HEATR1	METRC	RCYC1
SEP3	ORGANIC	METOL ETHYLSTY
HEATR3	WATER	WATER1
HEATR4	STYRENE	STYRENE1
HEATR5	ETHYLBZ	ETHYLBZ1

Appendix 3.2 (continued)

OVERALL MASS AND ENERGY BALANCE

	IN	OUT
CONVENTIONAL COMPONENTS (KMOL/HR)		
HYDROGEN	0.	351.182
METHANOL	495.105	7.350
WATER	0.	487.743
TOLUENE	489.927	2.178
ETHYLBNZ	0.	136.572
STYRENE	0.	351.185
TOTAL BALANCE		
MOLE (KMOL/HR)	985.031	1336.21
MASS (KG/HR)	61006.3	61007.0
ENTHALPY (MMKCAL/H)	-27.1782	-24.9057

ID	TYPE	FORMULA	COMPONENT'S NAME OR ALIAS	REPORT NAME
H2	C	H2	H2	HYDROGEN
METH	C	CH4O	CH4O	METHANOL
H2O	C	H2O	H2O	WATER
TOL	C	C7H8	C7H8	TOLUENE
EB	C	C8H10-4	C8H10-4	ETHYLBNZ
STY	C	C8H8	C8H8	STYRENE

PROPERTY OPTION SETS: SYSOP4 PENG-ROBINSON EQUATION OF STATE SYSOP18 REDLICH-KWONG-UNIFAC EQUATION OF STATE

SIMULATION BLOCK RESULTS

BLOCK: COND MODEL: FLASH3 (SYSOP18)

OUTLET TEMPERATURE	C	38.000
OUTLET PRESSURE	KPA	374.00
HEAT DUTY	MMKCAL/HR	-30.176
VAPOR FRACTION		0.23376
1ST LIQUID/TOTAL LIQUID		0.53349

BLOCK: HEATR1 MODEL: HEATER (SYSOP4)

OUTLET TEMPERATURE	C	525.00
OUTLET PRESSURE	KPA	570.00
HEAT DUTY	MMKCAL/HR	1.1283

BLOCK: HEATR2 MODEL: HEATER (SYSOP4)

OUTLET TEMPERATURE	C	525.00
OUTLET PRESSURE	KPA	570.00
HEAT DUTY	MMKCAL/HR	3.8387

BLOCK: HEATR3 MODEL: HEATER (SYSOP4)

OUTLET TEMPERATURE	C	38.000
OUTLET PRESSURE	KPA	365.00
HEAT DUTY	MMKCAL/HR	-1.0757

Appendix 3.2 (continued)

BLOCK: HEATR4 MODEL: HEATER (SYSOP4)

OUTLET TEMPERATURE	C	38.000
OUTLET PRESSURE	KPA	60.000
HEAT DUTY	MMKCAL/HR	-1.3645

BLOCK: HEATR5 MODEL: HEATER (SYSOP4)

OUTLET TEMPERATURE	C	38.000
OUTLET PRESSURE	KPA	55.000
HEAT DUTY	MMKCAL/HR	-0.49078

BLOCK: HTMEOH MODEL: HEATER (SYSOP4)

OUTLET TEMPERATURE	C	525.00
OUTLET PRESSURE	KPA	570.00
HEAT DUTY	MMKCAL/HR	8.5476

BLOCK: HTTOL MODEL: HEATER (SYSOP4)

OUTLET TEMPERATURE	C	525.00
OUTLET PRESSURE	KPA	570.00
HEAT DUTY	MMKCAL/HR	14.786

BLOCK: REACMIX MODEL: MIXER (SYSOP4)

BLOCK: REACTOR MODEL: RSTOIC (SYSOP4)

OUTLET TEMPERATURE	C	525.00
OUTLET PRESSURE	KPA	400.00
HEAT DUTY	MMKCAL/HR	3.9334

BLOCK: SEP1 MODEL: DSTWU (SYSOP18)

DISTILLATE TEMP. (C)	101.633
BOTTOM TEMP. (C)	140.245
MINIMUM REFLUX RATIO	2.54034
ACTUAL REFLUX RATIO	3.30244
MINIMUM STAGES	7.60642
ACTUAL EQUILIBRIUM STAGES	14.3032
NUMBER OF ACTUAL STAGES ABOVE FEED	8.77268
DIST. VS FEED	0.13990
NET CONDENSER DUTY (MMKCAL/H)	-2.77586
NET REBOILER DUTY (MMKCAL/H)	3.83161

BLOCK: SEP2 MODEL: DSTWU (SYSOP18)

DISTILLATE TEMP. (C)	114.733
BOTTOM TEMP. (C)	126.578
MINIMUM REFLUX RATIO	11.8984
ACTUAL REFLUX RATIO	19.0374
MINIMUM STAGES	46.4981
ACTUAL EQUILIBRIUM STAGES	66.7338
NUMBER OF ACTUAL STAGES ABOVE FEED	25.1392
DIST. VS FEED	0.28826
NET CONDENSER DUTY (MMKCAL/H)	-24.9833
NET REBOILER DUTY (MMKCAL/H)	24.3539

Appendix 3.2 (continued)

BLOCK: SEP3 MODEL: DSTWU (SYSOP18)

DISTILLATE TEMP. (C)	42.5056
BOTTOM TEMP. (C)	147.540
MINIMUM REFLUX RATIO	1.23711
ACTUAL REFLUX RATIO	1.73195
MINIMUM STAGES	13.4586
ACTUAL EQUILIBRIUM STAGES	24.6131
NUMBER OF ACTUAL STAGES ABOVE FEED	11.2035
DIST. VS FEED	0.24005
NET CONDENSER DUTY (MMKCAL/H)	-3.72595
NET REBOILER DUTY (MMKCAL/H)	6.44413

STREAM SECTION

STREAM ID	MEOH	METFEED	TOLFEED	RCYC1	RCYC2
FROM :	----	HTMEOH	HTTOL	HEATR1	HEATR2
TO :	HTMEOH	REACMIX	REACMIX	REACMIX	REACMIX
COMPONENTS:					
KMOL/HR					
HYDROGEN	0.0	0.0	0.0	0.0014	0.8302
METHANOL	495.1048	495.1048	0.0	73.5366	26.3954
WATER	0.0	0.0	0.0	4.8676	20.3439
TOLUENE	0.0	0.0	489.9266	1.9733-02	104.8707
ETHYLBNZ	0.0	0.0	0.0	4.9343-03	1.3738
STYRENE	0.0	0.0	0.0	1.3576-02	0.1092
TOTAL FLOW:					
KMOL/HR	495.1048	495.1048	489.9266	78.4438	153.9234
STATE					
VARIABLES:					
TEMP C	20.0000	524.0000	524.0000	524.0000	524.0000
PRES KPA	570.0000	570.0000	570.0000	570.0000	570.0000
PHASE	LIQUID	VAPOR	VAPOR	VAPOR	VAPOR
ENTHALPY					
MMKCAL/HR	-28.6271	-20.0795	16.2353	-3.2412	1.3713

STREAM ID	RXFEED	RXOUT	H2RICH	AQUEOUS	ORGANIC
FROM :	REACMIX	REACTOR	COND	COND	COND
TO :	REACTOR	COND	----	SEP1	SEP3
COMPONENTS:					
KMOL/HR					
HYDROGEN	0.8316	352.0122	351.1822	1.3731-03	0.8302
METHANOL	595.0338	107.2831	6.9806	73.9061	26.3954
WATER	25.2117	512.9624	5.8483	486.7626	20.3439
TOLUENE	594.8180	107.0672	1.1187	1.9733-02	105.9300
ETHYLBNZ	1.3787	137.9490	0.5610	4.9343-03	137.3846
STYRENE	0.1228	351.3034	0.9724	1.3576-02	350.3214
TOTAL FLOW:					
KMOL/HR	1217.3969	1568.5775	336.6634	560.7084	641.2056
STATE					
VARIABLES:					
TEMP C	523.2819	524.0000	38.0000	38.0000	38.0000
PRES KPA	400.0000	400.0000	374.0000	374.0000	374.0000
PHASE	VAPOR	VAPOR	VAPOR	LIQUID	LIQUID
ENTHALPY:					
MMKCAL/HR	-5.7139	-1.7805	-0.5872	-37.3462	5.9773

Appendix 3.2 (continued)

STREAM ID	ETHYLSTY	METOL	STYRENE	ETHYLBZ	METRC
FROM :	SEP3	SEP3	SEP2	SEP2	SEP1
TO :	SEP2	HEATR2	HEATR4	HEATR5	HEATR1
COMPONENTS:					
KMOL/HR					
HYDROGEN	1.0430-11	0.8302	2.5753-21	1.0430-11	1.3731-03
METHANOL	6.6374-18	26.3954	0.0	0.0	73.5366
WATER	5.6018-20	20.3439	0.0	0.0	4.8676
TOLUENE	1.0593	104.8707	9.5016-17	1.0593	1.9733-02
ETHYLBNZ	136.0107	1.3738	0.1088	135.9019	4.9343-03
STYRENE	350.2121	0.1092	346.7100	3.5021	1.3576-02
TOTAL FLOW:					
KMOL/HR	487.2822	153.9234	346.8188	140.4633	78.4438
STATE					
VARIABLES:					
TEMP C	147.5395	42.5056	126.5779	114.7325	101.6332
PRES KPA	115.0000	110.0000	60.0000	55.0000	360.0000
PHASE	LIQUID	MIXED	LIQUID	LIQUID	MIXED
ENTHALPY:					
MMKCAL/HR	11.1629	-2.4673	10.2371	0.2963	-4.3695

STREAM ID	WATER	WATER1	STYRENE1	ETHYLBZ1	TOLUENE
FROM :	SEP1	HEATR3	HEATR4	HEATR5	----
TO :	HEATR3	----	----	----	HTTOL
COMPONENTS:					
KMOL/HR					
HYDROGEN	8.3105-15	8.3105-15	2.5753-21	1.0430-11	0.0
METHANOL	0.3695	0.3695	0.0	0.0	0.0
WATER	481.8950	481.8950	0.0	0.0	0.0
TOLUENE	3.2334-13	3.2334-13	9.5016-17	1.0593	489.9266
ETHYLBNZ	5.7968-15	5.7968-15	0.1088	135.9019	0.0
STYRENE	6.9772-14	6.9772-14	346.7100	3.5021	0.0
TOTAL FLOW:					
KMOL/HR	482.2645	482.2645	346.8188	140.4633	489.9266
STATE					
VARIABLES:					
TEMP C	140.2451	38.0000	38.0000	38.0000	20.0000
PRES KPA	365.0000	365.0000	60.0000	55.0000	570.0000
PHASE	LIQUID	LIQUID	LIQUID	LIQUID	LIQUID
ENTHALPY:					
MMKCAL/HR	-31.9209	-32.9966	8.8726	-0.1944	1.4489

Appendix 3.3 Energy management information

BLOCK: COND:	CAL/SEC	-0.83948E+07
BLOCK: HEATR1:	CAL/SEC	0.31382E+06
BLOCK: HEATR2:	CAL/SEC	0.10682E+07
BLOCK: HEATR3:	CAL/SEC	-0.29880E+06
BLOCK: HEATR4:	CAL/SEC	-0.37902E+06
BLOCK: HEATR5:	CAL/SEC	-0.13632E+06
BLOCK: HTMEOH:	CAL/SEC	0.23769E+07
BLOCK: HTTOL:	CAL/SEC	0.41149E+07
BLOCK: REACTOR:	CAL/SEC	0.10929E+07

Appendix 3.3 (continued)

BLOCK: SEP1

DISTILLATE TEMP. (C)	101.633
BOTTOM TEMP. (C)	140.245
NET CONDENSER DUTY (CAL/SEC)	-771,061.
NET REBOILER DUTY (CAL/SEC)	1,064,320.

BLOCK: SEP2

DISTILLATE TEMP. (C)	114.733
BOTTOM TEMP. (C)	126.578
NET CONDENSER DUTY (CAL/SEC)	-6,939,720.
NET REBOILER DUTY (CAL/SEC)	6,764,860.

BLOCK: SEP3

DISTILLATE TEMP. (C)	42.5056
BOTTOM TEMP. (C)	147.540
NET CONDENSER DUTY (CAL/SEC)	-1,034,970.
NET REBOILER DUTY (CAL/SEC)	1,790,010.

STREAM ID	METFEED	TOLFEED	RCYC1	RCYC2	RXFEED
KMOL/HR	495.1048	489.9266	78.4438	153.9234	1217.3969
TEMP C	525.0000	525.0000	525.0000	525.0000	524.2842
CAL/SEC	-5.5749+06	4.5174+06	-8.9992+05	3.8278+05	-1.5747+06

STREAM ID	RXOUT	H2RICH	AQUEOUS	ORGANIC	ETHYLSTY
KMOL/HR	1568.5775	366.6634	560.7084	641.2056	487.2822
TEMP C	525.0000	38.0000	38.0000	38.0000	147.5395
CAL/SEC	-4.8178+05	-1.6312+05	-1.0374+07	1.6604+06	3.1008+06

STREAM ID	METOL	STYRENE	ETHYLBZ	METRC	WATER
KMOL/HR	153.9234	346.8188	140.4633	78.4438	482.2645
TEMP C	42.5056	126.5779	114.7325	101.6332	140.2451
CAL/SEC	-6.8537+05	2.8436+06	8.2308+04	-1.2137+06	-8.8668+06

STREAM ID	WATER1	STYRENE1	ETHYLBZ1
KMOL/HR	482.2645	346.8188	140.4633
TEMP C	38.0000	38.0000	38.0000
CAL/SEC	-9.1656+06	2.4646+06	-5.4016+04

4

Thermophysical properties for design simulations

RONALD P. DANNER
Pennsylvania State University
CALVIN F. SPENCER and MANOJ NAGVEKAR
The M. W. Kellogg Company

4.1 Introduction

The keystone of the design of a thermal system or any other chemical process is the process simulation program. Quoting Cox (1993), "All correlations, theories, and physical property data must be translated into computer terms before they truly become useful for chemical process design applications. Data banks and associated physical property models form the heart of any computer simulator calculation." The quality of these data modules can have extensive effects. Inaccurate data may lead to costly errors in judgment whether it is to proceed with a new process or modification or not to go ahead. Inadequate or unavailable data may cause a potentially attractive and profitable process to be delayed or rejected because of the difficulty or impossibility of properly simulating it. Even the most sophisticated software will not lead to a cost-effective solution if it is not backed up by an accurate database. The ability to effectively conserve energy in many processes is related directly to the accuracy of the physical and thermodynamic data available.

A number of projects, frequently funded by consortia of companies and government agencies, exist for the collection and evaluation of data concerning particular areas of interest: chemical processing, electric and gas production and distribution, biology, medicine, geology, meteorology, music, demographics, etc. This chapter will address the phase equilibrium, thermodynamic data, and physical properties generally required in the design of thermal and chemical processing systems.

Unfortunately, as reported by Selover and Spencer (1987) much of the work in the area of thermophysical properties goes unnoticed, especially by the industrial user community. The result of this observation is to reinvent the wheel and repeat work that has already been done. An even more damaging consequence is for industry to rely on outdated, incorrect, or misleading data that could cause serious problems in process design.

In this chapter we consider the different kinds of data that are used in simulators. What kinds of data are needed? How important are they? How accurate must they be? Where can one find such data? How is the quality of the data judged? When should correlation and prediction methods be used? How do simulation programs use these data and prediction models?

4.2 What can the design engineer expect to find in the literature?

Several types of resources are available for the engineer to use to deal with data needs. Frequently the engineer will first reach for a publication containing a database. Finding the values of interest can be a tedious task, given the large volume of data contained in many of these publications. It is becoming much more common for the user to have access to a computerized database. Search routines accompanying such databases provide an efficient way to determine if they contain the needed values.

If contradictory data are found for a particular property, as is often the case, the user must then decide which value should be used. If the property varies with an independent variable such as temperature or pressure, the user may want to develop an analytical function that accurately reflects this dependency. Hopefully, when the data are stored in a computer, also available will be a data analysis program that will provide programs for determining the statistical variance of the data and for fitting an analytical equation if needed.

Often, however, the user will find that the cupboard is bare. Despite the long history of measuring and organizing data, one will usually find only a tiny fraction of the data required for any process design. Most of the data must be predicted from generalized correlations, estimated, inferred, or extrapolated. The odds of locating mixture data are much smaller than of finding pure component data. One can hope that data for all the pure components can be found and then invoke a predictive method (mixing rule) to calculate the mixture property. In some cases, however, it is the interaction between different compounds that is predominant, and no approach based on simply combining the properties of the pure compounds can be used.

What the engineer needs is a universal, comprehensive property system. That is, a software package that has methods for the calculations of the relevant thermophysical properties of pure compounds and mixtures interfaced with an extensive reliable databank of pure component data. The system should have methods that can handle highly nonideal systems in terms of their chemical makeup and that can cover the region extending from vacuum to the supercritical region. The user should be able to use data from the database, enter new data as needed, predict values from certified methods as required, access regression procedures to obtain model parameters, and interface with some of the more popular spreadsheet and graphics software.

4.3 Importance of properties

The properties required for the design of a thermal or chemical processing system vary, of course, depending on the specific case. The general types of data involved can be categorized as shown in Table 4.1.

The relative importance of various properties in the design of a process necessarily varies with the type of process under consideration. In cases where vapor–liquid equilibrium occurs, vapor pressure is important. Although phase equilibria are especially important for design of separation equipment, they are also necessary for heat exchanger and compressor design to establish which phases are present. Critical constants are needed for any equipment that operates close to the critical temperature of a major component, or when generalized equations of state (for example, Soave–Redlich–Kwong or Peng–Robinson) or corresponding states are used. The equations of state, in turn, are used for calculation of fugacities, heat capacities, enthalpies, entropies, and densities.

Mixture densities must be known to convert between volume and mol units and, thus, are important for sizing equipment. Thermal properties are most important for heat exchangers and compressors but are also needed for reactors and separation equipment. The increased attention to energy recovery and heat

Table 4.1. *Properties needed for design*

Property type	Specific properties
Phase equilibria	Boiling and melting points, vapor pressure, fugacity and activity coefficients, solubility (Henry's constants, Ostwald or Bunsen coefficients)
P-*V*-*T* behavior	Density, volume, compressibility, critical constants
Thermal	Specific heat, latent heat, enthalpy, entropy
Transport	Viscosity, thermal conductivity, ionic conductivity, diffusivity
Boundary	Surface tension
Acoustic	Velocity of sound
Optical	Refractive index, polarization
Safety	Flammability limits, toxicity, autoignition temperature, flash point
Molecular	Virial coefficients, binary interaction parameters
Chemical equilibrium	Equilibrium constants, dissociation constants, enthalpies of formation, heat of reaction, Gibbs energy of formation, reaction rates

integration leads to an increased occurrence of heat interchange between process streams. This interchange frequently gives tighter temperature approaches in the exchangers, thus leading to an increased demand for more precise enthalpy and vapor–liquid equilibria data. Transport property data are needed for heat transfer computations for heat exchangers, for line sizing (thermal conductivities of mixtures), and for pressure drop calculations (viscosities of mixtures). Reaction data are, of course, important in the design of any type of reactor or equipment where reactions tend to occur.

Zudkevitch (1982) identified the most influential variables in the design of compressors and expanders as density, phase equilibrium, and heat capacity. The critical constants and entropy are next in importance. For heat exchangers, the primary variables are phase equilibrium and heat capacity; somewhat less significant are critical constants, entropy, and heat transfer coefficients.

4.4 What accuracy is required?

In the past, lack of accurate data was not a serious problem for many industrial applications. Frequently the fuel and materials required were readily available and inexpensive. Thus, efficient use of energy and materials was not a major concern and there was no economic driving force to obtain more accurate data. This situation has rapidly changed with the augmented demands for energy conservation; increasingly stringent environmental, waste-reduction, and safety regulations; and intense global competition. Before any plant can be constructed a detailed statement must be prepared outlining the environmental implications. These environmental concerns have extended the need not only for more accurate data but for data on a much wider range of compounds.

When data are taken in response to a pressing design need, they are frequently minimal in quantity and low in accuracy. This problem has produced a literature that contains sparse data on many substances and is often of low quality. High-quality data sponsored by industry are often proprietary and out of the public domain. Unfortunately, support for collection of such data in the United States has become scarce, and relatively few new experimental data are being measured.

When assessing the quality of data, their end-use in technology and processing must be kept in mind, especially in areas where off-design performance would place process guarantees at risk. For example, an error of plus or minus ten percent in the predicted relative volatility can be tolerated in many tower designs. On the other hand, an error of only one percent in the estimated relative volatility of a close-boiling system, such as propane-propylene, translates to approximately a ten percent variation in the number of trays or reflux ratio in a C_3-splitter design. Likewise, an error of two to five percent in liquid density is acceptable for most process design computations but certainly not for custody transfer applications.

Over the years errors have been uncovered in the basic physical constants for some of the most common components. For example, in a conventional olefins plant, the depropanizer is designed so as to meet the purity specifications on the MAPD (methyl acetylene/propadiene) in the bottom product. The predicted reflux ratio or the number of theoretical trays is therefore quite sensitive to the estimated phase equilibrium for these key C_3 trace impurities. Using early-1980s data, one would obtain a reflux ratio of 1.59 for the design scenario summarized in Table 4.2. Using current data available from a commercial simulation program, such as ASPEN PLUS™, the reflux ratio is 1.13. This forty percent difference in the reflux ratio was traced to inconsistencies in the critical pressure and the acentric factor of propadiene as shown in Table 4.3. As indicated, Daubert and Danner (1994) updated these values as part of their data compilation effort for the Design Institute for Physical Property Data (DIPPR®). Their recommendations were subsequently adopted by the American Petroleum Institute and other data compilers. The DIPPR values are

Table 4.2. *Basis for depropanizer design*

Temperature: 60 to 180°F
Pressure: 120 to 130 psia

Component	Feed rate kmol/hr
Propylene	493.6
Propane	16.7
Propadiene	30.0
Methyl acetylene	19.3
Butenes	240.8
n-Butane	7.6
C_{5+}	102.6

Column specifications: 62 trays
Total condenser
Distillate rate: = 559.4 kmol/hr
Design specification: 0.5 kmol MAPD in bottoms
Design variable: Reflux ratio

Table 4.3. *Evolution of propadiene physical constants*

Source	P_c, psia	ω
Reid and Sherwood (1966)	640.9	0.086
Reid, Prausnitz, and Sherwood (1977)	640.2	0.313
API[a] (1980)	640.7	0.313
ASPEN PLUS (1982)	640.7	0.313
Daubert and Danner (extant 1986)	793.4	0.159
API[a] (1987)	793.4	0.160

[a] American Petroleum Institute Technical Data Book as of indicated date.

now accessible in most of the simulation programs. With the revised constants, methyl acetylene is now correctly identified as the controlling component in the design of this unit.

A number of similar studies highlight the impact of uncertainties in the physical and thermodynamic property values on the design of various types of equipment. Table 4.4 lists some of these cases.

Erroneous properties can lead to substantial increases in the capital or operating costs for equipment such as heat exchangers, compressors, and expanders. The sizing of compressors depends on the density of the fluid. Errors in the heat transfer coefficient are especially important in the case of expensive exchangers such as found in the cryogenic industry. A number of detailed examples of the effects of inaccuracies in physical properties are given by Zudkevitch (1982).

Baldwin, Hardy, and Spencer (1987) examined the prediction of the transport properties of a water–methane–inorganic gas vapor as it was cooled from 550 to 295 K to remove the bulk of the water. They found deviations between the predicted and experimental values as large as seventy percent for the thermal conductivity, ninety seven percent for the viscosity, and forty three percent for the surface tension. This finding is particularly surprising for a condensing liquid that was essentially pure water. As indicated in Table 4.4, such large deviations can have a serious impact on equipment design.

Starling, Kumar, and Savidge (1982) investigated the thermophysical property requirements relevant to the syngas industry. To compete with the petroleum and natural gas industry, syngas design methods have to be very efficient. This efficiency requires design and operation of process equipment within close tolerances, which, in turn, require reliable property estimation methods. The accuracy requirements specified by Starling et al. are given in Table 4.5.

Table 4.4. *Effects of property estimate errors*

Property	Error in estimate	Resulting effect	Reference
Critical pressure Critical temperature	10% 5%	25% inaccuracy in specific heat near the critical point (for a light gas mixture)	Zudkevitch (1982)
Thermal conductivity	20%	9.2% error in boiling coefficient of reboilers	Squires and Orchard (1968)
Thermal conductivity	18%	12.3% error in heat transfer coefficient	Mani and Venart (1973)
Density	20%	10% error in boiling coefficient of reboilers	Squires and Orchard (1968)
Density	20%	18% error in pressure drop calculations	Williams and Albright (1976)
Viscosity	20%	4% error in pressure drop calculations	Williams and Albright (1976)
Density Viscosity Thermal conductivity Heat capacity	50% 50% 50% 50%	Heat transfer coefficient 110% high if all estimates are on high side; 60% too low if all estimates are on low side.	Najjar, Bell, and Maddox (1981)

Table 4.5. *Accuracy requirements for syngas processes*

Property	Specification
Density	0.1%
Temperature	0.02 K
Pressure	0.01%
Mol percent	Larger of 0.005 or 0.1% of mol%
Enthalpy	2.5 kJ/kg
Viscosity	1%
Thermal conductivity	1%

4.5 The computer – its advantages, its pitfalls

In the past, data were usually tabulated in books, which were updated from time to time. These books frequently took the form of loose-leaf notebooks. Today this type of hard copy is often only of secondary interest. The primary objective is to develop an electronic form of the database. For the storage of large databases and for providing the computational power to calculate properties at many different conditions of temperature, pressure, and composition, the computer is clearly the correct and necessary tool. Ten years ago this kind of software was available only on a mainframe computer because of the extensive memory and CPU requirements. With the introduction of more powerful PCs and workstations, however, many databases and prediction algorithms were transferred to these platforms. In the next few years the CD-ROM will emerge as the preferred method for distributing large databases and software packages as many potential users add disk readers to their PCs, providing ready access to thermophysical data packages for a much wider group of users in an economical and efficient form.

The interface between the computer database and the user frequently serves also as an information barrier. Whereas in previous times the user may have searched out the data directly, thus seeing the deviations in experimental values or contending with the approximations that were required to use a prediction method, the computer can effectively insulate the user from these considerations. The novice user is therefore susceptible to using data that are seriously in error. It is important for the computer system to provide warnings of the potential pitfalls and thus protect the inexperienced or unskilled user.

Kistenmacher and Burr (1983) warn that there is a general lack of confidence in the quality of the data predicted by computer models. The important question is therefore not the availability of data, but whether the data are accurate enough to enable a satisfactory process design to be made. In the past, many companies maintained a staff of highly qualified data analysts to serve their engineering needs. Recently, however, these data groups have been severely downsized or eliminated. On the other hand, computer software specialists are increasing in number. They will be responsible for incorporating databases into the simulation packages. This situation is also found in some of the companies that are developing commercial simulation packages. The pitfall is that while these individuals can do a good job of incorporating databases they will not likely detect erroneous data (Selover and Spencer 1987). In the past, an engineer searching for a piece of data would access the most recent values as they might be found on a chart or in a table specific for the component of interest. The incorporation of new data into a large electronic data system is a significant effort requiring advanced scheduling and careful oversight. Thus, the inertia is large. It is important that the database be brought up to date frequently and comprehensively if the newest information is to be used in the design.

Whether it is contained within a process simulator or is used as a stand-alone data package, a smart data program is clearly what the system designer needs. It should provide warnings when the calculations enter a region of questionable accuracy or applicability. It should be able to do a sensitivity analysis and let the user know which of the properties are the most critical to the design and what the estimated accuracy of these properties is. The ultimate goal is a computer with artificial intelligence. It would learn the best method to apply under different conditions and relieve the engineer from making many decisions along the way. It must, however, advise the user of the decisions that it has made and the basis for these decisions. Unfortunately, the development of a simulation program of this caliber is unlikely any time soon, even for the simpler, conventional systems.

Along with the emergence of faster and high-capacity computers at bargain basement prices has come the use of molecular modeling to calculate properties. More and more calculations are being made using Monte Carlo or molecular dynamic methods. These techniques provide an excellent tool for increasing our understanding of the relationships of thermophysical properties but are far too time-consuming and limited in their applicability to be generally useful in process simulation (Albright 1985). On the other hand, as these techniques improve, they will be applied by the data specialist to generate data difficult or impossible to obtain experimentally (Boston, Britt, and Tayyabkhan 1993).

4.6 Data compilations

As plants have become large, the need for better data also becomes more important. Overdesign or a faulty design have the potential to produce enormous costs in capital investment, feedstock requirements, byproduct disposal, and energy requirement. Thus, the astute engineer will seek accurate data and reliable prediction methods to meet the data requirements for design simulation.

The time scale required for the measurement of all experimental data pertinent to a process is generally orders of magnitude longer than the time frame allotted for its design. Thus, the engineer must rely on available data and prediction methods.

The selection of reliable numerical data is not an easy task. The accuracy claimed by many researchers is frequently significantly better than can be justified when the data are juxtaposed with data from other laboratories. Although individual data sets are said to have accuracies of plus or minus two or three percent, the scatter among these sets may be as high as plus or minus twenty five percent (Shaw 1984) (Dean 1977). Phase equilibrium measurements often show a wide variation and fail to meet standard thermodynamic consistency tests. Thus, the evaluation of the data is an important ancillary to the work and has lead to the development of a number of data centers that have taken on the responsibility of collecting, evaluating, and publishing databases. In Table 4.6 some of the major data centers are listed.

Table 4.6. *Some major data compilations*

Full name	Abbreviation	Address	Type of data	Output
American Petroleum Institute	API	American Petroleum Institute, 1220 L St., NW, Washington, DC 20005	Pure and mixture data relating to hydrocarbons and petroleum fluids	Hard copy and electronic formats. Catalog[a]
Beilstein Institute for Organic Chemistry	BEILSTEIN	Beilstein Institute, Varrentrappstrasse 40-42, D-60486 Frankfurt am Main, Germany	Structural, synthetic, chemical, thermophysical, spectral, electromagnetic, and mechanical data for wide range of organic chemicals	Beilstein handbook – hard copy and electronic on-line format Catalog
Center for Applied Thermodynamic Studies	CATS	University of Idaho, Department of Mechanical Engineering, Moscow, ID 83811-1101	Tables, charts, and computer programs for determination of fluid thermophysical properties including equations of state for pure fluids and mixtures	Published in *J. Phys. Chem. Ref. Data*, *Int. J. Thermophys.*, *Fluid Phase, Equilib.*, ASHRAE handbooks
Center for Information and Numerical Data Analysis and Synthesis	CINDAS	Purdue University, Purdue Industrial Research Park, 2595 Yeager Rd., West Lafayette, IN 47906	Mechanical and thermophysical properties of metals, alloys, ceramics, composites, aerospace structural materials, and IR detector sensors	Reference publications in open literature. Listing available from CINDAS
Design Institute for Physical Property Data	DIPPR®	American Institute of Chemical Engineers, 345 E. 47th St., New York, NY 10017	Thermophysical properties of pure chemicals (elements, inorganic and organic compounds) and mixtures	Loose-leaf notebooks, books, electronic formats Catalog[a]

Table 4.6 (Cont.)

Table 4.6. *(Continued)*

Full name	Abbreviation	Address	Type of data	Output
Deutsche Gesellschaft für Chemisches Apparatewesen, Chemische Technik und Biotechnologie e.V.	DECHEMA	Theodor-Heuss-Allee 25, D-60486 Frankfurt am Main, Germany	Pure component and mixture thermophysical properties for common chemicals, coal chemicals, and electrolytes; phase equilibria. Safety properties for flammable substances and mixtures	Handbooks, electronic format for in-house license, on-line. Catalog[a] (See Tables 4.10 and 4.11.)
Environmental Chemicals Data and Information Network	ECDIN	Environment Institute, Joint Research Center, 21020 Ispra (Varese), Italy	Physical properties, health, safety, environment, transport, storage data, and regulations	Electronic format on CD-ROM and on-line
Engineering Sciences Data Unit, Ltd.	ESDU	251-259 Regent St., London, W1R 7AD, UK	Thermophysical properties of common organic chemicals	Hard copy as data sheets
Gmelin Institute for Inorganic Chemistry	GMELIN	Varrenstrappstrasse 40-42, D-60486, Frankfurt am Main, Germany	Chemical and thermophysical properties of elements, inorganic and organic metallic compounds, and solid, liquid, and gaseous states	Published in *GMELIN Handbook of Inorganic and Organometallic Chemistry*, electronic on-line format
Landolt-Börnstein		Springer-Verlag New York, Inc., 175 Fifth Ave./13th Floor, New York, NY 10010	Particle physics, atomic, and molecular physics, crystal data, thermophysical properties, phase equilibria, mechanical and electrical properties	Landolt-Börnstein handbook with cumulative index and search software. Catalog
National Institute of Standards and Technology	NIST	Standard Reference Data, 221/A320 Physics Building, Gaithersburg, MD 20899	Analytical data, thermophysical properties for pure components and mixtures. Data for electrolytes, ceramics, safety, and kinetics	Published in *J. Phys. Chem. Ref. Data*, and monographs. Electronic format, on-line. Catalog[a]

Table 4.6 (Cont.)

Table 4.6. *(Continued)*

Physical Property Data Service	PPDS	National Engineering Laboratories, East Kilbride, Glasgow, Scotland G750QU, UK	Thermodynamic, transport, and phase equilibrium data for pure components and mixtures. Special data files on equations of state, petroleum fractions, steam, refrigerants, and aqueous solutions	Computer package available for license. On-line[a]
STN International	STN	2540 Olentangy River Rd., P.O. Box 3012, Columbus, OH 43210-0012	Numeric databases covering analytical, structural, chemical, thermophysical, electrical, and safety properties	On-line access to vendor databases including DIPPR® pure component database. Catalog[a]
Technical Database Services, Inc.	TDS	135 W. 50th St., Suite 1170, New York, NY 10020-1201	Numerica™ service providing numeric data for chemicals including thermophysical, phase equilibria, safety, and health properties as well as data on electrolytes and plastics	On-line access to vendor databases. Computer packages available for license including PC version of DIPPR® Data Compilation. Catalog[a]
Thermodynamics Research Center	TRC	TRC Data Distribution, TEES Fiscal Office, the Texas A&M University System, College Station, TX 77843-3124	Pure components and mixtures – thermophysical and spectroscopic data	Loose-leaf data sheets, electronic databases, and research publications. Catalog[a]

[a] These sources also incorporate correlation or prediction methods in their products.

Some of these databases were prepared many years ago and have not been updated primarily because of lack of adequate funding. Other databases provide values but give little or no information as to where the numbers came from. Quoting from Boston et al. (1993), "Two large compilations of pure species data have emerged as standards for process engineering calculations, particularly for vapor/liquid systems: the DIPPR compilations, ... and the PPDS® compilation from the Physical Property Data Service of the National Engineering Laboratory in the U.K. The most comprehensive compilation of physical property data for mixtures is the Dortmund Data Bank (DDB®), developed in Germany under the sponsorship of DECHEMA, the German counterpart of AIChE."

4.6.1 Pure component data

Many pure component properties may be of interest to the design engineer. In Table 4.7 are listed the data contained in the DIPPR Data Compilation (Daubert and Danner 1994). The properties can be divided into three major categories: descriptive information, universal constants, and temperature-dependent properties. The first category contains ways of identifying the components such as molecular formulas, various names, and CAS registry number. Providing such identifiers is important, because a compound may be specified in many ways, and for complex molecules in particular, numerous naming conventions exist. The universal constants include molecular weight, critical properties, dipole moment, etc. Temperature-dependent properties cover such things as vapor pressure, ideal gas heat capacity, density, thermal conductivity, etc. No pressure dependency has been provided in this database for the properties. For most applications the pressure dependency is much less important than the variation with temperature. Although this example does not cover all the properties that may be of interest, the selection was made by a group of data experts from industry, thus indicating their priorities.

4.6.2 Mixture data

A primary source of mixture data in the western scientific world is DECHEMA, a German technical nonprofit society founded in 1926 with the object of promoting chemical engineering and biotechnology. The Chemistry Data Series has become the depository of many types of mixture data. The stated primary purpose of the series is to provide chemists and engineers with the data needed for process design and optimization. In 1994 the series exceeded forty volumes with 25,000 pages and several thousand compounds. Heat capacity, enthalpy and entropy, phase equilibria, PVT, and transport properties have been emphasized. Most of the compilation covers low to moderate-boiling industrially

Table 4.7. *Properties included in the DIPPR® pure component data compilation*

Type of data	General category	Specific form
Descriptive information	Molecular formula	
	Molecular structure	
	Names	Common
		Chemical abstracts
		IUPAC
		Synonyms
	CAS registry number	
Constants	Molecular weight	
	Critical properties	Temperature
		Pressure
		Volume
		Compressibility factor
	Melting point at 1 atm	
	Triple point	Temperature
		Pressure
		Enthalpy of fusion
	Normal boiling point	
	Liquid molar volume at 298.2 K	
	Ideal gas (298.2 K)	Enthalpy of formation
		Gibbs energy of formation
	Enthalpy of combustion at 298.2 K	
	Molecular characterization factors	Acentric factor
		Radius of gyration
		Solubility parameter
		van der Waals volume
		van der Waals area
	Dipole moment	
	Refractive index at 293.2 K	
Temperature-dependent properties	Vapor pressure	
	Heat of vaporization	
	Second virial coefficient	
	Surface tension	
	Density	Solid
		Liquid
	Heat capacity	Solid
		Liquid
		Ideal gas
	Viscosity	Liquid
		Vapor
	Thermal conductivity	Liquid
		Vapor

important organic compounds in the fluid phase. Volumes on polymer-solvent systems and salt solutions have recently been added. Each book contains a succinct discussion of the properties in and use of the tables. Table 4.8 summarizes the contents of the DECHEMA Chemistry Data Series.

The DECHEMA Chemistry Data Series is just a subset of the complete physical property data available from DECHEMA. In addition to the numerical data, nonnumerical facts and chemical safety data are provided in the CHEMSAFE® software package. These data together with many other databases are available on electronic media. A data retrieval system, DETHERM®-SRD, accompanies

Table 4.8. *Contents of the DECHEMA Chemistry Data Series*

Vol.	Title	Description
1	Vapor–liquid equilibria	More than 15,300 isothermal or isobaric VLE data sets. Constants of the UNIQUAC, NRTL, Wilson, van Laar, and Margules correlations are given. Thermodynamic consistency is tested by two methods. Vapor pressure constants for the Antoine equations are listed.
2	Critical data of pure substances	Critical temperature, pressure, volume, and compressibility for more than 850 and compounds are given. Literature through 1993 is covered.
3	Heats of mixing	Data are presented in tables and graphs for binary and ternary mixtures. The corresponding regression parameters are given.
4	Selected compounds and binary mixtures	Key data of 14 binary test mixtures and their 20 pure components are presented. For the binary mixtures, the authors present VLE, volumes and heats of mixing, liquid viscosity, thermal conductivity, surface tension, and refractive index data.
5	Liquid–liquid equilibria	LLE data for more than 2,000 binary, ternary, and quaternary systems are given. The data are correlated using the NRTL and UNIQUAC models.
6	Vapor–liquid equilibria: mixtures of low-boiling substances	Data for more than 200 systems containing two or more components, all of which have boiling points lower than that of *n*-decane, are tabulated and presented in graphical form. A comprehensive literature survey is included. Parameters have been fitted to four generalized equations of state.
8	Solid–liquid equilibria	Data for more than 180 binary solid–liquid systems are given. Also listed are the parameters for activity coefficient models or equations of state.
9	Activity coefficients at infinite dilution	Includes approximately 29,000 sets of activity coefficients at infinite dilution.

Table 4.8 (Cont.)

Table 4.8. *(Continued)*

Vol.	Title	Description
10	Thermal conductivity and viscosity	Recommended values of the thermal conductivity and viscosity of binary systems consisting of a noble gas with another noble gas or mixtures of water with the lower alcohols are given. A literature survey listing sources of the relevant data for approximately 1,800 systems is included.
11	Phase equilibria of electrolytes	VLE, solubility, and enthalpy data for binary electrolyte solutions are presented in tables and diagrams.
12	Electrolyte solutions	A comprehensive collection of properties of electrolyte solutions in which methanol or ethanol is the solvent is given.
13	VLE of electrolyte solutions	A compilation of VLE data for electrolyte solutions obtained from the open literature together with a large collection of values measured at the Martin Luther University, Halle-Wittenberg, is given.
14	Polymer solutions	Data for binary polymer-solvent systems have been compiled. Included are solvent activities and critical solution temperatures.

Titles and Authors of DECHEMA Chemistry Data Series, Theodor-Heuss-Allee 25, D-60486 Frankfurt am Main, Germany.

(1) Vapor–liquid equilibrium data collection. Gmehling, J., Onken, U., Arlt, W., Grenzheuser, P., Weidlich, U., Kolbe, B. and Rarey, J.
(2) Critical data of pure substances. Simmrock, K. H., Janowsky, R. and Ohnsorge, A.
(3) Heats of mixing data collection. Christensen, C., Gmehling, J., Rasmussen, P., Weidlich, U. and Holderbaum, T.
(4) Recommended data of selected compounds and binary mixtures. Stephan, K. and Hildwein, H.
(5) Liquid–liquid equilibrium data collection. Arlt, W., Macedo, M.E.A., Rasmussen, P. and Sørenson, J. M.
(6) Vapor–liquid equilibria for mixtures of low boiling substances. Knapp, H., Döring, R., Oellrich, L., Plöcker, U., Prausnitz, J. M., Langhorst, R. and Zeck, S.
(8) Solid–liquid equilibrium data collection. Knapp, H., Teller, M. and Langhorst, R.
(9) Activity coefficients at infinite dilution. Gmehling, J., Tiegs, D., Medina, A., Soares, M., Bastos, J., Alessi, P., Kikic, I., Schiller, M. and Menke, J.
(10) Thermal conductivity and viscosity data of fluid mixtures. Stephan, K. and Heckenberger, T.
(11) Phase equilibria and phase diagrams of electrolytes. Engels, H.
(12) Electrolyte data collection. Barthel, J. and Neueder, R.
(13) Vapor-liquid-equilibrium-data for electrolyte solutions. Figurski, G.
(14) Polymer solution data collection. Hao, W., Elbro, H. S. and Alessi, P.

Table 4.9. *Electronic databases available from DECHEMA*

Data Type	Amount of Data Included
VLE (normal-boiling substances)	15,000 sets
VLE (low-boiling substances)	250 systems
LLE (normal-boiling substances)	8,450 sets
SLE (normal-boiling substances)	6,500 sets
Activity coefficients (infinite dilution)	29,000 points
Eutectic data	6,000 sets
Solubilities	48,000 sets
Azeotropic compositions	36,000 points
Excess enthalpies	9,600 sets
Excess heat capacities	720 sets
Diffusion coefficients	9,000 sets; 1,600 mixtures
Thermal conductivities	4,200 sets
Viscosities	7,200 sets
Safety parameters	17,000 sets; 2,080 compounds and mixtures
Anthracite coals	8,700 sets; 2,100 pit coal samples
Coal oils and chemicals	31,800 sets
Electrolyte data	54,000 systems

these computer programs. Table 4.9 is a listing of the available data collections from DECHEMA.

An extensive bibliographic vapor–liquid equilibria database has been compiled by Wichterle et al. (1991). Almost 9,000 references are cited covering more than 12,000 systems. The citations cover the period from 1900 to 1991. Both organic and inorganic compounds are covered, including aqueous solutions of electrolytes. Mixtures containing up to nine components are listed. Accompanying the book is a diskette containing the bibliographic data and a retrieval program. A review of this package has been provided by Lide (1994).

4.7 Correlation and prediction methods

In many cases one will not be able to find, or the process simulation program will not be able to access, the physical property data for the components at the temperature and pressure conditions of interest. Even if the required pure component data can be found, the need to treat mixtures of these components will almost certainly require the use of correlation of prediction methods. Correlation refers to the use of a limited amount of data to develop regression constants for a model, which then permits the interpolation of data at intermediate conditions and cautious extrapolation to conditions outside the range of the available data. Prediction – the calculation of data of a particular type without any data of that type being available, at least for the component or mixture of interest – would include methods based on corresponding states or group contribution, for example.

Correlation and prediction methods can be classified into three groups: theoretical, empirical, or semitheoretical. In a theoretical method, the equations are based on sound physical theory. Thus, they can be used reliably for interpolation and extrapolation as long as no assumptions in their development are violated.

An empirical model has no basis in theory. The available experimental data are simply fit to some arbitrary function. Interpolation can generally be made between the available data, but such models should not be extrapolated to other physical conditions or to other compounds or mixtures. Furthermore, any attempt to calculate derivative properties, those based on the slope of any empirical model, are likely to be significantly in error even within the range of the data.

The last class of correlation and prediction methods is the semitheoretical. In this case a model is developed based on rigorous principles. At some point in the development, however, either simplifying assumptions and approximations have to be made or a functional form has to be developed that contains parameters that cannot be measured. The functional form is maintained, but the unavailable parameters are replaced with regression constants. This is the realm of thermodynamics that has become known as molecular thermodynamics. It is perhaps best espoused by Prausnitz, Lichtenthaler, and de Azevedo in their book *Molecular Thermodynamics of Fluid Phase Equilibria* (1986). The goal is to get as much information from as little data as possible. Using these principles, models have been developed that can handle many different constituents (Evans 1990). The parameters, although not theoretical, often have physical significance that allows them to be evaluated at least qualitatively. The resulting correlations can be extrapolated within limits with much more confidence than empirical equations.

Correlation and prediction methods can also be classified as specific or generalized. Specific models are unique to a particular compound or mixture. They can be very accurate. For example equations have been derived to describe the PVT behavior of water, which contains more than thirty constants. Accurate and reliable thermodynamic derivative properties (enthalpy, entropy, expansivity, etc.) can be obtained from these equations. These are a valuable predictive tool for the engineer who is doing a steam balance. If, however, significant amounts of some other component are mixed with the steam, such as in a furnace quench system, the pure water model equations are of little value. Furthermore, one cannot know how to blend the numerous constants to account for the additional components.

The second class of methods are called generalized. In this case the constants of the model have not been regressed specifically for individual compounds. Rather, they can be predicted from some other property of the compound. The largest group of methods in this class are the corresponding states models. According to the concept of corresponding states, all fluids will behave identically if they are analyzed in terms of their reduced properties, T_r and P_r.

$$T_r = T/T_c \qquad P_r = P/P_c \tag{4.1}$$

This simple two-parameter corresponding states principle can be applied with reasonable accuracy to nonpolar fluids that are relatively small and symmetrical. Within these limitations, however, numerous properties of a compound or mixture can be predicted knowing only the critical constants of the compounds involved. For larger molecules, a third parameter, such as the acentric factor or radius of gyration, must be added to account for the size and shape of the molecules. And for polar compounds a fourth parameter is needed to account for the polar interactions. These methods, although not as accurate as the specific methods, are much more generally applicable and can usually provide a reasonable estimate when no other method may be applicable. For this reason, they have been incorporated in a wide range of simulation and other process design programs.

One other type of prediction method has become quite successful: the group contribution method. In this approach a molecule is looked at in terms of its constituent groups. For example, *n*-propyl alcohol, CH_3-CH_2-CH_2-OH, is represented as 1 CH_3 group, 2 CH_2 groups, and 1 OH group. Each group is assumed to contribute a unique amount to the total property of the molecule. This amount is the same for every molecule in which that group is found. By regressing extensive amounts of data, the contribution of each group to any property (or to the numerical value of a parameter in a model) can be determined. A value can be calculated for the property of interest for any molecule that can be constructed from the available groups. The major pitfall of this approach is the assumption that a group always contributes the same amount no matter what other groups are in the molecule or what the group's location is on the molecule. The deficiency of this assumption can be diminished by defining more complex groups that take into account the adjoining atoms. How complex the groups must be depends on the property and the accuracy required. Numerous examples exist of successful group-contribution methods being used for property prediction.

4.7.1 Compendiums of correlation and prediction methods

Many publications appear in the scientific literature describing methods of correlating or predicting various thermophysical properties. Judging the value of these methods without further testing is often difficult. Sometimes the authors test the model against a limited set of compounds or mixtures or examine only a small range of temperature and pressure. Thus, the individual engineer finds selecting a calculation procedure difficult. A number of research groups have undertaken the evaluation and recommendation of correlation and prediction methods. The primary ones are listed in Table 4.10. In addition, Mullins, Radecki, and Rogers (1994) discuss the basics of data correlation and

Table 4.10. *Sources of correlation and prediction methods*

Reference	Coverage
The Properties of Gases and Liquids, Reid, R. C., Prausnitz, J. M., and Poling, B. E., (1987)	Estimation of pure component constants, PVT behavior, departure functions, heat capacities, critical points, vapor pressure, enthalpy of vaporization, phase equilibria, viscosity, thermal conductivity, diffusion coefficients, surface tension. Includes many available methods. Covers all types of compounds.
Technical Data Book: Petroleum, Refining, American Petroleum Institute (loose-leaf format, continuously updated)	Contains 15 chapters on characterization of fractions, critical properties, vapor pressure, density, thermal properties, VLE, water solubility, surface tension, viscosity, thermal conductivity, diffusivity, and combustion. Emphasis on petroleum fluids. Extensive evaluation of methods. Only the recommended procedures are included. Available in loose-leaf notebook or in computerized form with subroutines for each of the recommended procedures.
Data Prediction Manual, Danner, R. P. and Daubert, T. E., (1987)	Emphasizes methods for petrochemicals and organometalllics. Covers physical, thermodynamic, and transport properties. Extensive evaluation of methods. Only the recommended procedures are included. Loose-leaf notebook form.
Thermodynamic Analysis of Vapor–Liquid Equilibrium, Gess, M. A., Danner, R. P. and Nagvekar, M. (1991)	Discusses activity coefficient models and equations of state for the calculation of VLE. Lists 104 systems that have been shown to be thermodynamically consistent. Presents guidelines for the selection of models for a particular type of system. Computer diskette that accompanies book allows the regression of VLE data by a number of activity coefficient models and two equations of state using four different mixing rules.
Ullmann's Encyclopedia of Industrial Chemistry, U. Onken and H.-I. Paul (1990)	Estimation methods are provided for critical properties, PVT behavior of gases and liquids, caloric properties, phase equilibria, enthalpies and Gibbs energies of formation, and the transport properties.

cite references to methods of predicting many thermophysical properties. Also, many of the sources listed in Table 4.6 provide correlation or prediction methods as indicated by the footnote to that table.

The references cited in Table 4.10 provide a means of predicting required properties for many types of compounds and systems. For systems containing highly polar, hydrogen-bonding, associating, or electrolyte components or for systems with multiple phases, the design engineer will quickly run into limitations and numerous questions as to how to best proceed. A number of organizations have combined the databases and the calculation methods to produce packages that will simulate the design of processes. In the final sections of this chapter, we will focus on the thermophysical property and phase equilibrium framework of some of the more popular commercial simulation programs.

4.8 Simulation programs

A wide array of commercial simulation programs is available today. A partial list of some of the more general-purpose simulation packages is given in Table 4.11. Broader-based overviews together with coverage of several special-purpose packages are given elsewhere (Simpson, 1994, Hydrocarbon Processing, 1996, and Chan et al., 1991). All of these programs, which are generally attainable

Table 4.11. *Partial list of commercial software packages available for process design or simulation*

Software	Vendor
ASPEN PLUS™	Aspen Technology, Inc. Cambridge, MA
CHEMCAD III™	Chemstations Houston, TX
HYSIM™	Hyprotech Calgary, Alberta, Canada
MAXSTILL PLUS SPAN	Kesler Engineering East Brunswick, NJ
PD-PLUS	Deerhaven Technical Software Burlington, MA
PRO/II™	Simulation Sciences Brea, CA
PROSIM®	Bryan Research & Engineering Bryan, TX

on a number of platforms, attempt, with varying degrees of success, to meet the diverse demands of the engineer. This section will concentrate on three of the programs with which we are most experienced: ASPEN PLUS^{TM}, HYSIM^{TM}, and PRO/II^{TM}.

The term *simulation*, as used here, indicates deterministic calculations for energy and mass balances occurring in a chemical process at the specified operating conditions. Phase equilibrium and thermal property data/estimates are of prime importance in calculations of this nature, but all of the properties discussed in this chapter have a role from conception to marketing of the final product, the operating plant.

All three of the foregoing programs are supported by extensive flexible primary pure component data banks as summarized in Table 4.12. Displaying many of the desirable ingredients of a high-quality database, as defined in this chapter, these packages supply certified universal constants and temperature-dependent parameters for 1,500 or more defined organic and inorganic compounds. All of these component libraries are essentially extensions of the previously mentioned DIPPR database with additional parameters and supplemental components furnished by the individual vendor. Properties are usually accessed during the setup phase of program execution using either a component ID or alias. For example, methanol is defined as CH4O in ASPEN PLUS, METHANOL in HYSIM, and MEOH in PRO/II. Along with the primary property database, additional pure component packages are provided, such as those listed in Table 4.13, for special applications such as solids handling and solution chemistry, and for user-supplied data.

For defined components not available in their databanks, all three programs present state-of-the-art estimation routes, primarily group contribution methodology, for calculating the universal constants and temperature-dependent coefficients required. This capability is important, especially when dealing with specialty chemicals, novel heat transfer fluids and solvents, and first-of-a-kind, emerging technologies.

For undefined components, namely petroleum fractions, several characterization (estimation) routes have been integrated in the respective programs, including the recommended API *Technical Data Book* procedures. PRO/II, widely used by industry for refinery design applications, also supplies a proprietary characterization route that extrapolates well for heavy petroleum fluids. These routes translate the available assay data on the fraction into pure component parameters needed by those property methods invoked by the simulator. Typically two of the following three properties must be defined for each fraction in the system: normal boiling point, specific gravity (API gravity) at 60°F, and molecular weight.

Table 4.12. *Overview of pure component data banks*

	ASPEN PLUS™ Version 9.21	PRO/II™ Version 3.30	HYSIM™ Version 2.53
Number of components	1557	1552	1499
Universal constants			
Molecular weight	X	X	X
Atomic structure	X		X
Normal boiling point	X	X	X
Normal freezing point	X	X	
Critical temperature	X	X	X
Critical pressure	X	X	X
Critical volume	X	X	X
Critical compressibility factor	X	X	
Acentric factor	X	X	X
Triple point temperature	X	X	
Triple point pressure	X	X	
Specific gravity, 60°F/60°F	X	X	
Standard liquid density, 60°F	X	X	X
Standard API gravity	X	X	
Liquid molar volume at NBP	X	X	
Heat of vaporization at NBP	X	X	
Heat of fusion	X	X	
Heat of combustion	X	X	X
Standard heat of formation, 77°F	X	X	X
Standard free energy of formation, 77°F	X	X	
Gross heating value		X	
Flash point	X		
Autoignition temperature	X		
Flammability limits	X		
Solubility parameter, 77°F	X	X	
Scatchard–Hildebrand volume parameter	X		
van der Waals volume/area parameters	X	X	
Dielectric constant	X		
Dipole moment	X	X	X
Radius of gyration	X	X	X
Refractive index	X		
Aniline point	X		
Octane number	X		
Truncated virial parameter		X	
UNIFAC functional groups	X	X	X

Table 4.12 (Cont.)

Table 4.12. *(Continued)*

	ASPEN PLUS™ Version 9.21	PRO/II™ Version 3.30	HYSIM™ Version 2.53
Number of components	1557	1552	1499
Temperature-dependent parameters			
Vapor pressure	X	X	X
Liquid density – Rackett	X	X	
Liquid density – Costald	X		X
Latent heat – Watson	X	X	
Latent heat – Cavett	X		X
Ideal gas heat capacity	X		X
Ideal gas enthalpy		X	X
Liquid heat capacity	X		
Liquid enthalpy		X	
Ideal gas free energy of formation			X
Vapor viscosity	X	X	
Liquid viscosity	X	X	X
Kinematic viscosity		X	
Vapor thermal conductivity	X	X	
Liquid thermal conductivity	X	X	
Surface tension	X	X	
Water solubility	X		
Virial equation	X		

4.9 Phase equilibrium

The most important industrial application of process simulation is, and will continue to be, establishing the phase equilibrium of a given system or, more precisely, the liquid–vapor flash computation. All such calculations are based on the conditions of thermodynamic equilibrium, that is, that the temperature, pressure, and fugacities of each component are equal in both phases. Fugacities are related to temperature, pressure, and mixture composition using equations of state. Within this framework the isofugacity criterion is usually written as

$$x_i \gamma_i f_i^{o,L} = y_i \phi_i^V P \tag{4.2}$$

Here, P is the pressure, and for each component i, x_i and y_i are the liquid and vapor phase mole fractions, γ_i is the liquid phase activity coefficient, $f_i^{o,L}$ is the pure component liquid fugacity, and ϕ_i^V signifies the vapor phase fugacity coefficient, $(f_i^V / y_i P)$.

In the approach using a single equation of state Equation (4.2), with some

Table 4.13. *Specialty pure component data banks*

Simulation program	Databank	Number of components	Description
ASPEN PLUS™	AQUEOUS	262	Used for electrolyte applications; key parameters are aqueous heat of and Gibbs free energy of formation and heat capacity at infinite dilution.
ASPEN PLUS™	INORGANIC	2,457	Used for solids, pyrometallurgical applications, and electrolyte applications; key data are the enthalpy, entropy, Gibbs free energy, and heat capacity correlation coefficients. Taken primarily from compilation of Knacke, Kubaschewskim, and Hesselmann (1991).
ASPEN PLUS™	SOLID	122	Used for solids and electrolyte applications; key parameters included solid heat capacity coefficients, standard free energy of and heat of formation, and ideal gas heat capacity coefficients.
ASPEN PLUS™ PRO/II™ HYSIM™	IN-HOUSE/ USER DATABANK	Varies with Simulation program	Used to store in-house data specific to a company's technologies or when data are of a proprietary nature not intended for all users of the program.

rearrangement, reduces to

$$K_i = \frac{y_i}{x_i} = \frac{\phi_i^L}{\phi_i^V} \tag{4.3}$$

where K_i is the liquid–vapor equilibrium constant of component i. The same equation of state is then used to estimate both the liquid and the vapor fugacity coefficients through standard thermodynamic relationships.

Among the equations of state that have been applied by most simulation programs are the traditional cubic forms, such as the Peng–Robinson (1976) and the Soave (1972) modification of the Redlich–Kwong equation of state. These equations are normally valid over a wide range of temperatures and pressures for nonpolar and moderately polar hydrocarbon–hydrocarbon and hydrocarbon–

inorganic gas systems and enjoy widespread use in industry, especially for gas processing, refinery design, and reservoir simulation.

Cubic equations require little input information – only the critical properties and acentric factor for the generalized parameters. Additionally, they employ mixing rules with a single adjustable binary interaction parameter, k_{ij}. The interaction parameter may be taken as zero for most hydrocarbon–hydrocarbon pairs, especially at or above ambient temperature. Nonzero k_{ij}'s are required, however, when the hydrocarbons are close-boiling, widely different in size, or paired with inorganic compounds. Finite k_{ij}'s may also be needed for cryogenic applications, critical region computations, and interactions with strongly polar components, such as water. With numerous improvements in the mixing rules over the last fifteen years, including Huron and Vidal (1979), Mathias (1983), Stryjek and Vera (1986), Panagiotopoulos and Reid (1986), and Wong and Sandler (1992), computations based on equations of state have now also been shown to be highly reliable for polar, nonideal chemical systems.

Equations of state offered by ASPEN PLUS^{TM}, HYSIM^{TM}, and PRO/II^{TM} are listed in Table 4.14. Binary interaction parameter data banks, validated to varying degrees, are also furnished for several of these equations.

In the two-equation approach, Equation (4.2) is restated as

$$K_i = \frac{\gamma_i f_i^{o,L}}{\phi_i^V P} \tag{4.4}$$

where the standard state liquid fugacity, $f_i^{o,L}$, is defined as

$$f_i^{o,L} = \phi_i^{(o,V)} P_i^o \exp \int_{P_i^o}^{P} \frac{V_i dP}{RT} \tag{4.5}$$

Here, for each component i, $\phi_i^{(o,V)}$ is the fugacity coefficient at saturation, P_i^o is the vapor pressure, V_i is the saturated liquid molar volume, and R is the gas constant.

In essence, this so-called two-equation approach imposes three corrections on the system: one for the nonideality of the liquid (activity coefficient), one for the nonideality of the vapor (fugacity coefficient), and one for the standard state liquid fugacity to which the activity coefficient must be referred. The fugacity coefficients (liquid and vapor) are, for the most part, calculated from the aforementioned equations of state. Several liquid phase activity coefficient methods are suitable in this framework, such as Wilson (1964), NRTL (Renon and Prausnitz 1968) and UNIFAC (Fredenslund, Gmehling, and Rasmussen 1977). The models integrated by ASPEN PLUS^{TM}, HYSIM^{TM}, and PRO/II^{TM} are enumerated in Table 4.15.

Table 4.14. *Phase equilibrium framework – equations of state*

Equation	ASPEN PLUS™ Version 9.21	PRO/II™ Version 3.30	HYSIM™ Version 2.53
Cubic equations			
Redlich–Kwong[a]	X		X
Redlich–Kwong–Soave	X	X	X
Redlich–Kwong–ASPEN	X		
Peng–Robinson	X	X	X
Graboski–Daubert	X		
Zudkevitch–Joffe			X
Schwartzentruber–Renon	X		
Modified mixing rules			
Huron–Vidal	X	X	
Stryjek–Vera			X
Panagiotopoulos–Reid		X	
Kabadi–Danner		X	X
Holderbaum–Gmehling	X		
Wong–Sandler	X		
Virial equations[a]			
Nothnagel	X		
Hayden–O'Connell	X		X
Generalized			
BWRS	X	X	
Lee–Kesler–Plocker	X		
Braun K 10 (graphical)	X	X	X

[a] Applicable to vapor phase only

The two-equation approach is generally applied for highly nonideal systems, particularly those containing polar and associated components, and for two-phase (liquid–liquid) and three-phase (vapor–liquid–liquid) equilibrium. Because different models are used for each phase, this route is not internally consistent, a drawback that becomes serious at high pressures, especially in the critical region.

The heart of the two-equation framework is the liquid phase activity coefficient model. Using the nonrandom, two-liquid (NRTL) model, an equation common to all the simulation programs, the activity coefficient for a component is given by

$$\ln \gamma_i = \frac{\sum_{j=1}^{m} \tau_{ji} G_{ji} x_j}{\sum_{l=1}^{m} G_{li} x_l} + \sum_{j=1}^{m} \frac{G_{ij} x_j}{\sum_{l=1}^{m} G_{lj} x_l} \left[\tau_{ij} - \frac{\sum_{r=1}^{m} \tau_{rj} G_{rj} x_r}{\sum_{l=1}^{m} G_{lj} x_l} \right] \tag{4.6}$$

Table 4.15. *Phase equilibrium framework – activity coefficient–based*

Equation	ASPEN PLUS™ Version 9.21	PRO/II™ Version 3.30	HYSIM™ Version 2.53
Ideal	X	X	X
Margules		X	X
van Laar	X	X	X
Wilson	X	X	X
NRTL	X	X	X
UNIQUAC	X	X	X
Regular solution	X	X	X
Chien-null	X		X
Generalized			
UNIFAC	X	X	X
Modified UNIFAC	X	X	
Chao–Seader types			
Chao–Seader	X	X	X
Grayson–Streed	X	X	X
Polymers			
Flory		X	
Ionic species			
Extended NRTL	X		

where, $\tau_{ji} = (g_{ji} - g_{ii})/RT$, $G_{ji} = \exp(-\alpha_{ji}\tau_{ji})$, and i, j, l, r are component indices. Parameter g_{ji} is an energy parameter characteristic of the j-i interaction, whereas the parameter α_{ji} is related to the nonrandomness in the mixture. This parameter normally ranges between 0.2 and 0.6 (Prausnitz et al. 1986). In essence, for each possible binary pair in a mixture, three NRTL constants, τ_{ij}, τ_{ji}, and α_{ji}, are required. These parameters can be obtained from binary or multicomponent phase equilibrium data. For an m-component mixture, there are $m(m-1)/2$ unlike binary pairs. Ten parameters would be needed, therefore, for complete coverage of a five-component system through the NRTL equation.

If the model parameters are not available and ideal solution behavior is assumed ($\gamma = 1.0$), the activity coefficient methodology can lead to serious error. A case in point is the first methyl tert-butyl ether (MTBE) plant designed by Kellogg in the mid-1980s. ASPEN PLUS™ was the proper simulation tool, but the lack of adequate NRTL parameters had the following consequences. First of all, methanol, a key reactant, never entered the reaction zone, because the MTBE-methanol azeotrope was predicted incorrectly, and methanol rather than MTBE went out the bottom of the MTBE reactive distillation column.

The water wash column likewise ceased to function, because the model failed to predict an adequate liquid–liquid split on methanol. Furthermore, the lights column could not achieve dimethyl ether separation at a level even close to that suggested by the licensor, because its volatility with respect to the C3 and C4 hydrocarbons was grossly underpredicted.

In effect, the core of the correlation, the model interaction parameters, is mandatory for accurate representation of the key components in the processes. All three of the simulation programs furnish model parameters for a fairly broad range of common organic pairs, but in many cases, especially for first-of-a-kind processes, the parameters are missing. Moreover, even when supplied, the parameters should be tested for applicability for the case at hand.

Model parameters are conventionally estimated through regression using phase equilibrium data from references, such as the previously mentioned DECHEMA publications. When pertinent data are unavailable, they can be calculated from widely used techniques, such as UNIFAC. UNIFAC correlates the activity coefficient in terms of size and energy contributions. These factors have been generalized in the form of the group contribution method discussed earlier. UNIFAC together with its many subsequent modifications provides coverage for numerous components and their mixtures, but deficiencies remain for many areas of prime importance in industry.

Summarizing, all three of the simulation programs provide a rational thermodynamic framework for modeling nonideal chemical processes. Development (selection) of the parameters, as well as identification of those needed for a particular case remains the job of a seasoned data specialist.

4.10 Thermal properties

Process simulation as well as a stand-alone energy analysis of a system requires reliable thermal properties, particularly enthalpy. HYSIM™, ASPEN PLUS™, and PRO/II™ compute system enthalpy in a similar fashion.

When the one-equation methodology is adopted for phase equilibrium, system enthalpy is normally predicted as follows:

$$H_T^P(m) = \sum_i^n y_i \Delta H_{f,i} + \sum_i^n y_i H_{i,T}^o - \left(H_T^o - H_T^P\right) \qquad (4.7)$$

Here $H_T^P(m)$ represents the mixture enthalpy at stream conditions, $\Delta H_{f,i}$ is the ideal gas heat of formation of component i, $H_{i,T}^o$ is the ideal gas enthalpy of component i at system temperature, and $H_T^o - H_T^P$ equals the isothermal enthalpy departure (pressure correction). Heat-of-formation data are usually directly retrieved from the pure component data bank(s) supplied by the respective programs. Ideal gas enthalpy is related to system temperature (T) as shown here.

$$H_{i,T}^{o} = \int_{T_{\text{ref}}}^{T} C_{Pi}^{o} dT \tag{4.8}$$

where T_{ref} is an arbitrary reference temperature and C_{Pi}^{o} is given by some functional form such as

$$C_{Pi}^{o} = A + BT + CT^2 + DT^3 + ET^4 + FT^5 \tag{4.9}$$

A through F are component-specific parameters usually furnished by the program for all data bank components. These coefficients, for the most part, have been taken from previously mentioned sources, such as Reid et al. and the API *Technical Data Book*, or regressed from data provided by TRC, DIPPR, and other related compilations. The ideal gas enthalpy is generally one of the most reliable properties available from a simulation program.

Using this framework, the isothermal enthalpy departure may be calculated from any one of the equations of state listed in Table 4.14 through the following thermodynamic relationship:

$$\left(\frac{\partial H}{\partial P}\right)_T = \frac{P}{\rho} - RT + \int_0^{\rho} \left[R - T\left(\frac{\partial P}{\partial T}\right)\right]\frac{d\rho}{\rho^2} \tag{4.10}$$

where ρ is the system density.

Enthalpy departures are also approximated from corresponding states theory, such as Curl–Pitzer (1958) or Lee–Kesler (1975), which relate the overall departure to a simple fluid term and a size correction term that accounts for molecular acentricity as shown here:

$$\left(\frac{H_T^o - H_T^P}{RT_c}\right) = \left(\frac{H_T^o - H_T^P}{RT_c}\right)^{(0)} + \omega\left(\frac{H_T^o - H_T^P}{RT_c}\right)^{(1)} \tag{4.11}$$

Both the simple fluid and size correction terms are functions of T_r and P_{rj}; the reduced temperature, T/T_c; pressure, P/P_c; and the acentric factor, ω. The Lee–Kesler method is accessible through all three of the simulation programs, whereas PRO/II also offers the Curl–Pitzer approach.

Departures estimated from these equations are typically valid for hydrocarbon, inorganic gas, and slightly polar organic systems. The approach is widely applied for gas processing, syn-fuel, and refinery-related design. Enthalpies estimated in this fashion are reliable, with deviations seldom exceeding plus or minus twelve kilojoules per kilogram. Overall, corresponding states tends to predict more reliable liquid phase enthalpies and, therefore, slightly more accurate latent heats of vaporization.

When an activity coefficient route is invoked for phase equilibrium, vapor enthalpy is normally predicted as just discussed. The following route, however, is followed for liquid enthalpy:

$$H^L(m) = \Sigma x_i \left(H_i^V - \Delta H_i^{\text{VAP}}\right) + H^{\text{MIX}} \tag{4.12}$$

Here $H^L(m)$ is the liquid phase enthalpy, H_i^V is the saturated vapor phase enthalpy of component i, ΔH_i^{VAP} is the latent heat of vaporization of component i, and H^{MIX} is the liquid phase heat of mixing. The component vapor enthalpy is predicted from Equations (4.8)–(4.10). Latent heat is determined from one of many variations of the Watson (1943) equation. Watson-based latent heats are usually reliable within plus or minus four and a half kilojoules per kilograms.

The heat-of-mixing term in Equation (4.12) is calculated rigorously as shown here:

$$H^{\text{MIX}} = -RT^2 \Sigma x_i \left(\frac{\partial \ln \gamma_i}{\partial T}\right) \tag{4.13}$$

where γ_i is the liquid phase activity coefficient. Heat-of-mixing data may be correlated using one of the activity coefficient correlations discussed earlier. To achieve an optimum fit of the data, use of a model other than that adopted for phase equilibrium is recommended. Liquid enthalpies estimated from Equation (4.12) are usually accurate within five to ten percent.

4.11 Design examples

In this section, application of ASPEN PLUS™, HYSIM™, and PRO/II™ is illustrated for four typical process design scenarios. Example 1 exemplifies the flash computation used to establish the operating conditions of a distillation column. As mentioned earlier, this is one of the most important industrial applications of process simulation. A bubble point flash is used to anchor the operating pressure based on the available refrigeration. Simulation programs are also widely employed to calculate the required heating or cooling duties for a given system. Example 2 calculates the overall duty together with the stream properties for a gas effluent from the final stage of a water–gas shift converter. Example 3 demonstrates the potential pitfalls associated with the use of improper thermodynamic models for a design. The need for precertification of the appropriate property route is emphasized. Pinch technology, a rapidly evolving methodology for the design of process heat and power systems, is highlighted in Example 4. The close marriage between process simulation and pinch technology is displayed.

Example 1 (Assessment of operating conditions) **Problem statement:** In a liquified natural gas fractionation train the vapor distillate off the deethanizer is cooled against propane refrigerant available at 277.6 K (40°F). This distillate contains 0.5 mol% methane, 95.5% ethane, and 4.0% propane. Assuming complete condensation of this distillate, at what pressure must the column operate to ensure a 2.78 K (5.0°F) approach temperature at the outlet of the condenser? The pressure drop across the condenser is 17.24 kilopascals (2.5 pounds per square inch absolute).

Solution To satisfy the given design criteria, the distillate must be at its bubble point, that is, a vapor fraction of 0.0 at $277.6 + 2.78 = 280.38$ K. Knowing the desired temperature and vapor fraction, the corresponding bubble point pressure may be readily obtained using the two-phase flash model available in any of the aforementioned programs. For mixtures of this nature, the traditional cubic equations of state can be used to estimate the K-values (Equation 4.3) required for this simulation. Predictions for the case at hand are summarized in Table 4.16. The calculated pressure concur well, varying by roughly 24 kilopascals. Although these differences are within the expected accuracy of the respective equations, the worst case (2,810 kilopascals) has been taken as the design basis. Accounting for the pressure drop across the condenser, the column operating pressure should be approximately 2,827 kilopascals. (Note: From an operating viewpoint, most heat exchange equipment is flexible enough to handle the uncertainties implicit in this computation).

Example 2 (Cooling curve duty) **Problem statement:** A typical water–gas shift conversion flow scheme is shown in Fig. 4.1. At a design pressure of 7,792 kilopascals, calculate the overall duty in cooling the effluent gas, which

Table 4.16. *Deethanizer overheads – predicted bubble point pressures*

	Bubble point pressure, kPa	
	SRK[a]	PR[b]
ASPEN PLUS™	2801	2786
PRO/II™	2807	2786
HYSIM™	2810	2800

[a] Soave–Redlich–Kwong equation of state (Soave 1972)
[b] Peng–Robinson equation of state (Peng and Robinson 1976)

Table 4.17. *Low temperature shift effluent*

Component	Flow, kmol/hr
Nitrogen	21.4
Hydrogen	8488.2
Methane	46.3
Argon	36.1
Carbon dioxide	4629.7
Carbon monoxide	210.5
Hydrogen sulfide	87.1
Water	9814.8
Total	23334.1

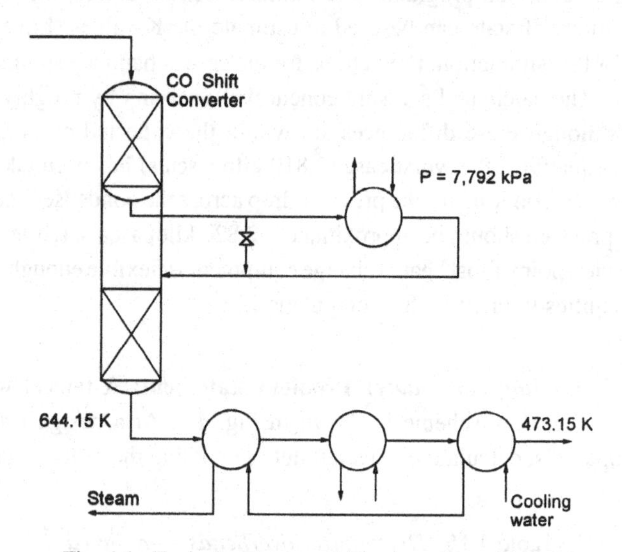

Fig. 4.1. Typical water–gas shift conversion process

exits the final CO shift stage at 644.15 K, to at least 30 K below its dew point through a downstream heat exchanger network. The shift effluent is defined in Table 4.17.

Solution Although all three programs can easily be used for computations of this type, many of the equilibrium routes furnished by these programs perform poorly for aqueous systems. Adjustments have been introduced by the respective vendors, such as the Free-Water methodology in ASPEN PLUS™,

Table 4.18. *CO shift effluent – predicted dew-point temperatures*

	Dew-point temperature, K	
	SRK[a]	PR[b]
ASPEN PLUS™	506.8	505.6
PRO/II™	512.1	512.1
HYSIM™	505.6	503.8

[a] Soave–Redlich–Kwong equation of state (Soave 1972)
[b] Peng–Robinson equation of state (Peng and Robinson 1976)

to alleviate some of the more serious deficiencies, but experience has shown that a few shortcomings remain. For simplicity, calculations for this example have likewise been made strictly within the context of the Soave–Redlich–Kwong and Peng–Robinson equations of state, even though an additional degree of user customization or use of a more sophisticated equation of state may be appropriate. The predicted dew points of the specified effluent are summarized in Table 4.18. Contrary to Example 1, the variations in these estimates are appreciable, particularly in comparison with those generated through PRO/II™. Lacking a fully definitive prediction, therefore, the worst case, 503.8 K, is again taken as the design basis. Using this temperature along with the desired 30 K additional cooling requirement, the effluent should be cooled to approximately 473.15 K (200°C). Duties predicted from the foregoing equations of state for this unit (644.15 → 473.15 K) are given in Table 4.19. Each program exhibits good internal agreement between the equations of state. On the other hand, PRO/II™ estimates a ten percent higher duty than HYSIM™ for this design, whereas ASPEN PLUS™ differs from each program by about five percent. These differences, as illustrated in the next example, can occasionally be resolved by comparison with pertinent experimental data.

In addition to the heating and cooling requirements and equilibrium conditions, complete stream properties can be generated along the entire heating and cooling path of the given system. An ASPEN PLUS™-based cooling curve printout for this design is shown in Table 4.20. The phase transition point is clearly marked, and liquid phase properties are predicted together with vapor properties at temperatures below this transition (dew point). All of the properties required for the design of this heat exchanger network can be obtained

Table 4.19. *CO shift effluent cooling curve duties*

	Overall duty, Megawatts	
	SRK[a]	PR[b]
ASPEN PLUS™	98.19	96.91
PRO/II™	103.83	103.89
HYSIM™	94.99	93.53

[a]Soave–Redlich–Kwong equation of state (Soave 1972)
[b]Peng–Robinson equation of state (Peng and Robinson 1976)

in this fashion using the HCURVE option of ASPEN PLUS™ or from similar features in either HYSIM™ or PRO/II™.

Example 3 (Tail gas drying) **Problem statement:** A drying step is common to almost all gas processing facilities. To properly size the molecular sieve drier or whatever medium may be used to dehydrate the system, accurately predicting the distribution of water throughout the system is critical, particularly the concentration in the hydrocarbon/inorganic-rich vapor phase.

A typical drying scheme is shown in Fig. 4.2. The water-saturated feed gas, most likely the effluent off a gas-treating facility where sour gases such as CO_2 and H_2S are removed, is cooled to about 294 K to remove bulk water and a heavy hydrocarbon liquid phase. The hydrocarbon and inorganic gas vapor is then dried to about one part per million H_2O in a mol sieve or equivalent unit before further downstream, usually cryogenic, processing. For the refinery tail gas defined in Table 4.21, compute the water concentration of the vapor feed to the drier and the amount of water condensed at 294 K. Assume that the feed is saturated with water at the given inlet conditions.

Solution Because the feed composition is given on a dry basis, its saturated water content must be firmly established. Based on a preliminary estimate, the dry gas is well below its dew-point pressure, and the water concentration can therefore be estimated through a two-phase vapor–liquid flash of the gas with excess water.

For computations of this type, it is tempting and unfortunately still a fairly common design practice to assume that the water vapor phase composition

Table 4.20. *CO shift effluent cooling curve from ASPEN PLUS™*

FLASH2 HCURVE:	FLSH21 HCURVE 1	
INDEPENDENT VARIABLE:	TEMP	
PROPERTY SET(S):	LIST1	
PRESSURE PROFILE:	CONSTANT	
PROPERTY OPTION SET:	SYSOP3	REDLICH-KWONG-SOAVE EQUATION OF STATE
FREE WATER OPTION SET:	SYSOP12	ASME STEAM TABLE
SOLUBLE WATER OPTION:	CORR SOLUBILITY DATA	

TEMP K	PRES KPA	DUTY KW	VFRAC	CPMX VAPOR J/KMOL-K	CPMX LIQUID J/KMOL-K	RHOMX VAPOR KMOL/CUM	RHOMX LIQUID KMOL/CUM
644.1500	7792.0000	0.0	1.0000	3.8670 + 04	MISSING	1.4605	MISSING
628.6045	7792.0000	−3895.5452	1.0000	3.8655 + 04	MISSING	1.5003	MISSING
613.0590	7792.0000	−7790.3023	1.0000	3.8654 + 04	MISSING	1.5425	MISSING
597.5136	7792.0000	−1.1686 + 04	1.0000	3.8670 + 04	MISSING	1.5875	MISSING
581.9681	7792.0000	−1.5584 + 04	1.0000	3.8704 + 04	MISSING	1.6356	MISSING
566.4227	7792.0000	−1.9486 + 04	1.0000	3.8760 + 04	MISSING	1.6871	MISSING
550.8772	7792.0000	−2.3396 + 04	1.0000	3.8843 + 04	MISSING	1.7426	MISSING
535.3318	7792.0000	−2.7315 + 04	1.0000	3.8957 + 04	MISSING	1.8024	MISSING
519.7863	7792.0000	−3.1248 + 04	1.0000	3.9109 + 04	MISSING	1.8673	MISSING
506.7818	7792.0000	−3.4551 + 04	DEW > 1.0000	3.9272 + 04	4.0920 + 04	1.9260	9.8414
504.2409	7792.0000	−4.1490 + 04	0.9698	3.9059 + 04	1.0181 + 05	1.9282	46.1076
488.6954	7792.0000	−7.4896 + 04	0.8324	3.7995 + 04	9.8498 + 04	1.9513	47.2273
473.1500	7792.0000	−9.8189 + 04	0.7471	3.7236 + 04	9.5869 + 04	1.9887	48.2616

Table 4.20 (Cont.)

Table 4.20. *(Continued.)*

TEMP K	PRES KPA	DUTY KW	VFRAC	KMX VAPOR WATT/M-K	KMX LIQUID WATT/M-K	MUMX VAPOR CP	MUMX LIQUID CP
644.1500	7792.0000	0.0	1.0000	0.1177	MISSING	2.7102 − 02	MISSING
628.6045	7792.0000	− 3895.5452	1.0000	0.1153	MISSING	2.6562 − 02	MISSING
613.0590	7792.0000	− 7790.3023	1.0000	0.1128	MISSING	2.6020 − 02	MISSING
597.5136	7792.0000	− 1.1686 + 04	1.0000	0.1104	MISSING	2.5478 − 02	MISSING
581.9681	7792.0000	− 1.5584 + 04	1.0000	0.1079	MISSING	2.4934 − 02	MISSING
566.4227	7792.0000	− 1.9486 + 04	1.0000	0.1055	MISSING	2.4389 − 02	MISSING
550.8772	7792.0000	− 2.3396 + 04	1.0000	0.1031	MISSING	2.3844 − 02	MISSING
535.3318	7792.0000	− 2.7315 + 04	1.0000	0.1007	MISSING	2.3298 − 02	MISSING
519.7863	7792.0000	− 3.1248 + 04	1.0000	9.8380 − 02	MISSING	2.2752 − 02	MISSING
506.7818	7792.0000	− 3.4551 + 04	DEW > 1.0000	9.6412 − 02	4.0796 − 02	2.2296 − 02	3.6005-02
504.2409	7792.0000	− 4.1490 + 04	0.9698	9.7444 − 02	0.1880	2.2290 − 02	0.2231
488.6954	7792.0000	− 7.4896 + 04	0.8324	0.1030	0.1975	2.2134 − 02	0.2461
473.1500	7792.0000	− 9.8189 + 04	0.7471	0.1072	0.2067	2.1810 − 02	0.2702

VFRAC ≡ vapor fraction, CP ≡ heat capacity, RHO ≡ density, K ≡ thermal conductivity, MU ≡ viscosity, MX ≡ mixture

Table 4.21. *Refinery tail gas – stream composition*

$T = 322.04$ K, $P = 5{,}516$ kPa,
Flow = 5,946 kmol/hr

Component	Mol%
Hydrogen	21.8
Nitrogen	2.4
Carbon monoxide	0.5
Methane	37.2
Ethylene	5.2
Ethane	15.0
Propylene	2.9
Propane	10.3
C4's	3.0
C5's	1.7

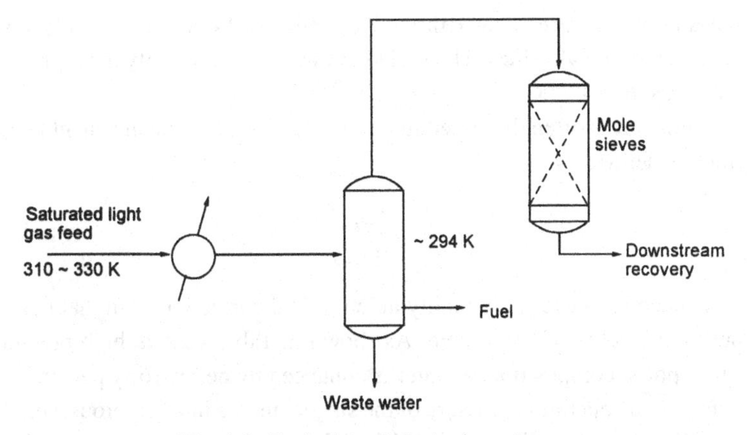

Fig. 4.2. Typical gas dehydration scheme

is simply

$$y_{H_2O} = \frac{P^o_{H_2O}}{P} \tag{4.14}$$

Here y_{H_2O} is the mol fraction of water in the vapor phase, $P^o_{H_2O}$ is the vapor pressure of water at the system temperature, and P is the system pressure. Although this assumption is adequate in many cases, it can give erroneous results especially at high system pressures. Consider the comparisons given in Table 4.22 for an aqueous natural gas at advanced pressures. The water

Table 4.22. *Calculated and observed vapor phase compositions in the water–natural gas system*

		Vapor-Phase Mol Fraction, $y_{H_2O} \times 10^3$		
Temp, °C	Press, kPa	McKetta–Wehe (1958)	Calc[a]	ϕ
37.8	6,895	1.075	0.950	1.13
	10,342	0.765	0.630	1.21
	13,790	0.610	0.475	1.28
	20,684	0.448	0.317	1.41
51.7	8,274	1.890	1.60	1.18
	10,342	1.550	1.29	1.20
	13,790	1.210	0.97	1.25
	20,684	0.890	0.645	1.38

[a]From the oversimplified relationship, Equation (4.14)

concentrations that are estimated through Equation (4.14) are consistently lower than the experimental values. The variation increases noticeably the higher the system pressure becomes.

This behavior is often discussed in terms of a so-called enhancement factor, ϕ, which is defined as

$$\phi = \frac{y_{exp} P}{P^o_{H_2O}}. \tag{4.15}$$

The enhancement factor is in reality an empirical correction term that defines the bulk nonideality of the system. As shown in Table 4.22, at high pressure, the vapor phase composition of water is enhanced by nearly forty percent.

In view of the operating pressure of the subject unit, a more rigorous, simulation program–based prediction is required. Water concentrations were predicted from the traditional cubic equations of state and from three more refined equations of state: the Kabadi–Danner (1985) SRK in PRO/II™, the HYSIM™ Stryjek-Vera Peng–Robinson, and a Soave–Redlich–Kwong equation of state as modified by the M. W. Kellogg Company. The results are enumerated in Table 4.23. As shown, the differences in the estimated water concentrations of the feed gas are clear-cut. More important, these variations translate to appreciable differences in the predicted drying capacity of the mol sieves unit. Comparing the extremes, Peng–Robinson in ASPEN PLUS™ and Soave–Redlich–Kwong in PRO/II™, the water concentration of the drier feed differs by more than 100%. Considering the well-documented inconsistencies of many of these

Table 4.23. *Process simulation of tail gas drying unit*

	Water concentration (kg/hr)		
Program	Saturated Feed Gas	Drier Feed	Waste Water
ASPEN PLUS^{TM}			
Soave–Robinson–Kwong	406.2	87.9	313.0
Peng–Robinson	424.2	93.0	326.2
Kellogg Modified SRK	276.9	56.6	215.0
HYSIM^{TM}			
Soave–Redlich–Kwong	311.6	65.3	240.3
Peng–Robinson	290.2	59.8	225.7
Stryjek–Vera PR	327.2	71.3	251.1
PRO/II^{TM}			
Soave–Redlich–Kwong	225.9	43.8	175.8
Kabadi–Danner SRK	273.1	56.3	212.3

equations for aqueous systems, as pointed out in Example 2, these variations are not surprising.

To assess its integrity for the subject design, each of the foregoing equations of state was tested against a few available experimental data commensurate with the stated operating conditions. The results are summarized in Table 4.24. Overall, the Kabadi–Danner, the HYSIM^{TM} Peng–Robinson, and the Kellogg modified Soave–Redlich–Kwong appear to give the most reliable estimates in the range of interest. These comparisons reinforce the trends in Table 4.23 showing excellent accord among the predicted drier requirements from these three equations. In essence, on the basis of the above comparisons for two of the key constituents in the feed to this unit, only these latter equations offer a fairly definitive route for system design.

This example demonstrates two key factors about the technical data that have been incorporated in simulation programs. First of all, a correlation, particularly an equation of state, can produce different results in different programs. Note the widespread differences in the estimates from the SRK in both Tables 4.23 and 4.24. Because of fine-tuning by the respective vendors, these differences are the rule rather than an exception. Secondly and more significantly, prudent application of the available property models is prerequisite to reliable and cost-effective design. All the vendors give an overview of the range or areas of application of the routes they offer, but these models should be closely examined and validated, particularly when the conditions have a major impact on the design of a given unit. The bottom line remains that a design is no better than the available data.

Table 4.24. *Comparison of predicted and experimental vapor phase concentrations of water*

	y_{H_2O}, ppm (mol) Exp.	Calc.	%Dev.	y_{H_2O}, ppm (mol) Exp.	Calc.	%Dev.
Water-methane	*298.15 K; 4,056 kPa*			*323.15 K; 6,205 kPa*		
ASPEN PLUS™						
Soave–Redlich–Kwong	915	1,216	32.9	2,450	3,422	39.7
Peng–Robinson	915	1,248	36.4	2,450	3,547	44.8
Kellogg Modified SRK	915	909	−0.7	2,450	2,398	−2.1
HYSIM™						
Soave–Redlich–Kwong	915	1,138	24.4	2,450	3,231	31.9
Peng–Robinson	915	904	−1.2	2,450	2,425	−1.0
Stryjek – Vera PR	915	1,208	32.0	2,450	3,377	37.8
PRO/II™						
Soave–Redlich–Kwong	915	774	−15.4	2,450	1,978	−19.3
Kabadi–Danner SRK	915	901	−1.5	2,450	2,376	−3.0
Water-hydrogen	*310.93 K; 3,103 kPa*			*310.93 K; 6,550 kPa*		
ASPEN PLUS™						
Soave–Redlich–Kwong	2,220	2,161	−2.7	1,160	1,050	−9.5
Peng–Robinson	2,220	2,227	0.3	1,160	1,116	−3.8
Kellogg Modified SRK	2,220	2,205	−0.7	1,160	1,120	−3.4
HYSIM™						
Soave–Redlich–Kwong	2,220	2,180	−1.8	1,160	1,086	−6.4
Peng–Robinson	2,220	2,376	7.0	1,160	1,290	11.2
Stryjek – Vera PR	2,220	2,431	9.5	1,160	1,306	12.6
PRO/II™						
Soave–Redlich–Kwong	2,220	2,097	−5.5	1,160	993	−14.4
Kabadi–Danner SRK	2,220	2,188	−1.4	1,160	1,103	−4.9

Example 4 (Process integration/pinch technology) Process integration or pinch technology, introduced in the early 1980s (Linnhoff and Vredeveld, 1984) is a methodology for the conceptual design of process heat and power systems. (See Chapter 5 for details.) In this framework the first and second laws of thermodynamics are systematically applied to process and utility systems. Using pinch technology tools, engineers can answer some fundamental questions about processes and the utility systems that surround them without committing to detailed process simulation.

All that are required are base-case heat and material balances to begin the analysis. These data can be generated from ASPEN PLUS™, HYSIM™, or PRO/II™. From these data, a combined hot composite curve (system temperature versus enthalpy) is constructed for all streams requiring cooling, and a cold composite curve is developed for all streams that must be heated. Plotting both curves on the same axes at a specific minimum approach temperature defines

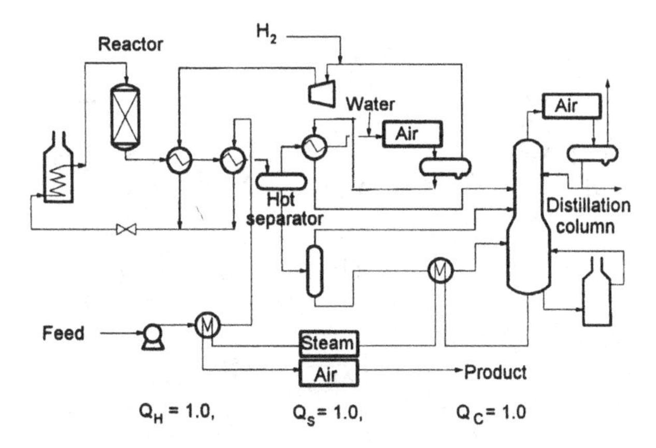

Fig. 4.3. Base case hydrotreater flow sheet

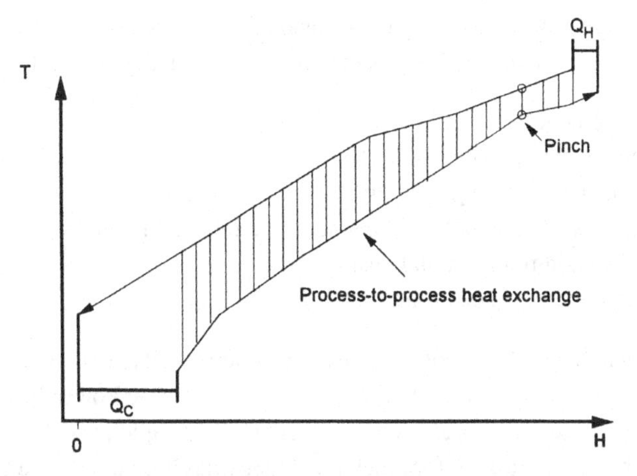

Fig. 4.4. Composite curves for hydrotreater

a target for minimum energy consumption. A representative pinch analysis is briefly detailed in the next few paragraphs.

For the base case diesel hydrotreater depicted in Fig. 4.3, it was desired to minimize utility consumption by optimizing the hot separator temperature. The heat requirements Q_H (fuel), Q_S (steam), and Q_C (cooling) have been normalized ($Q = 1.0$). From an ASPEN PLUS™ (SRK) simulation of the base unit, composite curves, as shown in Fig. 4.4, were constructed for the respective hot and cold streams. Employing these data together with state-of-the-art pinch software from Aspen Technology, ADVENT™, a revamped flow scheme illustrated in Fig. 4.5, was developed.

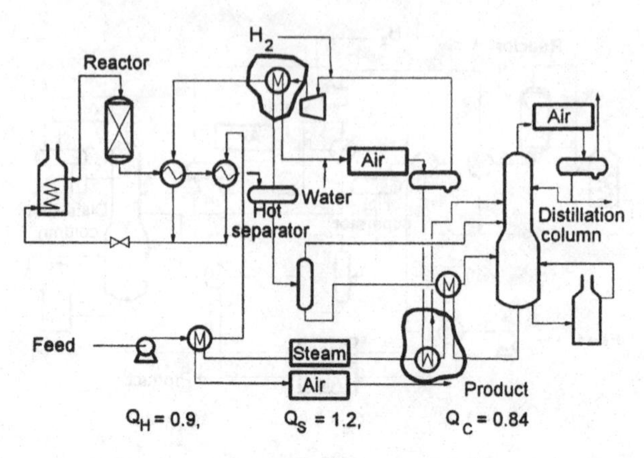

Fig. 4.5. Improved hydrotreater flow sheet

Note the areas of improvement, the enhanced heat interchange, in the revised process scheme. With these enhancements, several savings are realized:

- uses ten percent less fuel,
- generates twenty percent more steam,
- uses sixteen percent less cooling,
- saves approximately $400,000 per year in total operating costs
- requires no additional capital cost
- provides study payback around 1 week

Other pinch technology software packages are available including Super Target™ from Linnhoff March and Hpscan from Electric Power Research Institute. New technology is also being introduced at a rapid pace. For example, an expert system prototype for pinch technology has been developed by a joint venture between the University of Virginia and Linnhoff March (Samdani and Moore 1993).

Summarizing, a front-end pinch technology analysis allows the engineers to develop a more cost-effective conceptual design for the process and utility systems. Once the process topology has been finalized and key anchor point operating conditions specified, detailed heat and material balances are then generated from ASPEN™, HYSIM™, or PRO/II™ using pertinent equilibrium and enthalpy models.

References

Albright, M. A. (1985) An industrial's view of the thermophysical world. Paper 67e, AIChE Spring National Meeting, Houston, TX.

(extant 1993). *Technical Data Book: Petroleum Refining*, Washington, DC: American Petroleum Institute.

Baldwin, J. T., Hardy, B. L., & Spencer, C. F. (1987). The validation and maintenance of ASPEN PLUS at M. W. Kellogg. *Proceeding of the 1987 Summer Computer Simulation Conference*, 131–7. Society for Computer Simulation, San Diego, CA.

Boston, J. F., Britt, H. I. & Tayyabkhan, M. T. (1993). Software: tackling tougher tasks. *Chemical Engineering Progress*, Nov., 38–49.

Chan, W. K., Boston, J. F., & Evans, L. B (1991). Select the right software for modeling separation processes. *Chemical Engineering Progress*, Sept., 63–9.

Cox, K. R. (1993). Physical property needs in industry. *Fluid Phase Equilibria*, **82**, 15–26.

Curl, R. F. & Pitzer, K. S. (1958). Volumetric and thermodynamic properties of fluids: enthalpy, free energy and entropy: *Industrial and Engineering Chemistry*, **50**, 265–84.

Danner, R. P. & Daubert T. E. (1987). *Data Prediction Manual*. New York, NY: Design Institute for Physical Property Data, American Institute of Chemical Engineers.

Daubert, T. E. & Danner R. P. (1994) *Physical and Thermodynamic Properties of Pure Chemicals: Data Compilation*. Washington, DC: Taylor & Francis.

Dean, R. C. (1977). Truth in publication. *Journal of Fluids Engineering*, Vol 99, p. 270.

Evans, L. B. (1990). Process modeling: what lies ahead. *Chemical Engineering Progress*, May, 42–4.

Fredenslund, A., Gmehling, J. & Rasmussen P. (1977). *Vapor–Liquid Equilibria Using UNIFAC*. North Holland, Amsterdam: Elsevier.

Gess, M. A., Danner, R. P., & Nagvekar, M. (1991). *Thermodynamic Analysis of Vapor–Liquid Equilibrium*, New York: Design Institute for Physical Property Data, American Institute of Chemical Engineers.

Huron, M. J. & Vidal, J. (1979). New mixing rules in simple equations of state for representing vapor–liquid equilibrium of strongly non-ideal mixtures. *Fluid Phase Equilibria*, **3**, 255–71.

Parsons, H. C. (1996). Process simulation and design 196. *Hydrocarbon Processing*, 75(6): 51–96.

Kabadi, V. N. & Danner, R. P. (1985). A modified Soave-Redlich-Kwong equation of state for water–hydrocarbon phase equilibria. *Industrial Engineering Chemistry, Process Design and Development*, **24**, 537–41.

Kistenmacher, H. & Burr, P. (1983). The importance of data, its evaluation and consistency from an industrial point of view. Presented at the Third International Conference on Fluid Properties and Phase Equilibria for Chemical Process Design, Callaway Gardens, GA.

Knacke, O., Kubaschewskim O., & Hesselmann, K. (1991). *Thermochemical Properties of Inorganic Substances*, 2nd ed. New York, NY: Springer-Verlag.

Lee, B. I. & Kesler, M. (1975). A generalized thermodynamic correlation based on three-parameter corresponding states. *AIChE Journal*, **21**, 510–25.

Lide, D. R. (1994). Vapor–liquid equilibrium database (Computer software review). *Journal of Chemical Information and Computer Science*, **34**, 690.

Linnhoff, B. & Vredeveld, D. (1984). Pinch technology has come of age. *Chemical Engineering Progress*, July, 33–40.

Mani, P. & Venart, J. E. S. (1973). Thermal conductivity measurements of liquid and dense gaseous methane. *Advances in Cryogenic Engineering*, **18**, 280–8.

Mathias, P. M. (1983). A versatile phase equilibrium equation of state. *Industrial Engineering Chemistry, Process Design and Development*, **22**, 385–91.

McKetta, J. J. & Wehe, A. H. (1958). Use this chart for water content of natural gases. *Petroleum Refiner*, July, 153–6.

Mullins, M., Radecki, P. & Rogers, T. (1994). Engineering Chemical Data Correlation. In *Kirk-Othmer Encyclopedia of Chemical Technology*, 4th. ed. **9**: 463–524.

Najjar, M. S., Bell, K. J. & Maddox, R. N. (1981). The influence of improved physical property data on calculated heat transfer coefficients. *Heat Transfer Engineering*, **2**, 27–39.

Onken, U. E. & H.-I. Paul (1990). Estimation of physical properties. In *Ullmann's Encyclopedia of Industrial Chemistry*, 5th ed. H. Hoffman Vol. B1, chap. 6, pp. 6.1 to 6.59. New York: VCH Publishers.

Panagiotopoulos, A. Z. & Reid, R. C. (1986). New mixing rules for cubic equations of state for highly polar asymmetric mixtures. *ACS Symposia Series*, **300**, 571–82.

Peng, D. Y. & Robinson, D. B. (1976). A new two-constant equation of state. *Industrial Engineering Chemistry, Fundamentals*, **15**, 59–64.

Prausnitz, J. M., Lichtenthaler, R. N. & de Azevedo, E. G. (1986). *Molecular Thermodynamics of Fluid Phase Equilibria*. Englewood Cliffs, NJ: Prentice-Hall, Inc.

Reid, R. C. & Sherwood, T. K. (1966). *The Properties of Gases and Liquids*, 2nd. ed. New York: McGraw-Hill.

Reid, R. C., Prausnitz, J. M., & Poling, B. E. (1987). *The Properties of Gases and Liquids*, 4th ed. New York: McGraw-Hill.

Reid, R. C., Prausnitz, J. M., & Sherwood, T. K. (1977). *The Properties of Gases and Liquids*, 3rd ed. New York: McGraw-Hill.

Renon, H. & Prausnitz, J. M. (1968). Local compositions in thermodynamic excess functions for liquid mixtures. *AIChE Journal*, **14**, 135–44.

Samdani, G. & Moore, S. (1993). Pinch technology: Doing more with less. *Chemical Engineering*, July, 43–8.

Selover, T. B. & Spencer, C. F. (1987). An overview of international data projects and a proposal for quality control of computerized thermophysical property databases. In *Computer Handling and Dissemination of Data*, ed. P. S. Glaeser, North Holland: Elsevier Science Publisher B.V.

Shaw, J. A. (1984). Data for engineering design. In *Awareness of Information Sources*, ed. T. B. Jr. Selover, & Klein, M. AIChE Symp. Ser. #237, Vol. 80, 43–9.

Simpson, K. L. (1994). *1995 CEP Software Directory*. A supplement to *Chemical Engineering Progress*, Dec.

Soave, G. (1972). Equilibrium constants from a modified Redlich–Kwong equation of state. *Chemical Engineering Science*, **27**, 1197–203.

Starling, K. E., Kumar, K. H., & Savidge, J. L. (1982). Syngas thermodynamic properties prediction and data needs. *Proceedings 61st Annual GPA Convention*. Gas Properties Association, Tulsa.

Stryjek, R. & Vera, J. H. (1986). PRSV2: A cubic equation of state for accurate vapor–liquid equilibrium calculations. *Canadian Journal of Chemical Engineering*, **64**, 820–6.

Squires, E. W. & Orchard, J. C. (1968). How gas-plant performance compares with process designs. *Oil and Gas Journal*, 66, 136–40.

Watson, K. M. (1943). Thermodynamics of the liquid state, *Industrial Engineering Chemistry*, **35**, 398–402.

Wichterle, I., Linek, J., Wagner, Z., & Kehiaian, H. V. (1991). *Vapor–liquid equilibrium bibliographic database*. 81-83 re M Montreuil, France: ILDATA.

Williams, C. C. & Albright, M. A. (1976). Better data saves energy. *Hydrocarbon Processing*, May, 115–6.

Wilson, G. M. (1964). Vapor–liquid equilibrium. XI. A new expression for the excess free energy of mixing. *Journal of the American Chemical Society*, **86**, 127–33.
Wong, D. S. H. & Sandler, S. I. (1992). A theoretically correct mixing rule for cubic equations of state. *AIChE Journal*, **38**, 671–80.
Zudkevitch, D. (1982). Design data: importance of accuracy. In *Encyclopedia of Chemical Processing and Design* ed J. McKetta, Vol. 14, pp. 431–83. New York: M. Dekker.

5
Introduction to Pinch Analysis

BODO LINNHOFF
Linnhoff March International and University of Manchester Institute of Science and Technology

5.1 Introduction

In today's industrial practice, many plants are suffering from process designs that were completed in times when energy conservation and pollution prevention were not yet key issues. Such plants were built with little regard for reuse of utilities such as heat or water, because these came at low cost and could easily be discharged. Recently, however, environmental protection policies and the awareness that resources for heat and water are limited have led to considerable efforts in the fields of process design and process integration.

In a wide range of industries, Pinch Analysis is accepted as the method of choice for identifying medium- and long-term energy conservation opportunities. It first attracted industrial interest in the 1970s, against the background of rising oil prices. An engineering approach, but one incorporating scientific rigor, Pinch Analysis quantified the potential for reducing energy consumption through improved heat integration by design of improved heat-exchanger networks. Since then, Pinch Analysis has broadened its coverage to include reaction and separation systems in individual processes as well as entire production complexes. Today's techniques assess quickly the capital and operating-cost implications of design options that reduce intrinsic energy consumption, determine the potential for capacity debottlenecking, improve operation flexibility, and, most recently, quantify the scope for reduced combustion-related emissions. In addition, the same basic concepts have been extended to mass-transfer problems to give similar insights into waste minimization, with significant environmental implications.

In this chapter, the fundamentals and some extensions of Pinch Analysis are briefly reviewed.

5.2 The basics

Pinch Analysis is based on rigorous thermodynamic principles. These are used to construct diagrams and perform simple calculations that yield powerful insights into heat flows through processes. The key concepts are best explained by two types of graphical representation: composite curves and grand composite curves (Linnhoff et al. 1982). Both of these are stream-based *T-H* diagrams.

5.3 Stream-based *T-H* diagrams

The starting point is a stream-by-stream breakdown of the heat sources and sinks within a process. The breakdown shows the enthalpy change ΔH and temperatures T of all heat sinks (cold streams) and all heat sources (hot streams) in a given process. The first example has only two streams, *a* and *b* (Fig. 5.1a).

The interval boundary temperatures correspond to the start and end of a stream. In this case, four such temperatures exist, labeled T_1 through T_4. Next, the stream population between any two adjacent interval boundary temperatures is determined. In the example, stream a exists between T_1 and T_2; both streams a and b exist between T_2 and T_3; and only stream b exists between T_3 and T_4. This distribution quantifies the heating demand in each interval. Adding the heat load contribution each stream makes in each interval is simple. The results are shown in Fig. 5.1b. The single continuous curve represents all heat sinks of the process as a function of heat load versus temperature. This plot is the cold composite curve.

By an analogous procedure, the process heat sources (hot streams) can be combined to generate a hot composite curve, which represents all heat sources of the process as a function of heat load versus temperature.

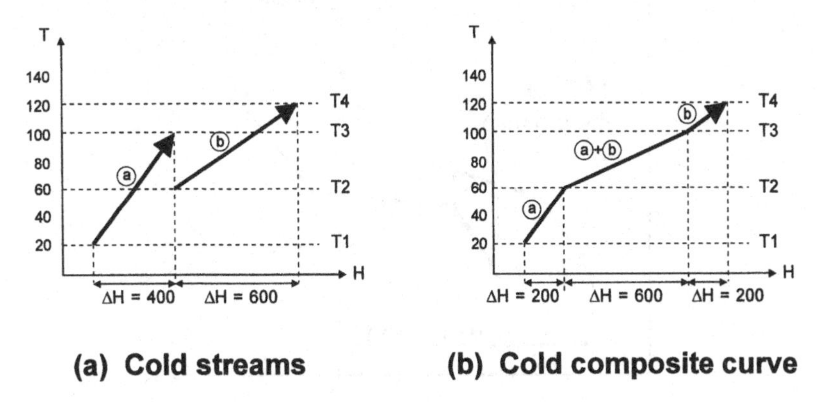

Fig. 5.1. Stream-based *T-H* plots.

In this example, the analysis is simplified because constant specific heats are assumed and the number of streams is small. In practice, variable specific heats are accommodated by segmented linearization of data. A large number of streams poses no practical problems with the computational power that is available from modern software.

5.4 Composite curves, energy targets, and the pinch

The two composite curves are conveniently plotted on the same axes in Fig. 5.2. Their positions relative to each other are set such that at the closest vertical approach between them, the hot composite curve is hotter by ΔT_{min} than the cold composite curve. ΔT_{min} is the minimum permissible temperature difference for heat transfer.

The combined plot is referred to as the composite curves and is a powerful tool for establishing energy targets. In the region of the diagram where the curves overlap, every point on the cold composite curve has a corresponding point on the hot composite curve directly above, which is at a temperature of at least ΔT_{min} higher. Therefore, heat sources exist within the process to satisfy all heating requirements represented by this overlapped portion of the cold composite curve, with temperature driving forces of at least ΔT_{min}. The projection of the overlap on the horizontal axis defines the scope for process-to-process heat recovery.

The section of the hot composite curve to the left of the overlap region represents the quantity of heat that cannot be recovered by cold process streams, because none of sufficiently low temperature is available. This heat surplus has

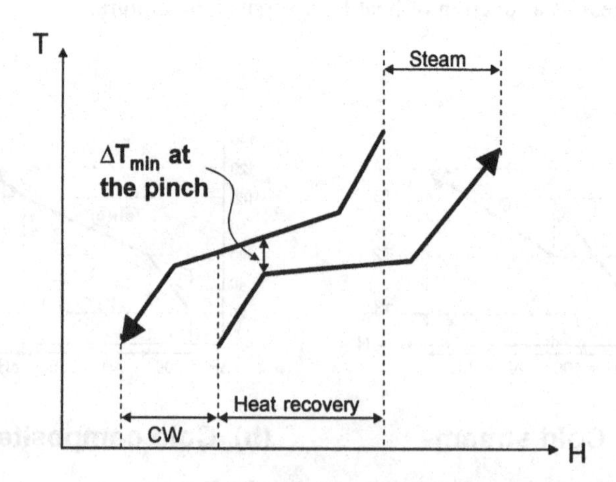

Fig. 5.2. Composite curves.

to be rejected to a cold utility, for example, cooling water, labeled *CW* in Fig. 5.2. Similarly, the section of the cold composite curve to the right of the overlap region defines the quantity of utility heating (such as steam) that is required. The quantity *CW* is the cold-utility target for the process at the specified ΔT_{min}. The quantity steam is the corresponding hot-utility target.

These targets represent attainable goals, based on thermodynamic principles. They give the minimum amounts of utility cooling and heating that are necessary to drive the process, subject to maintaining a temperature driving force of at least ΔT_{min} in any network of heat exchangers, heaters, and coolers – yet to be designed – that services all sources and sinks in the given process.

The point in Fig. 5.2 at which the curves come closest together is the pinch point (Linnhoff, Mason, and Wardle 1979). At the pinch, the vertical distance between the curves is equal to ΔT_{min}.

5.5 The pinch and heat-exchanger network design

Consider the characteristics of a heat-exchanger network (HEN) that is to achieve the energy targets, subject to a given ΔT_{min} requirement. By inspection of Fig. 5.2, we see that hot streams below the pinch cannot heat cold streams above the pinch. Next, if any hot stream above the pinch is used to heat a cold stream below the pinch, the overall hot and cold utility requirements will exceed their targets.

As we see in Fig. 5.3, any quantity *Y* of heat that is transferred across the pinch has to be replaced by an equal quantity of hot utility to satisfy the remaining heating requirements of cold streams above the pinch. Likewise, cold utility requirements increase an equivalent amount.

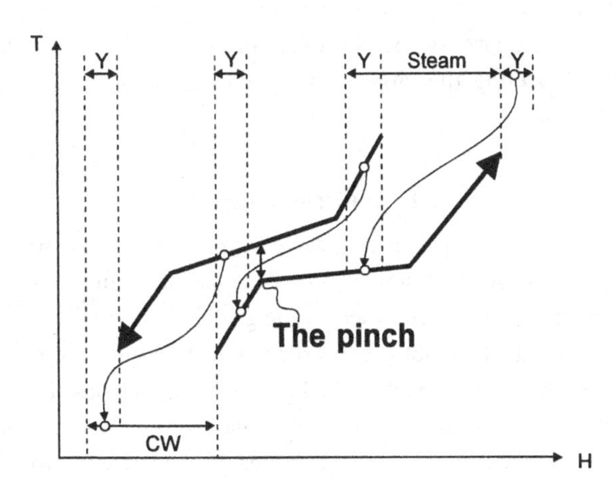

Fig. 5.3. Cross-pinch heat transfer.

This fact leads to a fundamental axiom:

- Do not transfer any heat across the pinch.

From this axiom are derived what have become known as the three golden rules of Pinch Analysis:

- Do not recover process heat across the pinch.
- Do not apply hot utilities to process streams below the pinch.
- Do not apply cold utilities to process streams above the pinch.

These rules are fundamental to the design of heat exchanger networks for maximum energy recovery and are summarized in the pinch heat-transfer equation:

$$Actual = target + XP_{\text{proc}} + XP_{\text{hot below}} + XP_{\text{cold above}} \quad (5.1)$$

where

actual	=	actual heat consumption of the process
target	=	minimum consumption as determined by the composite curves
XP_{proc}	=	heat transfer from hot process streams above the pinch to cold process streams below the pinch
$XP_{\text{hot below}}$	=	heat transfer from hot utilities to cold process streams below the pinch
$XP_{\text{cold above}}$	=	heat transfer from hot process streams above the pinch to cold utilities

Because the hot and cold composite curves are separated by ΔT_{min} at the pinch, it follows that heat exchangers in the pinch region must approach ΔT_{min} at the pinch. This requirement, together with the golden rules, provide a systematic design procedure for the placement of exchangers in overall networks that guarantees meeting the targets: the pinch design method (Linnhoff and Hindmarsh 1983).

5.6 The grid diagram

It is convenient for design purposes to have a representation of stream data and the pinch that also shows exchanger and utility placement in a simple graphical form. This representation is provided by the grid diagram in Fig. 5.4 (Linnhoff and Flower 1978), in which hot streams run from left to right at the top, whereas cold streams run from right to left at the bottom. The pinch position is indicated by a broken line at the division between the above- and below-pinch subsystems. Process exchangers are represented by circles on the streams involved, which are joined by vertical lines. Heaters are represented by circles on cold streams,

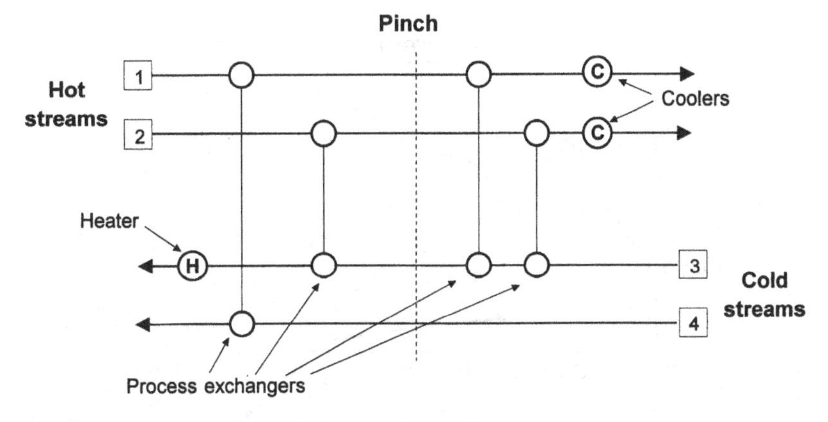

Fig. 5.4. Grid diagram.

and coolers by circles on hot streams. The grid diagram makes it obvious when cross-pinch transfer occurs, and the design task becomes straightforward. Note that the grid allows the design and evolution of any network structure without the need to reposition streams.

5.7 The grand composite curve, appropriate placement, and balanced composite curves

The hot and cold composite curves are representations of process heat sources and sinks. This information can also be amalgamated into a single plot. After an allowance is made for the $\Delta T_{\min}$ requirement by means of temperature shifting, the heat duties of the hot streams are offset against those of the cold streams in every interval. Each interval is thus found to be either in heat surplus, deficiency, or balance (Linnhoff et al. 1982). The numerical procedure is called the problem table (Linnhoff and Flower 1978), and the plot that results from it is the grand composite curve (Linnhoff et al. 1982) (Fig. 5.5).

The grand composite curve represents the net heating and cooling requirement of the process as a function of temperature. The curve depicts heating and cooling requirements at the process–utility interface and enables the engineer to screen feasible options for the hot utility (for example, furnace firing, hot oil, steam at various pressure levels) and the cold utility (for example, air cooling, water cooling, refrigeration at different levels) to satisfy the process requirements and to integrate heat engines and heat pumps with the process. This concept is called appropriate placement (Townsend and Linnhoff 1983).

The targeted utilities and the process streams are next integrated into one plot, the balanced composite curve (Fig. 5.6a), that shows the total heat balance

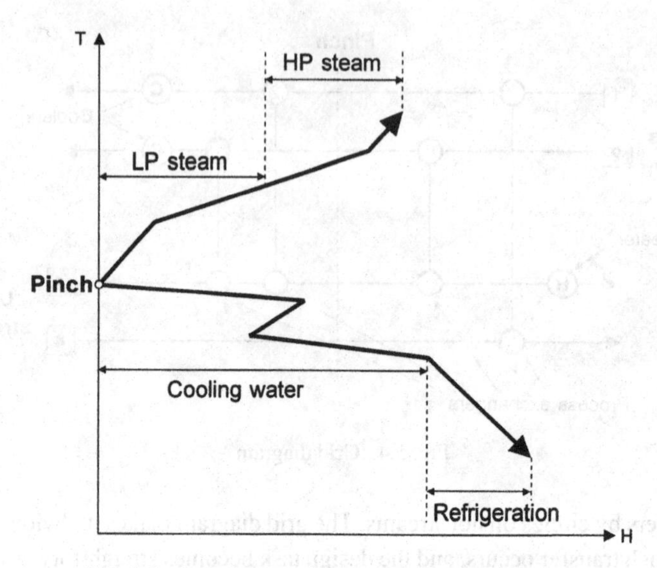

Fig. 5.5. Grand composite curve with utility targets.

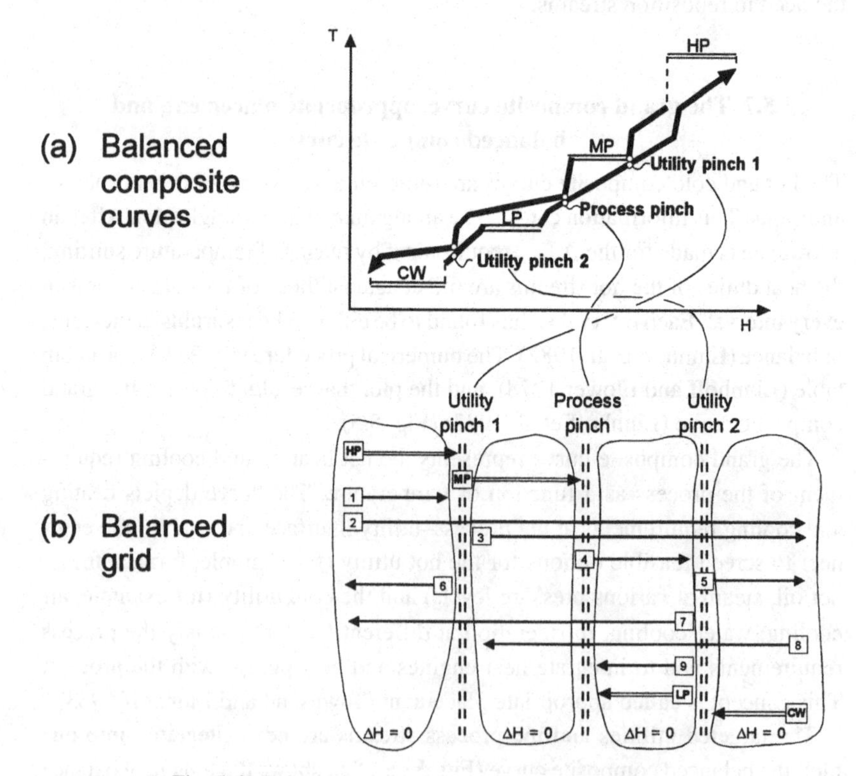

Fig. 5.6. Balanced composite curves and balanced grid.

Table 5.1. *Emissions reduction from BASF campaign*

Emission	*Saving*
Cooling duty to air and water	500 MW
Carbon dioxide	218 t/h
Sulphur dioxide	1.4 t/h
Nitrogen oxides	0.7 t/h
Soot	21 kg/h
Carbon monoxide	7 kg/h
Effluent from water treatment	70 t/h

of the system. The exchangers and utilities can now be designed as a merged system, using the balanced grid (Fig. 5.6b). (Linnhoff and de Leur 1988)

5.8 Additional benefits

Pinch technology as described saw industrial use during the late 1970s and early 1980s. Although the primary purpose was energy cost reduction, benefits for capital cost reduction, for the environment, and for design time savings soon became apparent. Combustion-related atmospheric emissions, such as SO_X, NO_X, and CO_X, are proportional to fuel firing rates, and consequently targets for these emissions could be established via hot-utility targets.

An early illustration of this analysis was a major campaign conducted by the German company BASF at its Ludwigshafen factory in the early 1980s (Körner 1988). Pinch methods were used to improve the energy efficiency of most of the processes on site. The objective was to save energy and reduce related operating costs. However, a review of the results clearly demonstrated that major environmental benefits were also obtained, with significant reductions in combustion-related atmospheric emissions, as listed in Table 5.1. Even wastewater emissions were reduced, as less water treatment was required for steam and cooling water.

In recent developments, outlined briefly in the following sections, pinch methods are used for the design of utility systems for total production sites as well as for individual processes.

5.9 Extensions of Pinch Analysis

Although the scope of Pinch Analysis has broadened considerably since the mid-1980s, it is still rooted firmly in thermodynamics, and most developments maintain the key strategy of setting targets before design. Pinch Analysis is now an overall methodology for what is sometimes referred to as process synthesis, including reactors, separators, furnaces, combined heat and power systems, etc. Pinch Analysis is the method of choice for identifying a wide range of process improvement options for the heart of chemical processes, or the heat and material balance. Objectives include emissions reduction, energy cost reduction, capital cost reduction, capacity increase, and yield improvement (Smith, Petela, and Spriggs 1990). Pinch Analysis is used equally for retrofits and for grassroots design and now for planning in both continuous and batch industries. An up-to-date overview was given recently (Linnhoff 1994).

The following specific developments have made Pinch Analysis a versatile and effective tool for the optimization of integrated process structures.

5.10 Supertargeting

It is possible to predict from the composite curves the overall target heat-transfer surface area, and from that the target capital cost. On the basis of the target energy cost and target capital cost, the target total cost can be predicted for any given ΔT_{min}. This concept of supertargeting (Linnhoff and Ahmad 1990) (Fig. 5.7) enables the designer to optimize ΔT_{min} before design. In other words, Pinch Analysis addresses not only the energy cost but also the capital cost and guides the overall optimization of the total cost of integrated systems at the targeting stage.

5.11 The plus/minus principle

Process changes are changes to the basic process heat and material balance that may or may not involve the chemistry of the process. The onion diagram

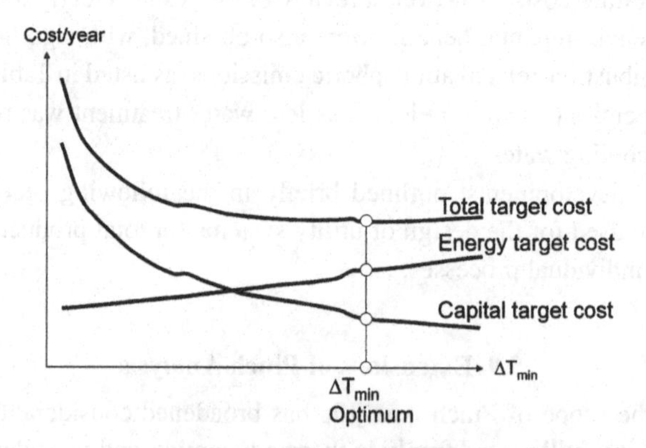

Fig. 5.7. Supertargeting.

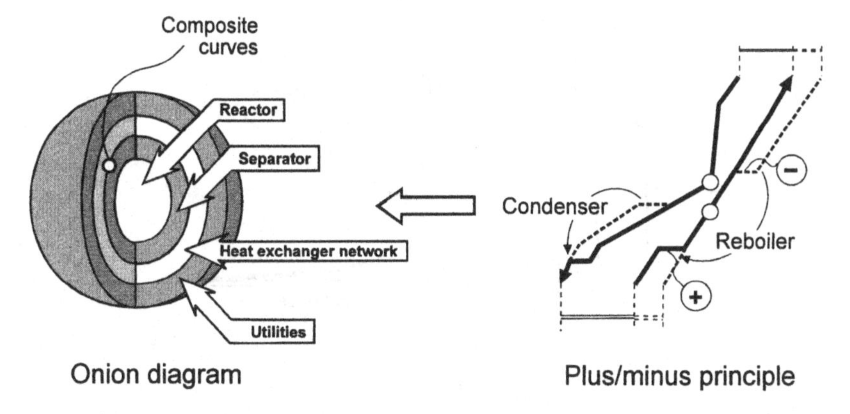

Fig. 5.8. Process changes.

(Linnhoff et al. 1982) is used to represent the hierarchy of chemical process design. The start is at the level of the reactors converting feeds to products. Once the reactor design is settled, the separation needs are known and the separators can be designed. Once the separator design is settled, and the mass flows, vapor–liquid equilibrium, temperatures, pressures, and flow rates are known, then the basic heat and material balance is defined and the composite curves representing the base-case process can be constructed.

The plus/minus principle (Linnhoff and Vredeveld 1984) (Fig. 5.8) uses the composite curves to identify process changes by going backward into the onion, based on the recognition that utility targets are reduced if the total hot-stream enthalpy change above the pinch is increased or the cold-stream enthalpy change is reduced. The principle gives the designer guidelines as to what changes are beneficial for energy or capital costs.

5.12 Column profiles

Conventional Pinch Analysis concerns heat exchange between streams within a process. Column composite curves and column grand composite curves enable a similar analysis inside distillation columns by using a plot of reboiling and condensing temperatures called the column profile (Dhole and Linnhoff 1993a) (Fig. 5.9). Similar to a grand composite curve, the column profile indicates at what temperatures heat needs to be supplied and rejected up and down the column. Not all heat needs to be supplied at the reboiling temperature; some can be supplied at lower temperatures using side reboilers. Not all heat needs to be removed at condensing temperature; partial heat removal through side condensers may be appropriate.

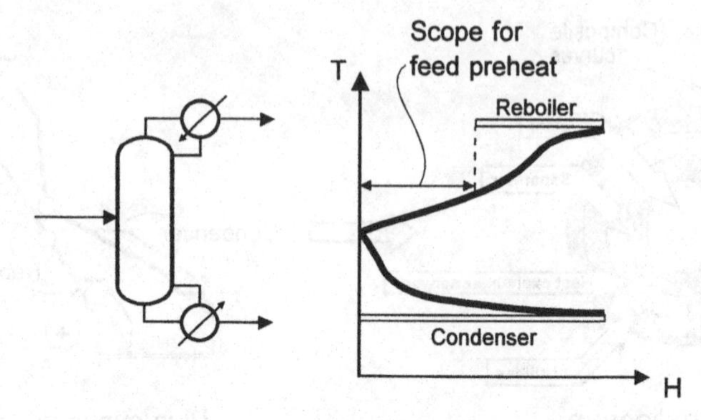

Fig. 5.9. Column profile.

5.13 Shaftwork targeting

Before Pinch Analysis emerged, the work of Carnot, Clausius, and others led to so-called second-law analysis of industrial energy systems. Over the last thirty years, this analysis has become better known as exergy analysis.

Exergy analysis allows calculation of irreversible exergy losses for individual process units and interprets these losses as a quantitative measure of the potential to generate shaft work following design improvements. This potential is lost in the analyzed system as a result of nonideal energy transfer and transformations. Exergy losses are thus caused by the thermodynamic inefficiencies of a real-life system when compared with an ideal system, which would contain only equilibrium processes. Thus far, this theory is the accepted one.

For practical process equipment, a certain minimum exergy loss is clearly unavoidable, and this minimum will be largely determined by economic considerations. In practical – optimized – heat exchanger networks, exergy losses are caused by heat transfer at finite temperature differences. To avoid these losses, we have two options:

- either substitute existing heat exchangers with equilibrium equipment by installing an infinitely large heat-transfer area to allow driving forces to become infinitely small,
- or accept existing exchangers, but introduce shaftwork recovery equipment, such as (ideal) heat engines or Carnot cycles.

Neither of these options would lead to improved practical systems. Exergy loss would be reduced, but capital cost would be increased. Therefore, although exergy analysis undoubtedly offers guidelines for design, it is up to the designer to differentiate between those exergy losses that are due to integration faults

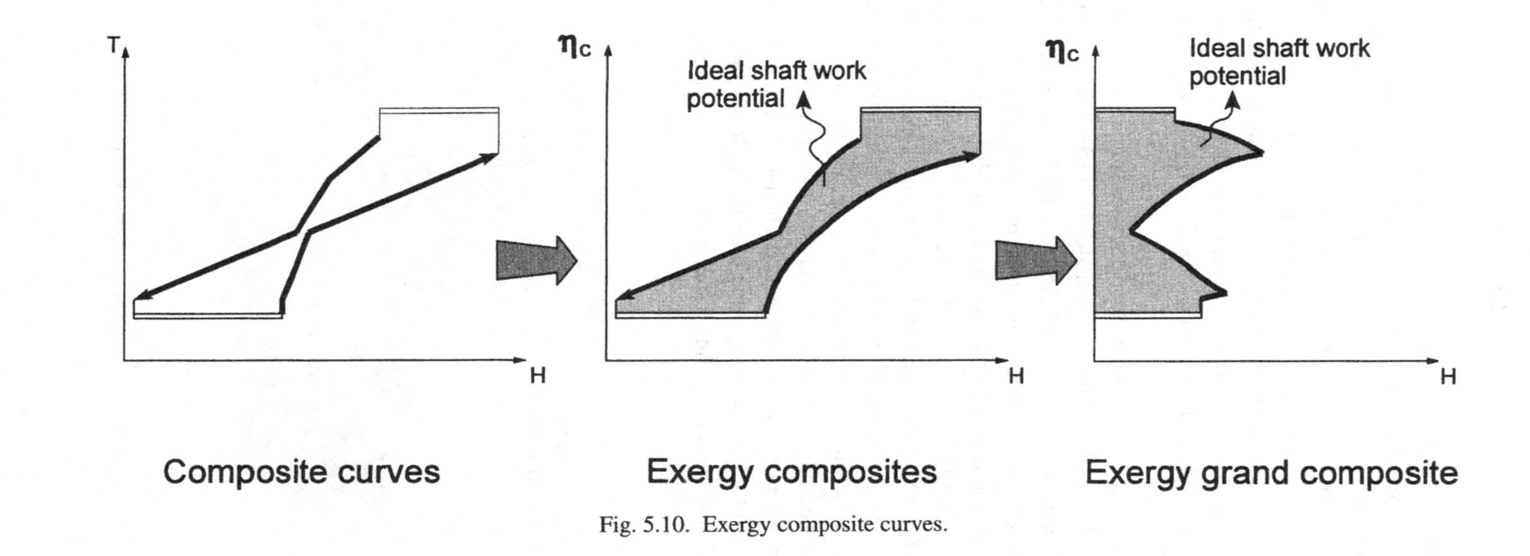

Fig. 5.10. Exergy composite curves.

and can be reduced without capital cost penalty, and those that are caused by realistic and optimized driving forces and cannot be reduced. Pinch Analysis, in essence, establishes targets for unavoidable exergy losses and offers specific information for the design of improved, practical systems that include only these unavoidable losses, while eliminating the avoidable ones. Pinches occur only when the designer departs from the concept of the ideal system and introduces irreversibilities. I have therefore suggested that Pinch Analysis can be best described as second-law analysis of nonideal energy systems (Linnhoff 1989) (Linnhoff 1993).

To represent exergy losses, Pinch Analysis uses composite curves on η_c-H diagrams (Fig. 5.10). These diagrams are analogous to the *T-H* diagrams described earlier. The variable η_c is the Carnot factor: $(1-T_o/T)$. On an η_c-H diagram, the area between the balanced composite curves or between the process grand composite curve and the utility profile exactly equals the overall exergy loss. This diagram can especially be of use in the design of combined process and refrigeration systems (Linnhoff and Dhole 1992) (Dhole and Linnhoff 1994) (Fig. 5.11).

As long as a network consisting of just heat exchangers is being designed following the basic pinch design rules, that is, no heat is transferred across the pinch, the method guarantees that only the minimum, unavoidable exergy loss occurs in the resulting design. A step-change reduction of this loss can then be achieved by incorporating additional utility levels, heat engines, heat pumps, etc., should the corresponding step-change increase in capital cost prove worthwhile.

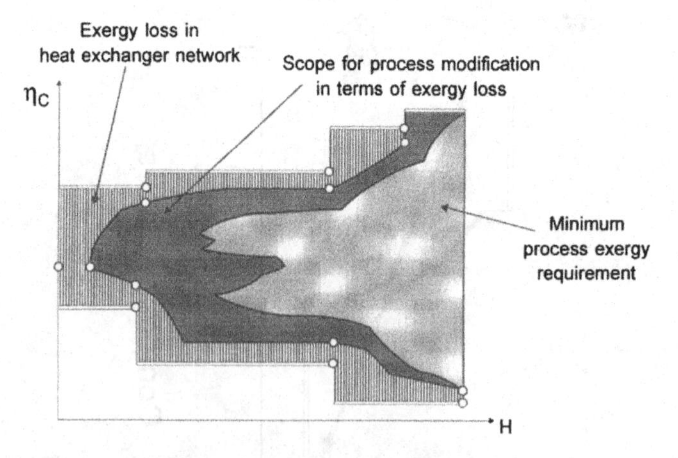

Fig. 5.11. Exergy grand composite curve.

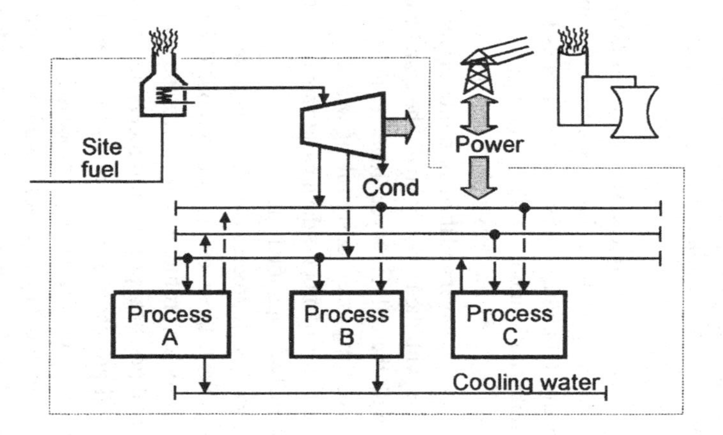

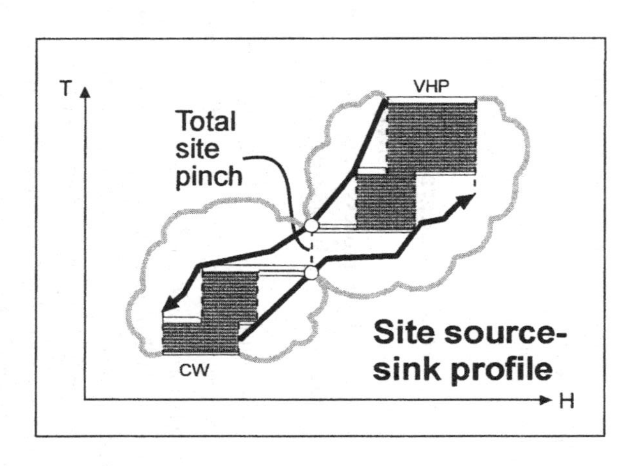

Fig. 5.12. Total site targeting.

In conclusion, Pinch Analysis concepts allow exergy analysis to be carried out in the targeting phase and for overall systems, taking into account the necessity of unavoidable exergy losses due to economic considerations.

5.14 Total site integration

The concepts just described have been used extensively in the last ten years in industrial design projects in individual process units. More recently, total site analysis has become important for better designs of heat and power systems for entire sites.

Boiler inefficiencies are often considered a local problem, related only to the boilers on a manufacturing site. In reality, these efficiencies are related to energy demands and losses throughout the facility.

On a typical site with several production processes, some processes have their own utilities, whereas others do not. A common set of central utilities usually exists, using imported fuel or power. Many industrial installations generate, import, or export power with steam turbines, gas turbines, or other prime movers; and load variations usually have an impact on the power required from electricity suppliers, thereby affecting the efficiency of public utilities and, in turn, power tariffs.

When a study of such a large, multiprocess production facility is required, traditional Pinch Analysis techniques for single processes become unwieldy. A new concept addresses this problem, setting sitewide targets for efficient heat and power systems (Dhole and Linnhoff 1993b) (Fig. 5.12).

On the onion diagram (Fig. 5.8), reactors and separators form the inner core, surrounded by the heat recovery layer and then the utilities layer. The composite curves link the core layers with the recovery layer; the grand composite curve links the core directly with the utilities layer. Together these concepts tell the engineer what the options are in process design along with the consequent utility choices. This simple translation of process designs into appropriate utility mixes provides the key to the analysis of a total site.

By using Pinch Analysis, each process can be represented by its optimized grand composite curve. Next, these curves are combined into a set of site source-sink profiles. The source profile, which depicts the heat to be exported from all processes after all practical process-to-process heat integration, is the net heat available from all processes as a function of temperature level and is analogous to the hot composite curve for a single process. Similarly, the sink profile represents the net heat required by all processes after integration within processes and is analogous to the cold composite curve. These total-site profiles give a simultaneous view of surplus heat from, and heat deficit for, all site processes.

Utilities are placed in between the total-site profiles, to give explicit cogeneration and fuel targets for the overall site. The heat loads for different site steam levels, a turbine system for cogeneration, central boiler capacity, and cooling demands are all visible at the targeting stage.

From the fuel targets and from fuel compositions, emission levels are targeted. Simultaneous targets can thus be set for site fuel, power export or import, cooling, and emissions.

Economic decisions and trade-offs are incorporated for process changes, fuel and steam demands, and infrastructure changes. Design and planning decisions are tested in the global context and revised as required, and the procedure is repeated to establish a set of planning scenarios meeting different criteria. As the targets meet specific conditions and constraints, the designer can with confidence generate designs for site development that take into account long- and short-term investment and benefit functions. Subject to the planning constraints applied, these designs will be optimum. Arguments for local and global control measures can be rationally assessed.

The approach has been successfully demonstrated in studies of both new and existing multiunit process plants (Linnhoff 1994).

References

Dhole, V. R. & Linnhoff, B. (1994). Overall design of subambient plants. *Computers in Chemical Engineering*, **18**, s105–11.

Dhole, V. R, & Linnhoff, B. (1993a). Distillation column targets. *Computers in Chemical Engineering*, **17**, 549–60.

Dhole, V. R. & Linnhoff, B. (1993b). Total site targets for fuel, co-generation, emissions, and cooling. *Computers in Chemical Engineering*, **17**, s101–9.

Körner, H. (1988). Optimal use of energy in the chemical industry. *Chemie Ingenieut Technik*, 60, 511–8.

Linnhoff, B. (1989). Pinch technology for the synthesis of optimal heat and power systems. *Transactions of the ASME, Journal of Energy Resources Technology*, **111**, 137–47.

Linnhoff, B. (1993). Pinch Analysis and exergy: a comparison. Paper presented at Energy Systems and Ecology Conference, Cracow, Poland, July 5 – 9.

Linnhoff, B. (1994). Pinch Analysis: Building on a decade of progress. *Chemical Engineering Progress*, August. pp. 32 – 57.

Linnhoff, B. & Ahmad, S. (1990). Cost optimum heat exchanger networks. I. Minimum energy and capital using simple models for capital cost. *Computers in Chemical Engineering*. **14**, 729–50.

Linnhoff, B. & Dhole, V. R. (1992). Shaftwork targets for low temperature process design. *Chemical Engineering Science*, **47**, 281 – 91.

Linnhoff, B. & de Leur, J. (1988). Appropriate placement of furnaces in the integrated process. Paper presented at IChemE Symposium No. 109 *Understanding Process Integration II*, March 22 – 3, University of Manchester Institute of Science and Technology, Manchester, UK.

Linnhoff, B. & Flower, J. R. (1978). Synthesis of heat exchanger networks. I. Systematic generation of energy optimal networks. *AIChE Journal*, 24, 633–42, and II. Evolutionary generation of networks with various criteria of optimality. *AIChE Journal*, 24, 642–54.

Linnhoff, B. & Hindmarsh, E. (1983). The pinch design method of heat exchanger networks. *Chemical Engineering Science*, 38, 745–63.

Linnhoff, B., Mason, D. R. & Wardle, I. (1979). Understanding heat exchanger networks. *Computers in Chemical Engineering*, 3, 295–302.

Linnhoff, B., Townsend, D. W., Boland, D., Hewitt, G. F., Thomas, B. E. A., Guy, A. R. & Marsland, R. H. (1982). *User Guide on Process Integration for the Efficient Use of Energy*. Rugby, UK: IChemE.

Linnhoff, B. & Vredeveld, D. R. (1984). Pinch technology has come of age. *Chemical Engineering Progress*, July, 33–40.

Smith, R., Petela, E. A. & Spriggs, H. D. (1990). Minimization of environmental emissions through improved process integration. *Heat Recovery Systems and CHP*, 10, 329–39.

Townsend D. W. & Linnhoff, B. (1983). Heat and power networks in process design. I. Criteria for placement of heat engines and heat pumps in process networks. *AIChE Journal*, 29, 742–48, and II. Design procedure for equipment selection and process matching. *AIChE Journal*, 29, 748–71.

6

Second Law applications in thermal system design*

MICHAEL J. MORAN

Ohio State University

The importance of developing thermal systems that effectively use energy resources such as oil, natural gas, and coal is apparent. Effective use is determined with both the First and Second Laws of thermodynamics. Energy entering a thermal system with fuel, electricity, flowing streams of matter, and so on is accounted for in the products and by-products. Energy cannot be destroyed – a First Law concept. The idea that something can be destroyed is useful. This idea does not apply to energy, however, but to exergy – a Second Law concept. Moreover, it is exergy and not energy that properly gauges the quality (usefulness) of, say, one kilojoule of electricity generated by a power plant versus one kilojoule of energy in the plant cooling water stream. Electricity clearly has the greater quality and, not incidentally, the greater economic value.

6.1 Preliminaries

In developed countries worldwide, the effectiveness of using oil, natural gas and coal has markedly improved over the last two decades. Still, use varies even within nations, and there is room for improvement. Compared with some of its principal international trading partners, for example, U.S. industry as a whole has a higher energy resource consumption per unit of output and generates considerably more waste. These realities pose a challenge for maintaining competitiveness in the global marketplace.

For industries where energy is a major contributor to operating costs, an opportunity exists for improving competitiveness through more effective use of energy resources. This is a well known and largely accepted principle today. A related but less publicized principle concerns the waste and effluent streams of plants. The waste of a plant is often not an unavoidable result of plant operation

* Adapted by permission of John Wiley & Sons, Inc., from Bejan, A., Tsatsaronis, G. & Moran, M. J. (1996). *Thermal Design and Optimization*. New York: Wiley.

but a measure of its inefficiency: the less efficient a plant the more unusable by-products it produces, and the converse. Effluents not produced owing to a more efficient plant require no costly cleanup, nor do they impose a burden on the environment. Such clean technology considerations provide another avenue for improving competitiveness.

Because economic, fuel-thrifty, environmentally benign plants have to be carefully engineered, an extensive body of literature has been developed to address various aspects of such demands on good engineering. See, for example, Douglas (1989), Kenney (1984), Peters and Timmerhaus (1991), Ulrich (1984), and other references to be cited.

Furthermore, in recognition of the link between efficient plants and the goal of better energy resource use with less environmental impact, the Second Law of thermodynamics is being used increasingly today in thermal system design. The principal objective of this chapter is to provide an introduction to such applications of the Second Law.

6.2 The Second Law in design: heat exchanger example

A critical early step in the design of a system is to pin down what the system is required to do and to express these requirements quantitatively, that is, to formulate the design specifications. A workable design is simply one that meets all the specifications. Among the workable designs is the optimal design: the one that is in some sense the best. Several alternative notions of "best" can apply: optimal cost, size, weight, reliability, etc. Although the term *optimal* can take on many guises depending on the context, the design that minimizes cost is usually the one of interest.

Owing to system complexity and uncertainties in data and information about the system, a true optimum is generally impossible to determine precisely. Accordingly, an acceptable design often is one that is close to optimal: a nearly optimal design. To illustrate, consider a counterflow heat exchanger. A key design variable for heat exchangers is $(\Delta T)_{\min}$, the minimum temperature difference between the two streams. Let us consider the variation of the total annual cost associated with the use of the heat exchanger as $(\Delta T)_{\min}$ varies, holding other variables fixed.

From engineering thermodynamics we know that the temperature difference between the two streams of a heat exchanger is a measure of nonideality (irreversibility) and that heat exchangers approach ideality as the temperature difference approaches zero. This source of nonideality exacts a penalty on the fuel required by the overall plant that has the heat exchanger as a component: part of the fuel provided to the overall plant is used to feed this source of

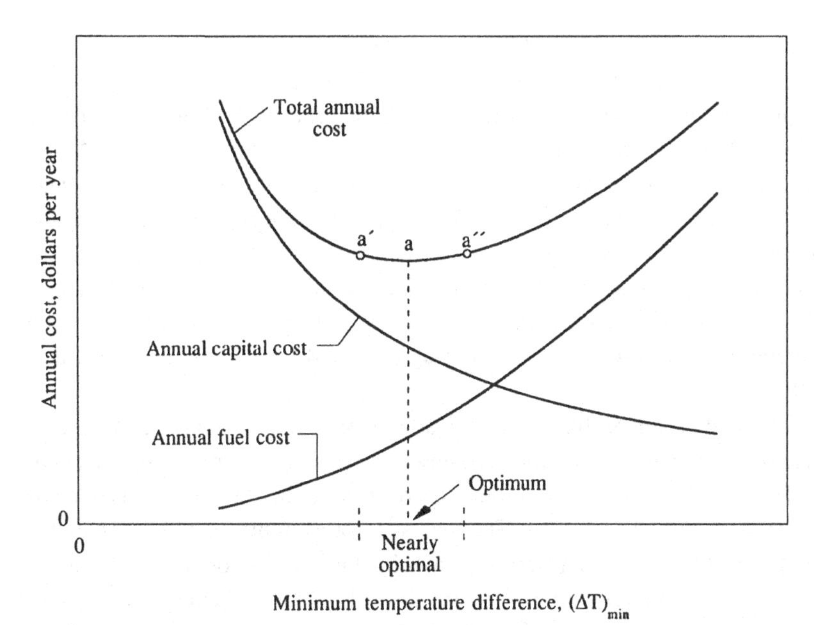

Fig. 6.1. Cost curves for a single heat exchanger.

nonideality. A larger temperature difference corresponds to a thermodynamically less ideal heat exchanger, and so the fuel cost increases with $(\Delta T)_{\text{min}}$, as shown in Fig. 6.1.

To reduce operating costs, then, we would seek to reduce the stream-to-stream temperature difference. From the study of heat transfer, however, we know that as the temperature difference between the streams narrows, more heat transfer area is required for the same rate of energy exchange. More heat transfer area corresponds to a larger, more costly heat exchanger. Accordingly, the capital cost increases with the heat transfer area and varies inversely with $(\Delta T)_{\text{min}}$, as shown in Fig. 6.1. In this figure the costs are on an annualized basis.

The total cost is the sum of the capital cost and the fuel cost. As shown in Fig. 6.1, the variation of the cost with $(\Delta T)_{\text{min}}$ is concave upward. The point labeled a locates the design with minimum total annual cost: the optimal design for a system consisting of a heat exchanger only. However, the variation of total cost with temperature difference is relatively flat in the portion of the curve whose end points are a' and a''. Heat exchangers with $(\Delta T)_{\text{min}}$ values giving total cost on this portion of the curve may be regarded as nearly optimal.

The design engineer would be free to specify, without a significant difference in total cost, any $(\Delta T)_{\text{min}}$ in the interval designated nearly optimal in Fig. 6.1.

For example, by specifying $(\Delta T)_{\text{min}}$ smaller than the economic optimum, near point a', the designer would save fuel at the expense of capital but with no significant change in total cost. At the higher end of the interval near point a'', the designer would save capital at the expense of fuel. Specifying $(\Delta T)_{\text{min}}$ outside this interval would be a design error for a stand-alone heat exchanger, however. With too low a $(\Delta T)_{\text{min}}$, the extra cost of the heat exchanger required would be much greater than the savings on fuel. With too high a $(\Delta T)_{\text{min}}$, the extra fuel cost would be much greater than the savings on the heat exchanger. As these conclusions stem from Second-Law reasoning and irreversibility considerations, any such design error may be called a Second-Law error (Sama 1992).

Thermal design problems are typically more complex than suggested by this example, of course. Costs are not known as precisely as implied by the curves of Fig. 6.1. Owing to uncertainties, cost data would be better represented on the figure as bands and not lines. The fuel cost, for example, requires a prediction of future fuel prices, and these vary with the dictates of the world market for such commodities. Moreover, the cost of a heat exchanger is not a fixed value but often the consequence of a bidding procedure involving many factors. Equipment costs also do not vary continuously as shown in Fig. 6.1 but in a stepwise manner with discrete sizes of the equipment.

Considerable additional complexity enters because thermal systems typically involve several components that interact with one another in complicated ways. Owing to component interdependency, a unit of fuel saved by reducing the nonideality of a heat exchanger, say, may be required elsewhere in the plant, resulting in no net change or even an increase in plant fuel consumption. Moreover, unlike the heat exchanger example just considered that involved a single design variable, several design variables usually must be considered and optimized simultaneously. Still further, the objective is normally to optimize an overall system of several components. Optimization of components individually, as was done in the example, usually does not guarantee an optimum for the overall system.

Thermal system design problems also may admit several alternative, fundamentally different solutions. Workable designs might exist, for example, that require one, two, or perhaps three or more heat exchangers. The most significant advances in design generally come from identifying the best solution from among several alternatives and not from merely polishing one of the alternatives. Accordingly, applying an optimization procedure to a particular design solution can consider trade-offs for only that solution (the trade-off between capital cost and fuel cost, for example). And no matter how rigorously formulated and solved an optimization procedure may be, such procedures are generally unable to identify the existence of alternative solutions or discriminate between fundamentally different solutions to the same problem.

6.3 Exergy fundamentals

This section briefly surveys exergy principles and illustrates their use. To explore the subject in greater depth, readers should refer to Bejan, Tsatsaronis, and Moran (1996), Kotas (1985), Moran (1989), Moran and Shapiro (1995), and Szargut, Morris, and Steward (1988).

6.3.1 Engineering thermodynamics principles

The exergy concept is developed using extensive property balances for mass, energy, and entropy, together with property relations. Here, we will consider properties of simple compressible substances exclusively.

Engineering applications involving exergy are generally analyzed on a control volume basis. Accordingly, the control volume formulations of the mass, energy, and entropy balances presented in this section play important roles. These are provided here in the form of overall balances assuming one-dimensional flow. Equations of change for mass, energy, and entropy in the form of differential equations are also available in the literature (Bird, Stewart, and Lightfoot 1960).

6.3.1.1 Mass rate balance

For applications in which inward and outward flows occur, each through one or more ports, the extensive property balance for mass expressing the conservation of mass principle takes the form

$$\frac{dm}{dt} = \sum_i \dot{m}_i - \sum_e \dot{m}_e, \tag{6.1}$$

where dm/dt represents the time rate of change of mass contained within the control volume, $\dot{m}_i$ denotes the mass flow rate at an inlet port, and $\dot{m}_e$ denotes the mass flow rate at an exit port.

6.3.1.2 Energy rate balance

Energy is a fundamental concept of thermodynamics and one of the most significant aspects of engineering analysis. Energy can be stored within systems in various macroscopic forms: kinetic energy, gravitational potential energy, and internal energy. Energy can also be transformed from one form to another and transferred between systems. Energy can be transferred by work, by heat transfer, and by flowing matter. The total amount of energy is conserved in all transformations and transfers.

On a time-rate basis, the extensive property balance for the energy of a closed system takes the form

$$\frac{d(U + KE + PE)}{dt} = \dot{Q} - \dot{W}, \tag{6.2}$$

where U, KE, and PE denote, respectively, the internal energy, kinetic energy, and gravitational potential energy. The terms $\dot{Q}$ and $\dot{W}$ account, respectively, for the net rates of energy transfer by heat and work. In Equation (6.2), energy transfer by heat to the system is considered positive and energy transfer by work from the system is considered positive.

The specific internal energy is symbolized by u or $\bar{u}$, respectively, depending on whether it is expressed on a unit mass or per-mole basis. The specific energy (energy per unit mass) is the sum of the specific internal energy, u, the specific kinetic energy, $V^2/2$, and the specific gravitational potential energy, gz. That is,

$$\textit{specific energy} = u + \frac{V^2}{2} + gz, \tag{6.3}$$

where V is the velocity, z is the elevation, and g is the acceleration of gravity.

Energy can enter and exit control volumes by work, by heat transfer, and by flowing matter. Because work is always done on or by a control volume where matter flows across the boundary, it is convenient to separate the work rate (or power) into two contributions. One contribution is the work rate associated with the force of the fluid pressure as mass is introduced at the inlet and removed at the exit. This is commonly referred to as flow work. The other contribution, denoted as $\dot{W}_{cv}$, includes all other work effects, such as those associated with rotating shafts, displacement of the boundary, and electrical effects. On a one-dimensional flow basis, the total work rate associated with a control volume is

$$\dot{W} = \dot{W}_{cv} + \sum_e \dot{m}_e(p_e v_e) - \sum_i \dot{m}_i(p_i v_i), \tag{6.4}$$

where p_i and p_e denote the pressures and v_i and v_e denote specific volumes at the inlets and exits, respectively. The terms $\dot{m}_i(p_i v_i)$ and $\dot{m}_e(p_e v_e)$ account for flow work at the inlets and exits, respectively. Energy also enters and exits control volumes with flowing streams of matter. Thus, on a one-dimensional flow basis, the rate at which energy enters with matter at inlet i is $\dot{m}_i(u_i + V_i^2/2 + gz_i)$.

Collecting results, the following form of the control volume energy rate balance evolves:

$$\frac{d(U + KE + PE)}{dt} = \dot{Q} - \dot{W}_{cv} + \sum_i \dot{m}_i\left(u_i + p_i v_i + \frac{V_i^2}{2} + gz_i\right)$$
$$- \sum_e \dot{m}_e\left(u_e + p_e v_e + \frac{V_e^2}{2} + gz_e\right). \tag{6.5}$$

Introducing the specific enthalpy h ($h = u + pv$), the energy rate balance becomes

$$\frac{d(U + KE + PE)}{dt} = \dot{Q} - \dot{W}_{cv} + \sum_i \dot{m}_i \left(h_i + \frac{V_i^2}{2} + gz_i \right) - \sum_e \dot{m}_e \left(h_e + \frac{V_e^2}{2} + gz_e \right). \quad (6.6)$$

6.3.1.3 *Entropy rate balance*

An extensive property balance also can be written for entropy. The closed system entropy balance reads

$$\underbrace{S_2 - S_1}_{\text{entropy change}} = \underbrace{\int_1^2 \left(\frac{\delta Q}{T} \right)_b}_{\text{entropy transfer}} + \underbrace{S_{gen}}_{\text{entropy generation}} \quad (6.7)$$

Because entropy is a property, the entropy change on the left side of Equation (6.7) can be evaluated independently of the details of the process, knowing only the end states. However, the two terms on the right side depend explicitly on the nature of the process and cannot be determined solely from knowledge of the end states.

The first term on the right side of Equation (6.7) is associated with heat transfer to or from the system during the process. This term can be interpreted as the entropy transfer accompanying heat transfer. The direction of entropy transfer is the same as the direction of heat transfer, and the same sign convention applies as for heat transfer. The second term as the right side, S_{gen}, accounts for entropy generated (produced) within the system owing to irreversibilities. Irreversibilities can be divided into two classes, internal and external. Internal irreversibilities are those occurring within the system, whereas external irreversibilities are those occurring within the surroundings, normally the immediate surroundings. The term S_{gen} is positive when internal irreversibilities are present during the process and vanishes when no internal irreversibilities are present. The Second Law of thermodynamics can be interpreted as specifying that entropy is generated by irreversibilities and conserved only in the limit as irreversibilities are reduced to zero. Because S_{gen} measures the effect of irreversibilities present within a system during a process, the value of S_{gen} depends on the nature of the process and not solely on the end states, and thus S_{gen} is not a property. When internal irreversibilities are absent during a process, no entropy is generated within the system and the process is said to be internally reversible.

A rate form of the closed system entropy balance that is frequently convenient is

$$\frac{dS}{dt} = \sum_j \frac{\dot{Q}_j}{T_j} + \dot{S}_{\text{gen}}, \tag{6.8}$$

where dS/dt is the time rate of change of entropy of the system. The term $\dot{Q}_j/T_j$ represents the time rate of entropy transfer through the portion of the boundary whose instantaneous temperature is T_j. The term $\dot{S}_{\text{gen}}$ accounts for the time rate of entropy generation due to irreversibilities within the system.

As for the case of energy, entropy can be transferred into or out of a control volume by streams of matter. This is the principal difference between the closed system and control volume forms. Accordingly, for control volumes the counterpart of Equation (6.8) is

$$\underbrace{\frac{dS}{dt}}_{\substack{\text{rate of}\\ \text{entropy}\\ \text{change}}} = \underbrace{\sum_j \frac{\dot{Q}_j}{T_j} + \sum_i \dot{m}_i s_i - \sum_e \dot{m}_e s_e}_{\substack{\text{rates of}\\ \text{entropy}\\ \text{transfer}}} + \underbrace{\dot{S}_{\text{gen}}}_{\substack{\text{rate of}\\ \text{entropy}\\ \text{generation}}}, \tag{6.9}$$

where dS/dt represents the time rate of change of entropy within the control volume. The terms $\dot{m}_i s_i$ and $\dot{m}_e s_e$ account, respectively, for rates of entropy transfer into and out of the control volume accompanying mass flow. $\dot{Q}_j$ represents the time rate of heat transfer at the location on the boundary where the instantaneous temperature is T_j, and $\dot{Q}_j/T_j$ accounts for the accompanying rate of entropy transfer. $\dot{S}_{\text{gen}}$ denotes the time rate of entropy generation due to irreversibilities within the control volume.

When applying the entropy balance, in any of its forms, the objective is often to evaluate the entropy generation term. However, the value of the entropy generation for a given process of a system usually does not have much significance by itself. The significance is normally determined through comparison: The entropy generation within a given component might be compared with the entropy generation values of the other components included in an overall system formed by these components. By comparing entropy generation values, the components where appreciable irreversibilities occur can be identified and rank-ordered, allowing attention to be focused on the components that contribute most heavily to inefficient operation of the overall system.

To evaluate the entropy transfer term of the entropy balance requires information about both the heat transfer and the temperature on the boundary where the heat transfer occurs. The entropy transfer term is not always subject to direct evaluation, however, because the required information is either unknown or not defined, such as when the system passes through states sufficiently far from equilibrium. In practical applications, it is often convenient, therefore, to enlarge the system to include enough of the immediate surroundings that the temperature on the boundary of the enlarged system corresponds to the ambient temperature, T_{amb}. The entropy transfer rate is then simply $\dot{Q}/T_{amb}$. However, as the irreversibilities present would not be just those for the system of interest but those for the enlarged system, the entropy generation term would account for the effects of internal irreversibilities within the system and external irreversibilities within that portion of the surroundings included within the enlarged system.

6.3.1.4 Steady-state rate balances

For control volumes at steady state, the identity of the matter within the control volume changes continuously, but the total amount of mass remains constant. At steady state, Equation (6.1) reduces to

$$\sum_i \dot{m}_i = \sum_e \dot{m}_e. \tag{6.10}$$

As steady state, the energy and entropy rate balances given by Equations (6.6) and (6.9) reduce, respectively, to read

$$0 = \dot{Q} - \dot{W}_{cv} + \sum_i \dot{m}_i \left(h_i + \frac{V_i^2}{2} + gz_i \right) - \sum_e \dot{m}_e \left(h_e + \frac{V_e^2}{2} + gz_e \right). \tag{6.11}$$

$$0 = \sum_j \frac{\dot{Q}_j}{T_j} + \sum_i \dot{m}_i s_i - \sum_e \dot{m}_e s_e + \dot{S}_{\text{gen}}. \tag{6.12}$$

When supplemented by appropriate property relations, Equations (6.10)–(6.12) allow control volumes at steady state to be investigated thermodynamically.

6.3.2 Defining exergy

An opportunity for doing work exists whenever two systems at different states are placed in communication, for in principle, work can be developed as the two are allowed to come into equilibrium. When one of the two systems is a suitably idealized system called an environment and the other is some system of interest, exergy (availability) is the maximum theoretical useful work (shaft work or electrical work) obtainable as the systems interact to equilibrium, heat transfer occurring with the environment only. Alternatively, exergy is the minimum theoretical useful work required to form a quantity of matter from substances present in the environment and bring the matter to specified state. As exergy is a measure of the departure of the state of the system from that of environment, exergy is an attribute of the system and environment together. Once the environment is specified, however, a value can be assigned to exergy in terms of property values for the system only, so exergy can be regarded as an extensive property of the system.

Exergy can be destroyed and generally is not conserved. A limiting case is when exergy would be completely destroyed, as occurs when a system comes into equilibrium with the environment spontaneously with no provision to obtain work. The capability to develop work existing initially would be completely wasted in the spontaneous process. Moreover, because no work need be done to effect such a spontaneous change, it may be concluded that the value of exergy (the maximum theoretical work obtainable) is at least zero and therefore cannot be negative.

6.3.2.1 Environment

Any system, whether a component in a larger system such as a steam turbine in a power plant or the larger system (power plant) itself, operates within surroundings of some kind. Distinguishing between the environment and system's surroundings is important. The term *surroundings* refers to everything not included in the system. The term *environment* applies to some portion of the surroundings, the intensive properties of each phase of which are uniform and do not change significantly as a result of any process under consideration. The environment is regarded as free of irreversibilities. All significant irreversibilities are located within the system and its immediate surroundings.

Because the physical world is complicated, models with various levels of specificity have been proposed for describing the environment. In the present discussion the environment is modeled as a simple compressible system, large in extent, and uniform in temperature, T_o, and pressure, p_o. In keeping with the idea that the environment has to do with the actual physical world, the values for p_o and T_o used throughout a particular analysis are normally taken as typical environmental conditions, such as one atmosphere and 25°C (77°F). The environment is regarded as composed of common substances existing in abundance within the atmosphere, the oceans, and the crust of the Earth. The substances are in their stable forms as they exist naturally, and developing work from interactions – physical or chemical – between parts of the environment is not possible. Although its intensive properties do not change, the environment can experience changes in its extensive properties as a result of interactions with other systems. Kinetic and potential energies are evaluated relative to coordinates in the environment, all parts of which are considered to be at rest with respect to one another. Accordingly, a change in the energy of the environment can be a change in its internal energy only.

6.3.2.2 *Dead state*

When the pressure, temperature, composition, velocity, or elevation of a system is different from the environment, an opportunity to develop work exists. As the system changes state toward that of the environment, the opportunity diminishes, ceasing to exist when the system and environment, at rest relative to one another, are in equilibrium. This state of the system is called the dead state. At the dead state, the conditions of mechanical, thermal, and chemical equilibrium between the system and the environment are satisfied: the pressure, temperature, and chemical potentials of the system equal those of the environment, respectively. In addition, the system has zero velocity and zero elevation relative to coordinates in the environment. Under these conditions, no possibility exists of a spontaneous change within the system or the environment, nor can they interact.

Another type of equilibrium between the system and the environment can be identified. This form of equilibrium is restricted – only the conditions of mechanical and thermal equilibrium must be satisfied. This state of the system is called the restricted dead state. At the restricted dead state, the fixed quantity of matter under consideration is imagined to be sealed in an envelope impervious to mass flow, at zero velocity and elevation relative to coordinates in the environment, and at the temperature T_o and pressure p_o.

6.3.2.3 *Exergy components*

In the absence of nuclear, magnetic, electrical, and surface tension effects, the exergy **E** can be divided into four components: physical exergy, $\boldsymbol{E}^{PH}$, kinetic exergy, E^K, potential exergy, E^P, and chemical exergy, E^{CH}, that is

$$\mathbf{E} = \boldsymbol{E}^{PH} + E^K + E^P + E^{CH}, \tag{6.13}$$

where the bold type here distinguishes the associated exergy concepts from exergy transfers introduced with Equation (6.21). The sum of the kinetic, potential, and physical exergies is also referred to in the literature as the thermomechanical exergy (availability) (Moran 1989) (Moran and Shapiro 1995).

Although exergy is an extensive property, it is often convenient to work with it on a unit mass or molar basis. The specific exergy on a mass basis, $\boldsymbol{e}$, is given by

$$\begin{aligned} \boldsymbol{e} &= \boldsymbol{e}^{PH} + e^K + e^P + e^{CH} \\ &= \boldsymbol{e}^{PH} + \frac{V^2}{2} + gz + e^{CH}. \end{aligned} \tag{6.14}$$

In Equation (6.14), V and z denote velocity and elevation relative to coordinates in the environment. When evaluated relative to the environment, the kinetic and potential energies of the system are, in principle, fully convertible to work as the system is brought to rest relative to the environment, and so they correspond to the kinetic and potential exergies, respectively.

Considering a system at rest relative to the environment ($e^K = e^P = 0$), the physical exergy is the maximum theoretical useful work obtainable as the system passes from its initial state, where the temperature is T and the pressure is p, to the restricted dead state, where the temperature is T_o and pressure is p_o. The chemical exergy is the maximum theoretical work obtainable as the system passes from the restricted dead state to the dead state, where it is in complete equilibrium with the environment. The use of the term *chemical* here does not necessarily imply a chemical reaction.

6.3.2.4 Physical exergy

The following expression for the physical exergy, E^{PH}, is readily derived using energy and entropy balances:

$$E^{PH} = (U - U_o) + p_o(V - V_o) - T_o(S - S_o) \tag{6.15}$$

where U_o, V_o, and S_o denote the energy, volume, and entropy of the system when at the restricted dead state. Various idealized devices can be invoked to visualize the development of work as a system passes from a specified state to the restricted dead state (Kotas 1985) (Moran 1989).

On a mass-unit basis, the physical exergy is

$$e^{PH} = (u - u_o) + p_o(v - v_o) - T_o(s - s_o). \tag{6.16}$$

6.3.2.5 Chemical exergy

When evaluating chemical exergy – the exergy component associated with the departure of the chemical composition of a system from that of the environment – the substances composing the system must be referred to the properties of a suitably selected set of environmental substances. For there to be no possibility of developing work from interactions, physical or chemical, between parts of the environment, it is essential that these environmental reference substances be in equilibrium, mutually and with the rest of the environment. The natural environment is not in chemical equilibrium, however, and a compromise between physical reality and the requirements of thermodynamic theory is necessary. Such considerations have led to alternative models for evaluating chemical exergy, and the term *exergy reference environment* is frequently used to distinguish the thermodynamic concept from the natural environment. The modeling of exergy reference environments is thoroughly discussed in the literature. For simplicity, the present development features the use of standard chemical exergies determined relative to a standard environment.

Standard chemical exergies are calculated on the basis that the environmental temperature and pressure have standard values and the environment consists of a set of reference substances with standard concentrations reflecting the chemical makeup of the natural environment. The reference substances generally fall into three groups: gaseous components of the atmosphere, solid substances from the lithosphere, and ionic and nonionic substances from the oceans. Alternative standard exergy reference environments that have gained acceptance for engineering evaluations are detailed by Ahrendts (1980) and Szargut et al. (1988). Readers should refer to these sources for the specific choices of the reference substances, the methods used to calculate the standard chemical exergies, tables of standard chemical exergies, and comparisons of the two approaches.

The use of standard chemical exergies greatly facilitates the application of exergy principles. The term *standard* is somewhat misleading, however, for no one specification of the environment suffices for all applications. Still, chemical exergies calculated relative to alternative specifications of the environment generally agree well. In particular, the effect owing to slight variations in T_o and p_o about their nominal values (25°C, one atmosphere) can be neglected. For a broad range of engineering applications, the simplicity and ease of use of standard chemical exergies generally outweighs any slight lack of accuracy that might result.

A common feature of standard exergy reference environments is a gas phase, intended to represent air, that includes N_2, O_2, CO_2, $H_2O(g)$, and other gases. Each gas i present in the gas phase is at temperature T_o and partial pressure $p_i^e = x_i^e p_o$, where the superscript e denotes the environment and x_i^e is the mole fraction of gas i in the gas phase. Referring to the device at steady state shown in Fig. 6.2, the standard chemical exergy for a gas included in the environmental gas phase can be evaluated as follows: Gas i enters at temperature T_o and pressure p_o, expands isothermally with heat transfer only with the environment, and exits to the environment at temperature T_o and partial pressure $x_i^e p_o$. The maximum theoretical work per mole of gas i would be developed when the expansion occurs without internal irreversibilities. Accordingly, with energy and entropy balances together with the ideal gas equation of state, the chemical exergy per mole of i is

$$\begin{aligned} \bar{e}_i^{\mathrm{CH}} &= -\bar{R}T_o \ln \frac{x_i^e p_o}{p_o} \\ &= -\bar{R}T_o \ln x_i^e. \end{aligned} \tag{6.17}$$

As an application of Equation (6.17), consider CO_2 with $x_{CO_2}^e = 0.00033$ and $T_o = 298.15$ K, as in Szargut et al. (1988). Inserting values into Equation (6.17) gives $\bar{e}_{CO_2}^{\mathrm{CH}} = 19{,}871$ kilojoules per kilomol, which is closely the value listed for CO_2 in the reference cited.

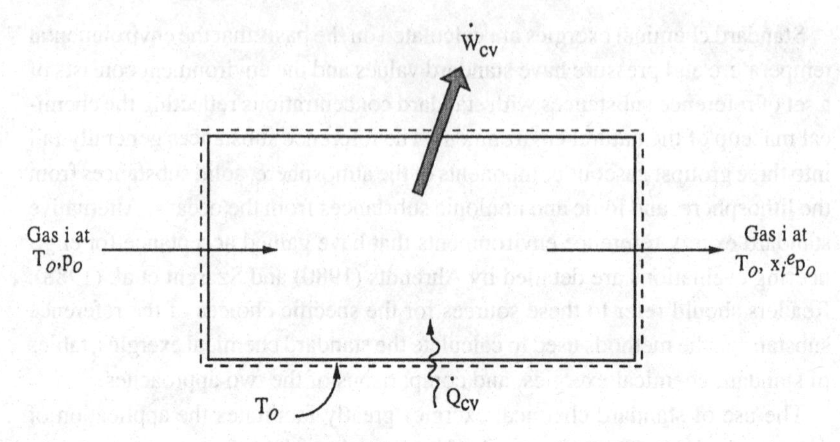

Fig. 6.2. Device for evaluating the chemical exergy of a gas.

The chemical exergy of a mixture of n gases, each of which is present in the environmental gas phase, can be obtained similarly. We may think of a set of n devices such as shown in Fig. 6.2, one for each gas in the mixture. Gas i, whose mole fraction in the gas mixture at T_o, p_o is x_i, enters at T_o and the partial pressure $x_i p_o$. As before, the gas exits to the environment at T_o and the partial pressure $x_i^e p_o$. Paralleling the previous development, the work per mole of i is $-\overline{R}T_o \ln(x_i^e/x_i)$. Summing over all components, the chemical exergy per mole of mixture is

$$\overline{e}^{\rm CH}_{mixture} = -\overline{R}T_o \sum x_i \ln \frac{x_i^e}{x_i}.$$

This expression can be written alternatively with Equation (6.17) as

$$\overline{e}^{\rm CH}_{mixture} = \sum x_i \overline{e}_i^{\rm CH} + \overline{R}T_o \sum x_i \ln x_i. \tag{6.18}$$

Equation (6.18) remains valid for mixtures containing gases, for example, gaseous fuels, other than those assumed present in the reference environment and can be extended to mixtures (and solutions) that do not adhere to the ideal gas model (Kotas 1985). In such applications, the terms $\overline{e}_i^{\rm CH}$ are selected from a table of standard chemical exergies.

As a further illustration of the evaluation of chemical exergy, consider the case of hydrocarbon fuels. In principle, the standard chemical exergy of a substance can be evaluated by considering an idealized reaction of the substance with other substances for which the chemical exergies are known. For the case of a pure hydrocarbon fuel C_aH_b at T_o, p_o, refer to the system shown in Fig. 6.3, where all substances are assumed to enter and exit at T_o, p_o. For operation at steady state, heat transfer only with the environment, and no internal irreversibilities,

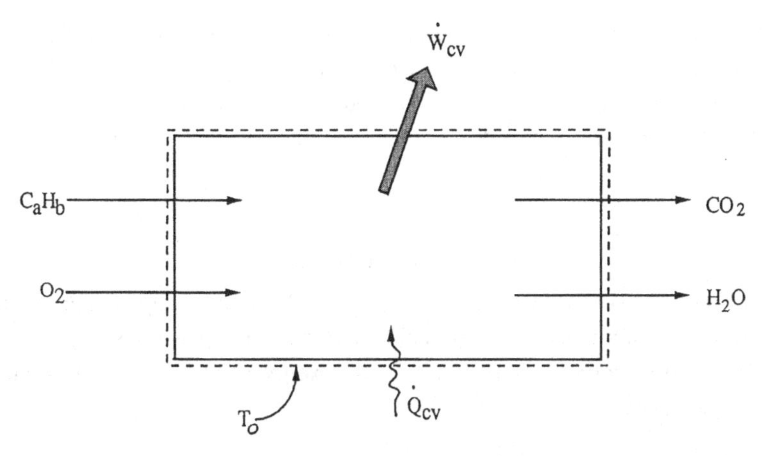

Fig. 6.3. Device for evaluating the chemical exergy of C_aH_b

a systematic application of mass, energy, and entropy balances results in the following alternative expressions:

$$\bar{e}_F^{CH} = \overline{HHV}(T_o, p_o) - T_o\left[\bar{s}_F + \left(a + \frac{b}{4}\right)\bar{s}_{O_2} - a\bar{s}_{CO_2} - \frac{b}{2}\bar{s}_{H_2O}\right](T_o, p_o)$$

$$+ a\bar{e}_{CO_2}^{CH} + \left(\frac{b}{2}\right)\bar{e}_{H_2O}^{CH} - \left(a + \frac{b}{4}\right)\bar{e}_{O_2}^{CH} \tag{6.19a}$$

$$\bar{e}_F^{CH} = \left[\bar{g}_F + \left(a + \frac{b}{4}\right)\bar{g}_{O_2} - a\bar{g}_{CO_2} - \frac{b}{2}\bar{g}_{H_2O}\right](T_o, p_o)$$

$$+ a\bar{e}_{CO_2}^{CH} + \left(\frac{b}{2}\right)\bar{e}_{H_2O}^{CH} - \left(a + \frac{b}{4}\right)\bar{e}_{O_2}^{CH}, \tag{6.19b}$$

where the subscript F denotes the fuel, $\overline{HHV}$ denotes the higher heating value, $\bar{s}$ denotes absolute entropy, and $\bar{g}$ denotes the Gibbs function.

With Equations (6.19a) and (6.19b), the chemical exergy of C_aH_b can be calculated using the standard chemical exergies of O_2, CO_2, and $H_2O(\ell)$, together with selected property data: the fuel higher heating value and absolute entropies, or Gibbs functions of formation. As an application, consider the case of carbon. With $a = 1$, $b = 0$, Gibbs function data from Moran and Shapiro (1995), and standard chemical exergies for CO_2 and O_2 as reported by Szargut et al. (1988). Equation (6.19b) gives $\bar{e}_C^{CH} = 410{,}280$ kilojoules per kilomol, which agrees closely with the value listed by Szargut et al. Similarly, for the case of hydrogen – $a = 0$, $b = 2$ – we have $\bar{e}_{H_2}^{CH} = 236{,}095$.

When tabulated chemical exergy data are lacking, as, for example, in the cases of coal, char, and fuel oil, the approach of Equation (6.19a) can be invoked using a measured heating value and an estimated value of the fuel absolute entropy determined with procedures discussed in the literature. For an illustration, see Bejan et al. (1996).

6.3.3 Exergy balance

As for the extensive properties mass, energy, and entropy, exergy balances can be written in alternative forms suitable for particular applications of practical interest. A frequently convenient form of the exergy balance for closed systems is the rate equation

$$\underbrace{\frac{d\mathbf{E}}{dt}}_{\text{rate of exergy change}} = \underbrace{\sum_j \left(1 - \frac{T_o}{T_j}\right)\dot{Q}_j - \left(\dot{W} - p_o\frac{dV}{dt}\right)}_{\text{rates of exergy transfer}} - \underbrace{\dot{E}_D}_{\text{rate of exergy destruction}}, \tag{6.20}$$

where $d\mathbf{E}/dt$ is the time rate of change of exergy. The term $(1 - T_o/T_j)\dot{Q}_j$ represents the time rate of exergy transfer accompanying heat transfer at the rate $\dot{Q}_j$ occurring at the location on the boundary where the instantaneous temperature is T_j. $\dot{W}$ represents the time rate of energy transfer by work, and the accompanying exergy transfer is $(\dot{W} - p_o dV/dt)$, where dV/dt is the time rate of change of system volume. $\dot{E}_D$ accounts for the time rate of exergy destruction due to irreversibilities within the system and is related to the rate of entropy generation within the system by $\dot{E}_D = T_o\dot{S}_{\text{gen}}$.

The counterpart of Equation (6.20) applicable to control volumes includes additional terms that account for exergy transfers into or out of a control volume where streams of mass enter and exit:

$$\underbrace{\frac{d\mathbf{E}}{dt}}_{\text{rate of exergy change}} = \underbrace{\sum_j \left(1 - \frac{T_o}{T_j}\right)\dot{Q}_j - \left(\dot{W}_{cv} - p_o\frac{dV}{dt}\right) + \sum_i \dot{m}_i e_i - \sum_e \dot{m}_e e_e}_{\text{rates of exergy transfer}} - \underbrace{\dot{E}_D}_{\text{rate of exergy destruction}}. \tag{6.21}$$

As for control volume rate balances considered previously, the subscripts i and e denote inlets and exits, respectively.

In Equation (6.21), the term $d\mathbf{E}/dt$ represents the time rate of change of the exergy of the control volume. $\dot{Q}_j$ represents the time rate of heat transfer at the location on the boundary where the instantaneous temperature is T_j, and the accompanying exergy transfer is given by $(1 - T_o/T_j)\dot{Q}_j$. As in the control volume energy rate balance, $\dot{W}_{cv}$ represents the time rate of energy transfer by work other than flow work. The accompanying exergy transfer is given by $(\dot{W}_{cv} - p_o dV/dt)$, where dV/dt is the time rate of change of volume of the control volume itself. The term $\dot{m}_i e_i$ accounts for the time rate of exergy transfer accompanying mass flow and flow work at inlet i. Similarly, $\dot{m}_i e_i$ accounts for the time rate of exergy transfer accompanying mass flow and flow work at exit e. In subsequent discussions, the exergy transfer rates at control volume inlets and exits are denoted, respectively, as $\dot{E}_i = \dot{m}_i e_i$ and $\dot{E}_e = \dot{m}_e e_e$. Finally, $\dot{E}_D$ accounts for the time rate of exergy destruction due to irreversibilities within the control volume.

At steady state, Equation (6.21) reduces to

$$0 = \sum_j \left(1 - \frac{T_o}{T_j}\right)\dot{Q}_j - \dot{W}_{cv} + \sum_i \dot{m}_i e_i - \sum_e \dot{m}_e e_e - \dot{E}_D. \qquad (6.22)$$

To complete the introduction of the control volume exergy balance, an equation for evaluating the terms e_i and e_e appearing in Equations (6.21) and (6.22) is considered next: When mass enters or exits a control volume, exergy transfers accompanying mass flow and flow work occur, that is,

$$\begin{bmatrix} \textit{time rate of exergy transfer} \\ \textit{accompanying mass flow} \end{bmatrix} = \dot{m}\mathbf{e}. \qquad (6.23)$$

On a time-rate basis the flow work is $\dot{m}(pv)$, and the accompanying exergy transfer is given by

$$\begin{bmatrix} \textit{time rate of exergy transfer} \\ \textit{accompanying flow work} \end{bmatrix} = \dot{m}[pv - p_o v]. \qquad (6.24)$$

A single expression consisting of the sum of these contributions is

$$\begin{bmatrix} \textit{time rate of exergy transfer} \\ \textit{accompanying mass flow and flow work} \end{bmatrix} = \dot{m}[\mathbf{e} + (pv - p_o v)]. \qquad (6.25)$$

The sum in square brackets on the right is symbolized by e, that is,

$$e = \mathbf{e} + (pv - p_o v). \qquad (6.26a)$$

(Although the first term of Equation (6.26a), $\mathbf{e}$, is never negative, the second term is negative when $p < p_o$, giving rise to negative values for e at certain states.) Introducing Equations (6.14) and (6.16), Equation (6.26a) gives the equation for evaluating e_i and e_e in Equations (6.21) and (6.22)

$$e = \underline{(h - h_o) - T_o(s - s_o)} + \frac{V^2}{2} + gz + e^{CH}, \qquad (6.26b)$$

where h_o and s_o denote the specific enthalpy and entropy, respectively, at the restricted dead state. The underlined term can be identified as the physical component of the exergy of a flow stream, written as

$$e^{PH} = (h - h_o) - T_o(s - s_o). \qquad (6.27)$$

6.3.4 Exergy analysis

Thermal systems are typically supplied with exergy inputs derived directly or indirectly from the consumption of fossil fuels or other energy resources. Accordingly, needless destruction and loss of exergy waste these energy resources. Exergy analysis aims at the quantitative evaluation of the exergy destructions and losses associated with a system. Exergy analysis additionally often involves the calculation of measures of performance called exergetic efficiencies, also known as Second Law efficiencies.

As an illustration of exergy analysis, refer to Tables 6.1 and 6.2, which give data for the cogeneration system shown in Fig. 6.4. Table 6.1 provides mass flow rates, state data, and exergy flow rates at key state points. Table 6.2 provides a rank-ordered listing of the exergy destruction within the principal components of the system.

The exergetic efficiency is calculated as the percentage of the fuel exergy supplied to a system that is recovered in the product of the system. Identifying the product of the cogeneration system as the sum of the net power developed and the net increase of the exergy of the feedwater,

$$\begin{aligned} \varepsilon &= \frac{\dot{W}_{net} + (\dot{E}_9 - \dot{E}_8)}{\dot{E}_{10}} \\ &= \frac{30\,\text{MW} + 12.75\,\text{MW}}{85\,\text{MW}} = 0.503(50.3\%), \end{aligned} \qquad (6.28)$$

where the exergy values are obtained from Table 6.1. The exergy carried out of the system of State 7, amounting to 3.27 percent of the fuel exergy, is regarded as a loss. A wide variety of alternative exergetic (Second Law) efficiency expressions is found in the literature (Bejan et al. 1996) (Kotas 1985) (Moran 1989) (Szargut et al. 1988).

The exergy destruction and loss data summarized in Table 6.2 clearly identify the combustion chamber as the major site of thermodynamic inefficiency. The

Table 6.1. *Mass flow rate, state data, and exergy rate data for the cogeneration system of Fig. 6.4*

		Mass Flow	T	p	Exergy (MW)		
State	Substance	Rate (kg/s)	(K)	(bar)	$\dot{E}^{PH}$	$+ \dot{E}^{CH}$	$= \dot{E}$
1	Air[a]	91.28	298.15	1.013	0.00	0.00	0.00
2	Air	91.28	603.74	10.130	27.54	0.00	27.54
3	Air	91.28	850.00	9.623	41.94	0.00	41.94
4	Combustion products[b,c]	92.92	1520.00	9.142	101.09	0.37	101.45
5	Combustion products	92.92	1006.16	1.099	38.42	0.37	38.79
6	Combustion products	92.92	779.78	1.066	21.39	0.37	21.76
7	Combustion products	92.92	426.90	1.013	2.41	0.37	2.78
8	Water	14.00	298.15	20.000	0.03	0.04	0.07
9	Water	14.00	485.57	20.000	12.78	0.04	12.82
10	Methane	1.64	298.15	12.000	0.63	84.37	85.00

[a]Molar analysis (%): 77.48 N_2, 20.59 O_2, 0.03 CO_2, 1.90 $H_2O(g)$
[b]Molar analysis (%): 75.07 N_2, 13.72 O_2, 3.14 CO_2, 8.07 $H_2O(g)$
[c]Heat transfer from the combustion chamber is estimated as two percent of the lower heating value of the fuel.

Table 6.2. *Exergy destruction data for the cogeneration plant Fig. 6.4*

	Exergy Destruction		
Component	MW	(%)[a]	(%)[b]
Combustion chamber[c]	25.48	64.56	29.98
Heat recovery steam generator	6.23	15.78	7.33
Gas turbine	3.01	7.63	3.54
Air preheater	2.63	6.66	3.09
Gas compressor	2.12	5.37	2.49
Overall plant	39.47	100.00	46.43[d]

[a]Exergy destruction within the component as a percentage of the total exergy destruction within the plant
[b]Exergy destruction within the component as a percentage of the exergy entering the plant with the fuel
[c]Includes the exergy loss accompanying heat transfer from the combustion chamber
[d]An additional 3.27 percent of the fuel exergy is carried out of the plant at State 7 and is charged as a loss.

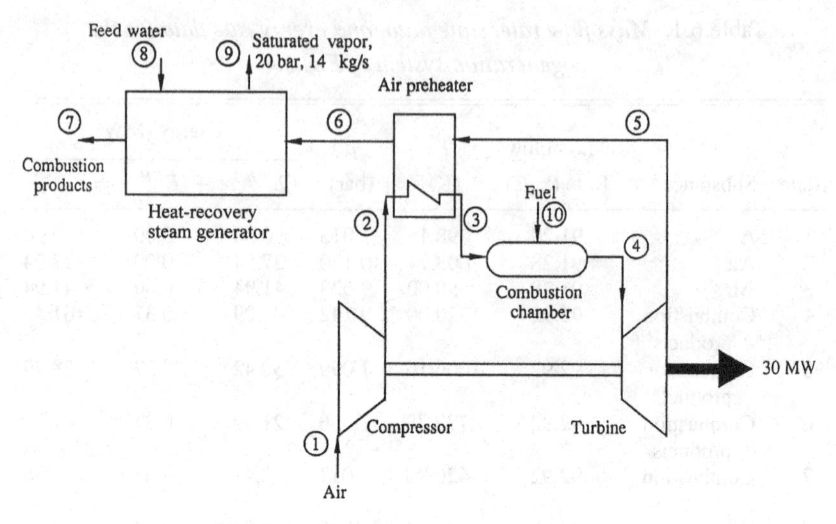

Fig. 6.4. Gas turbine cogeneration system.

next most prominent site is the heat recovery steam generator (HRSG). Roughly equal contributions to inefficiency are made by the gas turbine, air compressor, and the loss associated with Stream 7. The gas compressor is an only slightly smaller contributor.

The exergy destructions in these components stem from one or more of three principal internal irreversibilities associated, respectively, with chemical reaction, heat transfer, and friction. All three irreversibilities are present in the combustion chamber, where chemical reaction and heat transfer are about equal in significance and friction is of secondary importance. For the HRSG and preheater the most significant irreversibility is related to the stream-to-stream heat transfer, with friction playing a secondary role. The exergy destruction in the adiabatic gas turbine and air compressor is caused mainly by friction.

Although combustion is intrinsically a significant source of irreversibility, a dramatic reduction in its effect on exergy destruction cannot be expected using ordinary means. Still, the inefficiency of combustion can be ameliorated by preheating the combustion air and reducing the air–fuel ratio. The effect of the irreversibility associated with heat transfer tends to lessen as the minimum temperature difference between the streams, $(\Delta T)_{\min}$, is reduced; this reduction is achievable through the specification of the heat exchanger. The exergy destruction within the gas turbine and the air compressor decreases as the isentropic turbine and compressor efficiencies increase, respectively. The significance of the exergy loss associated with Stream 7 reduces as the gas temperature reduces.

These considerations are a basis for implementing practical engineering measures to improve the thermodynamic performance of the cogeneration system.

Such measures have to be applied judiciously, however. Measures that improve the thermodynamic performance of one component might adversely affect another, leading to no net overall improvement. Moreover, measures to improve thermodynamic performance invariably have economic consequences. The objective in thermal system design normally is to identify the cost-optimal configuration, requiring consideration of both thermodynamic and economic imperatives.

6.4 Sample case: preliminary design of a cogeneration system

Engineering design requires highly structured critical thinking and active communication among the group of individuals, the design team, whose responsibility is the development of the design of a product or system. The typical design project begins with an idea that something might be worth doing: a primitive problem proposed in recognition of a need or an economic opportunity. The development of concepts for implementing the idea takes place next. This stage is crucially important because decisions made here can determine up to eighty percent of the total project cost. The component parts and their interconnections are then detailed. A number of simultaneous, parallel activities occur at this stage. These might include further analysis, sizing and costing of equipment, optimization, and controls engineering. The objective is to combine several pieces of equipment into one smoothly running system. The design effort then moves into project engineering, or the detailed design stage, where a list of actual equipment to be purchased or fabricated is developed, piping and instrumentation diagrams are prepared, blueprints for required construction are drawn, and so on. The end result of the design process is the final stage: the commissioning and safe operation of the system.

A fundamental aspect of the design process makes it mandatory that knowledge about the project be generated as early as possible: At the outset, knowledge about the problem and its potential solutions is at a minimum, but designers have the greatest freedom of choice because few decisions have been made. As the design process evolves, however, knowledge increases but design freedom decreases because earlier decisions tend to constrain what is allowable later. Accordingly, most of the early design effort should focus on establishing an understanding of, and the need for, the problem.

6.4.1 Understanding the problem

Design engineers typically deal with ill-defined situations in which problems are not crisply formulated, the required data are not immediately at hand, decisions about one variable can affect a number of other variables, and economic considerations are of overriding importance. A design assignment also rarely takes the form of a specific system to design. The engineer often receives only a general statement of need or opportunity. This statement constitutes the primitive problem. The following provides an illustration.

Primitive problem: To provide for a plant expansion, additional power and steam are required. Determine how the power and steam are best supplied.

The first stage of the design process is to understand the problem. At this stage the object is to determine which qualities the system should possess and not how to achieve them. The question of how enters later. Specifically, at this initial stage the primitive problem must be defined, and the requirements to which the system must adhere must be determined. The key question now for design team members to be asking is "Why?". The team must examine critically every proposed requirement, seeking justification for each. Functional performance, cost, safety, and the environment are just a few of the issues that have to be considered in developing the requirements.

Not all requirements are equally important or even absolutely essential. The list of requirements normally can be divided into hard and soft constraints: musts and wants. The musts are essential and have to be met by the final design. The wants vary in importance and may be placed in rank order by the design team through negotiation. Although the wants are not mandatory, some of the top-ranked ones may be highly desirable and useful in screening alternatives later.

Next, the design team must translate the requirements into quantitative specifications. Each requirement must be expressed in measurable terms, for the lack of a measure often means that the requirement is not well understood. Moreover, only with numerical values can the design team evaluate the design as it evolves.

To illustrate, consider the *primitive problem* stated earlier. Let us assume that the following musts are identified: the power developed must be at least thirty megawatts. The steam must be saturated or slightly superheated at 20 bar and have a mass flow rate of at least fourteen kilograms per second. The system must adhere to all applicable environmental regulations. These environment-related musts also would be expressed numerically. Normally many other requirements – both musts and wants – would apply. Those listed are only representative.

6.4.2 Concept generation

The focus of the first stage of the design process is on building the design team's knowledge base. That knowledge is then used to develop alternative solutions to the primitive problem. Emphasis shifts from the what-to-do phase to the how-to-do-it phase. The design team considers specifically how the primitive problem can be solved.

Normally, a relation exists between the number of alternative solutions generated and their quality: To uncover a few good concepts, several have to be generated initially. Because the methods of engineering analysis and optimization can only polish specific solutions, alternatives must be generated having sufficient quality to merit polishing. Major design errors more often occur owing to a flawed alternative being selected than to subsequent shortcomings in analysis and optimization. And by the time a better-quality alternative is recognized, it is frequently too late to take advantage of the alternative without a costly revisiting of earlier stages of the design process.

The concept-generation stage of the design process opens by identifying as many alternative concepts as possible. New technology should be carefully considered of course, but concepts that may have been unusable in the past should not be passed over uncritically. They may now be viable owing to changes in technology or costs. Various group-interactive strategies such as brainstorming and analogical thinking have been found to be effective in sparking ideas at this stage.

The concept-creation phase can lead to a number, perhaps a great number, of plausible alternative solutions to the primitive problem differing from one another in basic concept and detail. Alternative concepts for the *primitive problem* might include the following:

- Generate all the steam required in a boiler, and purchase the required power from the local utility.
- Cogenerate steam and power. Generate all the steam required, and
  - generate the full electricity requirement,
  - generate a portion of the electricity requirement and purchase the balance of the electricity needed from the utility, or
  - generate more than the electricity requirement, and sell the excess electricity to the utility.

Each cogeneration alternative might be configured using a steam turbine system, a gas turbine system, or a combined steam and gas turbine system, and the type of fuel (coal, natural gas, oil) introduces still more flexibility.

Alternative concepts such as these should be screened to eliminate all but the most worthwhile. Perils are associated with screening, however. One is that an inferior concept will be retained only to be discarded later after considerable additional effort has been expended on it. Another more serious peril is that the best alternative is unwittingly screened out. Still, choices must be made among alternatives – carefully. Various means are available to assist the design team in screening concepts. These means include but are not limited to the use of a decision matrix or a decision tree. However, such methods do not make the decisions but only allow for orderly interaction and decision making by the design team.

For illustration, let us suppose that after preliminary screening the following alternatives have been retained for further screening and evaluation:

- Employ a coal-fired steam turbine cogeneration system.
- Employ a gas turbine cogeneration system using a natural gas as the fuel.
- Employ a combined steam and gas turbine cogeneration system using natural gas as the fuel.

These alternatives would be subject to further screening and evaluation in the concept development stage of design until the preferred design, the base-case design, emerges. The concept development stage concludes with the flow sheets serving as the point of departure for the detailed design stage.

6.4.3 Concept development: base-case flow sheet

The concept development stage is idealized in Fig. 6.5 in three interrelated steps: synthesis, analysis, and optimization. Synthesis is concerned with putting together separate elements into a whole. In this step the particular equipment making up the overall thermal system and its interconnections is specified. The schematics of Fig. 6.6 show possible outcomes of the synthesis step for the three alternatives under consideration. Synthesis is considered in more detail later in this section.

The analysis step of concept development generally entails thermal analysis (solving mass, energy, and exergy balances, as required), costing and sizing equipment on at least a preliminary basis, and considering other key issues quantitatively. The aim is to identify the preferred design approach from among the configurations synthesized. Referring again to Fig. 6.6, on the basis of analysis the gas turbine cogeneration system of Fig. 6.6(b) emerges as the preferred alternative. In this instance, the analysis is primarily economic. The same conclusion is reached for somewhat different requirements for power and process steam by Stoll (1989) using the investment payback period as a basis for comparison.

The optimization step of concept development can take two general forms: structural optimization and parameter optimization. In structural optimization the equipment inventory and interconnections between equipment are altered to

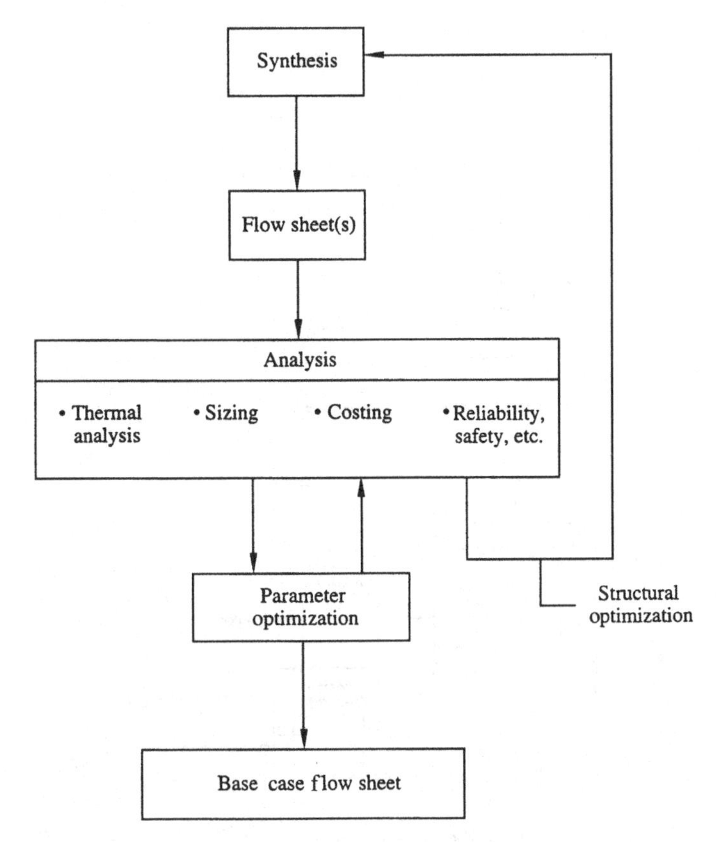

Fig. 6.5. Concept development steps: synthesis, analysis and optimization.

achieve a superior design. Structural optimization is indicated in Fig. 6.5 by the return arrow linking analysis to synthesis and can be illustrated by comparing Figs. 6.4 and 6.6(b): Fig. 6.4 is a thermodynamically improved version of the cogeneration configuration of Fig. 6.6(b), achieved by including an air preheater. In parameter optimization the pressures, temperatures, and chemical compositions at various state points and other key system parameters are determined, at least approximately, with the aim of satisfying some desired objective – minimum total cost of the system products, for example. At this juncture the term optimization generally means only improvement toward the objective, however, because a more thorough optimization effort may be undertaken at the detailed design stage, where the design is finalized.

Although the goal of the concept development stage is clear, means for achieving it are not, because no step-by-step procedure is generally accepted for arriving at a preferred flow sheet beginning from one or more alternative concepts.

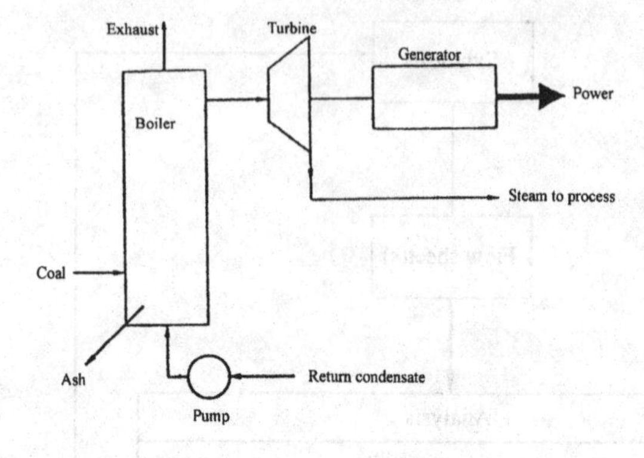

(a) Coal-fired steam turbine cogeneration system

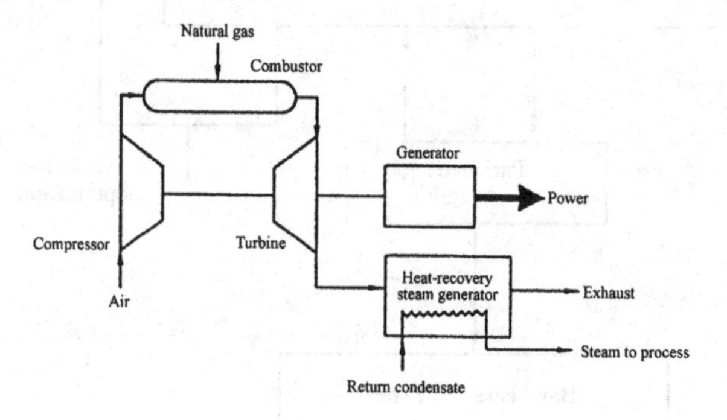

(b) Gas turbine cogeneration system

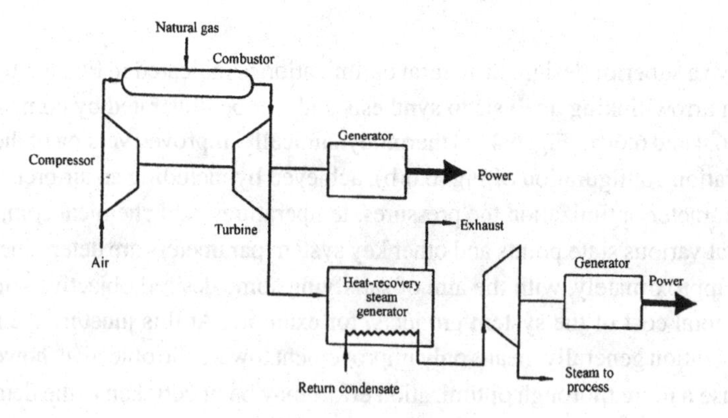

(c) Combined steam and gas turbine cogeneration system

Fig. 6.6. Cogeneration system alternatives.

However, several considerations apply broadly. When synthesizing a design, it is wise to start simply and add detail as decisions are made and avenues of inquiry open. Ideally a hierarchy of decision levels is traversed, with the amount of detail and the accuracy of the calculations increasing with each successive decision level. With this approach a flow sheet evolves in a step-by-step manner, beginning with the most important details and continuing in ever finer detail. Each decision level should involve an economic evaluation so that later decisions rest on, and are guided by, the economic evaluations at earlier levels.

Process synthesis is inherently combinatorial. Take, for example, the design of a thermal system in which several liquid or gas streams enter the system at respective states and several such streams exit at other states. Within the system the streams may react chemically, exchange energy by heat transfer, undergo pressure changes with an attendant development of, or requirement for, power, and so on. The number of possible flow sheets for such a system might be considerable (for complex systems, 10^6 flow sheets, for example). Each flow sheet would be described in terms of numerous – perhaps hundreds – of equations, and several dozen, or more, optimization variables may be associated with each alternative. For systems of practical interest, this combinatorial aspect soon becomes challenging, and an approach based on the complete enumeration and evaluation of all alternatives is not viable, even with the use of computers.

Engineers have traditionally approached such daunting design problems using experience, inventiveness, and extensive iteration as alternatives are evaluated against various criteria. To avoid the need to consider all possible alternatives, plausible but fallible screening rules are typically used to eliminate cases lacking merit and to determine feasible alternatives relatively quickly. Such design guidelines, or heuristics, are drawn from the experience of designers who have solved similar problems and recorded common features of their solutions. The use of design guidelines does not ensure the discovery of a satisfactory design, let alone an optimal design, but provides only a reasonable exploratory approach to design.

Table 6.3 shows a sampling of the numerous design guidelines that have been reported in the literature. The table entries are based on reasoning from the Second Law of thermodynamics. As combustion and heat transfer typically contribute most heavily to the inefficiency of thermal systems, two table sublists refer explicitly to them. A third, more general, list is also included. Finally, one of the table entries suggests the preheating-of-combustion-air feature embodied in the design of Fig. 6.4.

Although not a fully solved problem, the area of heat exchanger network synthesis has progressed substantially and currently stands as the most mature

Table 6.3. *Design guidelines*[a,b]

Combustion – Minimize the use of combustion. When combustion cannot be avoided

- minimize the use of excess air, and
- preheat the reactants.

Heat Transfer – Avoid unnecessary or cost-ineffective heat transfer. When using heat transfer, note the following:

- The higher the temperature T at which a heat transfer occurs when $T > T_o$, the more valuable the heat transfer and, consequently, the greater the need to avoid direct heat transfer to the ambient, to cooling water, or to a refrigerated stream.
- The lower the temperature T at which a heat transfer occurs when $T < T_o$, the more valuable the heat transfer and, consequently, the greater the need to avoid direct heat transfer with the ambient or a heated stream.
- The lower the temperature level, the greater the need to minimize the stream-to-stream temperature difference.
- Avoid the use of intermediate heat transfer fluids when exchanging energy by heat transfer between two streams.
- In the design of heat exchanger networks (HENs), consider using the pinch method (Chapter 5).

General

- Maximize the use of cogeneration where feasible.
- Some exergy destructions and exergy losses can be avoided, and others cannot. Efforts should be centered on those that can be avoided.
- For compressors, turbines, and motors consider the most thermodynamically efficient options.
- Minimize the use of throttling; check whether power-recovery expanders are a cost-effective alternative for pressure reduction.
- The greater the mass rate of flow, the greater the need to use the exergy of the stream effectively.
- The lower the temperature level, the greater the need to minimize friction.

[a] Based on the presentations of Bejan et al. (1996) and Sama (1992)
[b] Many additional design guidelines are provided in the engineering literature. See, for example, Appendix B of Ulrich (1984).

of the process synthesis subfields. A typical case concerning the synthesis of a heat exchanger network is to devise a network to exchange energy by heat transfer among a given set of process streams, each of which is to be heated or cooled from an inlet temperature to a desired outlet or target temperature, subject to the constraint that the annualized cost (cost of the equipment required plus the cost of the utility streams such as steam and cooling water) is minimized. One popular approach to heat exchanger network synthesis – the pinch method – which has led to significant savings in both capital and operating costs in industrial settings, is discussed in Chapter 5. The pinch method rests on reasoning from the Second Law of thermodynamics and thus has been included as an entry in Table 6.3.

6.5 Computer-aided process engineering

Because thermal system design involves considerable analysis and computation, including the mathematical modeling of individual components and the system as a whole, computers can be applied with significant effect. Although not realized uniformly in each instance, benefits of computer-aided process engineering (CAPE) may include increased engineering productivity, reduced design costs, and results exhibiting greater accuracy and internal consistency. Computer-aided process engineering falls within the purview of CAD and CAM (computer-aided design/computer-aided manufacturing). In thermal system design, however, the computer aids are more process-oriented than, as in CAD and CAM, product-oriented. An overview of computer-aided process engineering featuring aspects related to the Second Law of thermodynamics is provided in this section. The discussion is keyed to the concept development steps of Fig. 6.5.

6.5.1 Preliminaries

Information on computer-aided process engineering relevant to thermal system design is abundant in the engineering literature (see, for example, Peters and Timmerhaus [1991]). Current computer hardware and software developments are surveyed in the "Chemputer" and "Software" sections of the periodicals *Chemical Engineering* and *Chemical Engineering Progress*, respectively. The "CPI Software" and "Software Exchange" sections of *Chemical Processing* and *Mechanical Engineering*, respectively, also report on software of value to CAPE.

The extent to which computer-aided process engineering can be applied is limited by the availability of property data in suitable forms. Accordingly, such data are vitally important. One respected source of property data is the program DIPPR. Other noteworthy programs include PPDS and DETHERM. A survey of commercial on-line databases relevant to CAPE is provided by Anon. (1989). Chapter 4 further discusses properties for design simulations.

Computer-aided process engineering also relies heavily on suitable process-equipment-design programs. Libraries of programs are available for the design or rating calculations of one of the most common thermal system components: heat exchangers. Widely used heat exchanger programs stem from the large-scale research efforts of Heat Transfer Research Inc. (United States) and the Heat Transfer and Fluid Flow Service (United Kingdom). Software for numerous other types of equipment, including piping networks, is discussed in the literature. Many companies also have developed proprietary software.

6.5.2 Process synthesis software

In light of the explosively combinational nature of process synthesis noted previously, recent efforts have been directed to making process synthesis more systematic, efficient, and rapid. With the advent of very high-speed computers and rapidly improving software, engineers are increasingly resorting to computers for this purpose. A considerable body of literature has been developed (see Nishida, Staphanopoulos, and Westerberg (1981) for an introduction).

Although still in its infancy, one direction that this has taken is the development of expert systems for synthesizing flow sheets based on the experience of specialists and using principles from the field of artificial intelligence. Of special interest here are expert systems using the Second Law of thermodynamics to guide the search procedure, even if only implicitly.

The process invention procedure is a hierarchical expert system for the synthesis of process flow sheets for a class of petrochemical processes (Kirkwood, Locke, and Douglas 1988). It combines qualitative knowledge in the form of heuristics with quantitative knowledge in the form of design and cost models. The heuristics are used to select the unit operations and identify the interconnections between the units, to identify the principal design variables, and to identify process alternatives at each level of the hierarchy. For heat exchanger network synthesis, for example, the Second Law–based pinch method is employed. Design and cost models are used to determine process flows, equipment sizes and costs, feed stream and utility costs, and the process profitability at each level.

A knowledge-based approach to flow-sheet synthesis of thermal systems with heat–power–chemical transformations is presented by Kott, May, and Hwang (1989). The approach is based on a state-space search guided by heuristics and uses exergy analysis explicitly. Given a nearly arbitrary set of input and output flows, the method aims at generating an at least satisfactory flow sheet. The human designer is not required to input any predetermined equipment or processes, nor does the method rely on selecting among a particular set of predetermined alternatives. Rather, the method is guided by the need to eliminate sources of thermodynamic inefficiency while avoiding increases in equipment cost. Accordingly, both thermodynamic and economic considerations are used centrally in devising a flow sheet.

Another expert system for the design of thermal systems is discussed in Chapter 9. This procedure aims at synthesizing a flow sheet starting from a list of components stored in a database. The synthesis is conducted using a set of rules of interference, including rules of symbolic logic and a set of design guidelines intended to reproduce the expertise of a human designer. The design guidelines are drawn from the Second Law of thermodynamics and correspond to those listed in Table 6.3.

In their present states of development such expert systems provide plausible means for synthesizing flow sheets quickly. They aim at inventing feasible designs but not necessarily a final design. They allow for rapid screening of alternatives and obtaining first estimates of design conditions. The output of those procedures can be used as the input for one of the conventional simulators considered next.

6.5.3 Analysis and optimization: flow sheeting software

Greater success has been achieved thus far in applying computer aids to analysis and parameter optimization than to process synthesis. This type of application is commonly called flow sheeting or process simulation. Flow-sheeting software allows the engineer to model the behavior of a system, or components of it, under specified conditions and do thermal analysis, sizing, costing, and optimization. Spreadsheet software is a less sophisticated but still effective approach for a wide range of applications. Such software has become popular because of its availability for PCs at reasonable cost, ease of use, and flexibility.

Flow sheeting has developed along two lines: the sequential-modular approach and the equation-solving approach. In the sequential-modular approach, library models associated with the various components of a specified flow sheet are called in sequence by an executive program, using the output stream data for each component as the input for the next component. Access to physical property data is such that consistent data are used throughout. An advantage of the sequential-modular approach is simplicity. For example, efforts to upgrade program performance can be directed relatively independently to the three extant areas: model library, executive, physical properties. But although the approach is well suited to steady-state calculations, it is not readily extended to dynamic simulation. Most of the more widely used flow-sheeting programs, such as ASPEN PLUS, PROCESS, and CHEMCAD, are of the sequential modular type.

In the equation-solving approach, all the equations representing the individual flow-sheet components and the links between them are assembled as a set of equations for simultaneous solution. SPEEDUP is one of the more widely adopted programs of the equation-solver type. Flexibility is one of the advantages that this approach enjoys over the sequential-modular approach; for example, incorporating the capability of solving differential equations allows dynamic simulation. On the other hand, the need to solve simultaneously several hundred (or thousand) equations representing an industrial-scale system is considerably more demanding mathematically than is implementing the sequential-modular approach.

A survey of the capabilities of fifteen commercially available process simulators is reported by Biegler (1989). Of these, ten are sequential-modular and five are equation solvers. A brief description of the features of each simulator is given, including options for sizing and costing equipment, computer operating systems under which the simulator is available, and optimization capabilities (if any).

Optimization deserves special comment, because one of the main reasons for developing a flow-sheet simulation is system improvement. And for complex thermal systems described in terms of a large number of equations, including nonlinear equations and nonexplicit variable relationships, the term *optimization* implies improvement rather than calculation of a global mathematical optimum. Optimization considerations enter both at the design stage and in relation to plant operations. Optimization of plant operations involves timely adjustments owing to variations in feedstocks, changes between summer and winter operating conditions, effects of equipment fouling, and so on. Many of the leading sequential-modular programs have optimization capabilities, and this technology is rapidly improving. The newer equation-solving programs also offer flexible and relatively rapid optimization capabilities. Vendors should be contacted for up-to-date information on the features of flow-sheeting software, including optimization capabilities.

Conventional optimization procedures may suffice for relatively simple thermal systems, although even for such systems the cost and performance data are seldom in the form required for optimization. Moreover, with increasing system complexity, conventional methods can become unwieldy, time-consuming, and costly. The method of thermoeconomics, which combines exergy analysis with principles of engineering economics, may then provide a better approach, particularly when chemical reactions are involved, regardless of the amount of information on cost and performance. Thermoeconomics aims to facilitate feasibility and optimization studies during the design phase of new systems, and process improvement studies of existing systems. Thermoeconomics also aims to facilitate decisions on plant operation and maintenance. Knowledge acquired through thermoeconomics assists materially in improving system efficiency and reducing the product costs by pinpointing required changes in structure and parameter values. Chapter 8 of this book introduces thermoeconomics, and additional details are provided by Bejan et al. (1996) and Tsatsaronis (1993).

6.6 Closure

A close reading of the early literature of the Second Law of thermodynamics originating from the time of Carnot shows that Second Law reasoning in today's engineering sense – optimal economic use of resources – was present from the beginning. Interest in this theme occasionally flickered in the interim, but it was kept alive and now forms the basis for contemporary applications of engineering thermodynamics. The bonds that have been forged in recent years between thermodynamics and economics have made the theme of optimal use of resources more explicit and have allowed Second Law methods to be even more powerful and effective than ever before for the design and analysis of thermal systems.

The development of Second Law analysis owes considerably to the efforts of a great many towering figures in the history of science and technology, whose contributions have advanced our present understanding immeasurably. Still, the bulk of the literature of Second Law analysis in engineering has appeared in the last quarter-century, and most of those who have contributed remain active somewhere over the globe. Because of this activity, contemporary Second Law analysis may appear to some observers as a new field replete with growing pains, rather than the relatively mature product of an evolutionary process of some two centuries that it is. Some additional evolution surely will occur, but as we begin a new millennium we can regard Second Law analysis as established, and expect Second Law analysis to be used increasingly for the design and analysis of thermal systems.

References

Ahrendts, J. (1980). Reference states. *Energy*, **5**, 667–77.

Anon. (1989). A wealth of information online. *Chemical Engineering*, **96**, 112–27.

Bejan, A., Tsatsaronis, G. & Moran M. J. (1996). *Thermal Design and Optimization*. New York: Wiley.

Biegler, L. T. (1989). Chemical process simulation. *Chemical Engineering Progress*, October, 50–61.

Bird, R. B., Stewart, W. E. & Lightfoot, E. N. (1960). *Transport Phenomena*. New York: Wiley.

Douglas, J. M. (1989). *Conceptual Design of Chemical Processes*. New York: McGraw-Hill.

Kenney, W. F. (1984). *Energy Conservation in the Process Industries*. Orlando, FL: Academic Press.

Kirkwood, R. L., Locke, M. H. & Douglas, J. M. (1988). A prototype expert system for synthesizing chemical process flow sheets. *Computers in Chemical Engineering*. **12**, 329–43.

Kotas, T. J. (1985). *The Exergy Method of Thermal Plant Analysis*. London: Butterworths.

Kott, A. S., May, J. H. & Hwang, C. C. (1989). An autonomous artificial designer of thermal energy systems: part 1 and part 2. *Journal of Engineering for Gas Turbines and Power*, **111**, 728–39.

Moran, M. J. (1989). *Availability Analysis: A Guide to Efficient Energy Use*. New York: ASME Press.

Moran, M. J. & Shapiro, H. N. (1995). *Fundamentals of Engineering Thermodynamics*, 3rd ed. New York: Wiley.

Nishida, N., Staphanopoulos, G. & Westerberg, A. W. (1981). A review of process synthesis. *AIChE Journal*, **27**, 321–51.

Peters, M. S. & Timmerhaus, K. D. (1991). *Plant Design and Economics for Chemical Engineers*, 4th ed. New York: McGraw-Hill.

Sama, D. (1992). A common-sense 2nd law approach to heat exchanger network design. In *Proceedings of the International Symposium On Efficiency, Costs, Optimization and Simulation of Energy Systems* ed. A. Valero & G. Tsatsaronis, pp. 329–38. New York: ASME.

Stoll, H. G. (1989). *Least-Cost Electric Utility Planning*. New York: Wiley-Interscience.

Szargut, J., Morris, D. R. & Steward, F. R. (1988). *Exergy Analysis of Thermal, Chemical, and Metallurgical Processes*. New York: Hemisphere.

Tsatsaronis, G. (1993). Thermoeconomic Analysis and Optimization of Energy System. *Progress in Energy and Combustion Science*, **19**, 227–57.

Ulrich, G. D. (1984). *A Guide to Chemical Engineering Process Design and Economics*. New York: Wiley.

7

Thermodynamic optimization of heat transfer and fluid flow processes

ADRIAN BEJAN

Duke University

7.1 The method of entropy generation minimization or finite time thermodynamics

In the 1970s the method of Second Law optimization or entropy generation minimization (EGM) emerged as a distinct field of activity at the interface between heat transfer, engineering thermodynamics, and fluid mechanics. The position of the field is illustrated in Fig. 7.1, which is reproduced from the first book ever published on this method (Bejan 1982). The method relies on the simultaneous application of principles of heat and mass transfer, fluid mechanics, and engineering thermodynamics, in the pursuit of realistic models of processes, devices, and installations. By realistic models we mean models that account for the inherent irreversibility of engineering systems.

The objective in the application of the EGM method is to find designs in which the entropy generation is minimum. According to the Gouy–Stodola theorem (Bejan 1982),

$$W_{\text{lost}} = T_0 S_{\text{gen}} \tag{7.1}$$

a minimum entropy generation design characterizes a system with minimum destruction of available work (exergy). In the case of power plants, for example, minimum entropy generation rate is equivalent to maximum instantaneous power output. In the design of refrigeration plants, the minimum entropy generation rate design is exactly the same as the design with maximum refrigeration load, or minimum mechanical power input.

The method consists of dividing the system into subsystems that are in local (or internal) thermodynamic equilibrium. The subsystem properties are governed by the laws of engineering (classical) thermodynamics. Entropy is generated at the boundaries between subsystems, as heat and mass flow through the boundaries. These flows are accounted for by writing the rate (per unit time) equations of the disciplines of heat and mass transfer and fluid mechanics.

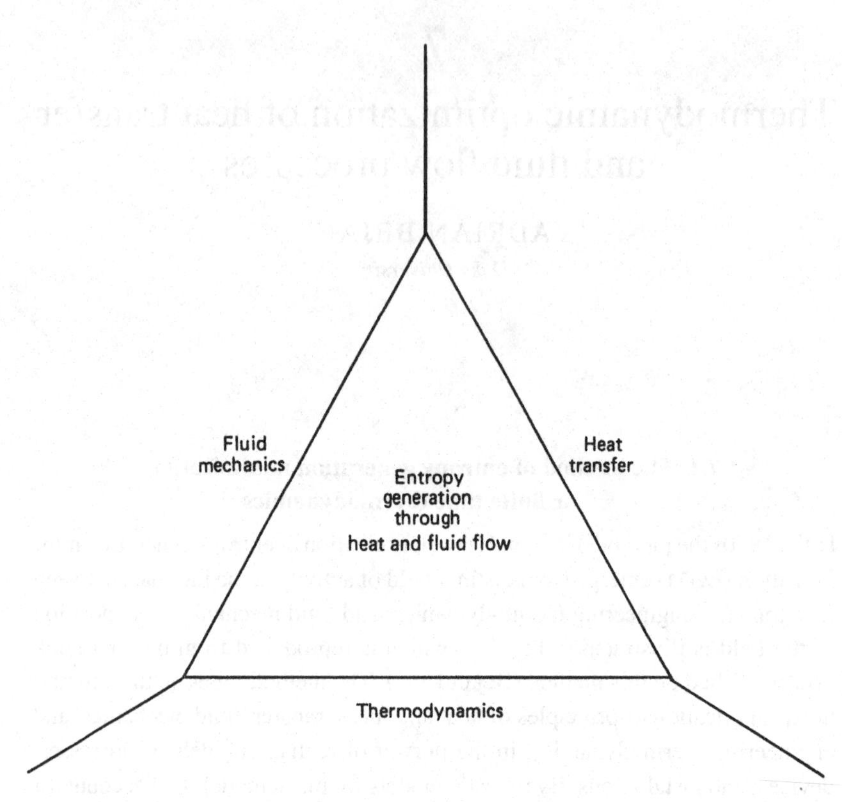

Fig. 7.1. The triangular domain occupied by the field of entropy generation minimization, or finite time thermodynamics (the book cover of Bejan, 1982).

In this way, the total rate of entropy generation is calculated in relation to the overall physical characteristics (finite size, finite time) of the greater system. Only then can the total entropy generation be monitored and minimized by properly varying the physical characteristics of the system. The subsystems can be macroscopic (two or more in a system), or infinitesimally small (an infinite number). In other words, system modeling and entropy generation analysis can be pursued at the macroscopic level or at the differential level: examples of both abound in Bejan (1982).

The growth and versatility of the field are illustrated further by more recent reviews (Bejan 1987) (Bejan 1988a). Physicists too recognized the usefulness of the method and expanded the field further under the name of finite time thermodynamics (FTT) (Andresen, Salamon, and Berry 1984). The FTT methodology relies on the same combined principles of heat transfer and engineering thermodynamics and seeks to identify designs in which the entropy generation rate is minimum or the instantaneous power output is maximum (in power applications).

The objective of this chapter is to review the most fundamental implications of thermodynamic optimization (EGM or FTT) in the design of heat transfer and fluid flow processes. The chapter shows how to construct simple but realistic models of systems that operate irreversibly. It also shows how to optimize their dimensions or time-dependent operation such that they generate minimum entropy or maximum power.

7.2 Balancing the heat transfer and fluid flow irreversibilities

7.2.1 Internal flow

We begin with the simplest illustrations of the EGM–FTT design optimization method. Later, as we review the expanding list of applications and literature, we will focus on more complex systems.

Consider first a heat exchanger passage, which is a duct of arbitrary cross section (A) and arbitrary wetted perimeter (p). The engineering function of the passage is specified in terms of the heat transfer rate per unit length (q') that is to be transmitted to the stream ($\dot{m}$); that is, both q' and $\dot{m}$ are fixed. In the steady state, the heat transfer q' crosses the temperature gap ΔT formed between the wall temperature ($T + \Delta T$) and the bulk temperature of the stream (T). The stream flows with friction in the x direction; hence, the pressure gradient $(-dP/dx) > 0$.

Taking as thermodynamic system a passage of length dx, the First Law and the Second Law state that

$$\dot{m}\, dh = q'\, dx \quad \text{and} \quad \dot{S}'_{\text{gen}} = \dot{m}\frac{ds}{dx} - \frac{q'}{T + \Delta T} \geq 0, \tag{7.2}$$

where $\dot{S}'_{\text{gen}}$ is the entropy generation rate per unit length. Combining these statements with $dh = T\, ds + v\, dP$, the design-important quantity $\dot{S}'_{\text{gen}}$ becomes (Bejan 1978a)

$$\dot{S}'_{\text{gen}} = \frac{q'\Delta T}{T^2(1 + \Delta T/T)} + \frac{\dot{m}}{\rho T}\left(-\frac{dP}{dx}\right) \cong \frac{q'\Delta T}{T^2} + \frac{\dot{m}}{\rho T}\left(-\frac{dP}{dx}\right) \geq 0. \tag{7.3}$$

The denominator of the first term on the right-hand side has been simplified by assuming that the local temperature difference ΔT is negligible compared with the local absolute temperature T.

The heat exchanger passage is a site for both flow with friction and heat transfer across a finite ΔT; the $\dot{S}'_{\text{gen}}$ expression thus has two terms, each accounting for one irreversibility mechanism. We record this observation by rewriting Equation (7.3) as

$$\dot{S}'_{\text{gen}} = \dot{S}'_{\text{gen},\Delta T} + \dot{S}'_{\text{gen},\Delta P}. \tag{7.4}$$

In other words, the first term on the right-hand side of Equation (7.3) represents the entropy generation contributed by heat transfer. The relative importance of the two irreversibility mechanisms is described by the irreversibility distribution ratio ϕ, which is defined by (Bejan 1982)

$$\phi = \frac{\text{fluid flow irreversibility}}{\text{heat transfer irreversibility}}. \tag{7.5}$$

Equation (7.4) can then be rewritten as

$$\dot{S}'_{\text{gen}} = (1+\phi)\dot{S}'_{\text{gen},\Delta T}. \tag{7.6}$$

The irreversibility distribution ratio ϕ has since been used in heat exchanger applications, where it was named the Bejan number (Be) (for example, Paoletti, Rispoli, and Sciubba 1989; Benedetti and Sciubba 1993).

A remarkable feature of the $\dot{S}'_{\text{gen}}$ expression (7.3) and of many like it for other simple devices is that a proposed design change (for example, making the passage narrower) induces changes of opposite signs in the two terms of the expression. An optimum trade-off then exists between the two irreversibility contributions, an optimum design for which the overall measure of exergy destruction ($\dot{S}'_{\text{gen}}$) is minimum while the system continues to serve its specified function (q', $\dot{m}$).

The trade-off between heat transfer and fluid flow irreversibilities becomes clearer if we convert Equation (7.3) into the language of heat transfer engineering, in which the heat exchange passage is usually discussed. For this purpose, we recall the definitions of friction factor, Stanton number, mass velocity, Reynolds number, and hydraulic diameter (for example, Bejan 1993a):

$$f = \frac{\rho D_h}{2G^2}\left(-\frac{dP}{dx}\right) \tag{7.7}$$

$$St = \frac{q'/(p\Delta T)}{c_P G} \tag{7.8}$$

$$G = \frac{\dot{m}}{A} \tag{7.9}$$

$$Re = \frac{GD_h}{\mu} \tag{7.10}$$

$$D_h = \frac{4A}{p}, \tag{7.11}$$

where $q'/(p\Delta T)$ of Equation (7.8) is the average heat transfer coefficient. The entropy generation rate formula (7.3) becomes

$$\dot{S}'_{\text{gen}} = \frac{(q')^2 D_h}{4T^2 \dot{m} c_P St} + \frac{2\dot{m}^3 f}{\rho^2 T D_h A^2} \tag{7.12}$$

Considering that both q' and $\dot{m}$ are fixed, we note that the thermodynamic design of the heat exchanger passage has two degrees of freedom, the perimeter p and the cross-sectional area A, or any other pair of independent parameters, such as (Re, D_h) or (G, D_h).

The competition between heat transfer and fluid flow irreversibilities is hinted at by the positions occupied by St and f on the right-hand side of Equation (7.12). The Reynolds and Colburn analogies regarding turbulent momentum and heat transfer teach us that St and f usually increase simultaneously (Bejan 1984a) as the designer seeks to improve the thermal contact between wall and fluid. Thus, what is good for reducing the heat transfer irreversibility is bad for the fluid flow irreversibility, and vice versa.

The tradeoff between the two irreversibilities and the minimum value of the overall $\dot{S}'_{\text{gen}}$ can be illustrated by assuming a special case of passage geometry, namely the straight tube with circular cross section. In this case, p and A are related through the pipe inner diameter, D, the only degree of freedom left in the design process. Writing

$$D_h = D, \qquad A = \pi\frac{D^2}{4}, \quad \text{and} \quad p = \pi D, \tag{7.13}$$

Equation (7.12) becomes

$$\dot{S}'_{\text{gen}} = \frac{(q')^2}{\pi T^2 k Nu} + \frac{32\dot{m}^3 f}{\pi^2 \rho^2 T D^5}, \tag{7.14}$$

where $Re = 4\dot{m}/(\pi\mu D)$ and $Nu = \bar{h}D_h/k = St\,Re\,Pr$. Invoking two reliable correlations for Nu and f in fully developed turbulent pipe flow, such as $Nu = 0.023\,Re^{0.8}Pr^{0.4}$ and $f = 0.046\,Re^{-0.2}$, and combining them with Equation (7.14), yields an expression for $\dot{S}'_{\text{gen}}$ that depends only on Re. Solving $d\dot{S}'_{\text{gen}}/d(Re) = 0$, we find that the entropy generation rate is minimized when the Reynolds number (or pipe diameter) reaches the optimum value (Bejan 1980):

$$Re_{\text{opt}} = 2.023\,Pr^{-0.071}B_0^{0.358}. \tag{7.15}$$

This compact formula allows the designer to select the optimum tube size for minimum irreversibility. Parameter B_0 is fixed as soon as q', $\dot{m}$, and the working fluid are specified:

$$B_0 = \dot{m}q'\frac{\rho}{\mu^{5/2}(kT)^{1/2}}. \tag{7.16}$$

The effect of Re on $\dot{S}'_{\text{gen}}$ can be expressed in relative terms as

$$\frac{\dot{S}'_{\text{gen}}}{\dot{S}'_{\text{gen,min}}} = 0.856\left(\frac{Re}{Re_{\text{opt}}}\right)^{-0.8} + 0.144\left(\frac{Re}{Re_{\text{opt}}}\right)^{4.8}, \tag{7.17}$$

where $\dot{S}'_{\text{gen,min}} = \dot{S}'_{\text{gen}}(Re_{\text{opt}})$.

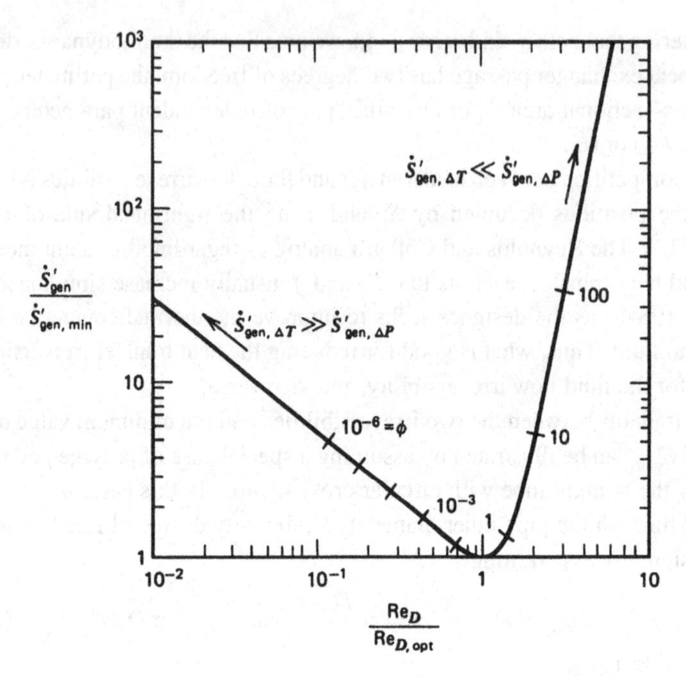

Fig. 7.2. The relative entropy generation rate for forced convection through a smooth tube (Bejan 1980).

Figure 7.2 shows that the entropy generation rate of the tube increases sharply on either side of the optimum. The irreversibility distribution ratio varies along the V-shaped curve, increasing in the direction of small D (that is, large Re, because $\dot{m}$ = constant), in which the overall entropy generation rate is dominated by fluid friction effects. At the optimum, the irreversibility distribution ratio assumes the value $\phi_{opt} = 0.168$.

This most fundamental issue of thermodynamic irreversibility at the heat exchanger passage level was reconsidered by Kotas and Shakir (1986). They took into account the temperature dependence of transport properties and showed that the operating temperature of the heat exchanger passage has a profound effect on the thermodynamic optimum. For example, the optimum Reynolds number increases as the absolute temperature T decreases. The minimum irreversibility corresponding to this optimum design also increases as T decreases.

7.2.2 *Heat transfer augmentation*

Another example of the competition between different irreversibility mechanisms occurs in connection with the general problem of heat transfer augmentation, in which the main objective is to devise a technique that increases the wall–fluid heat transfer coefficient relative to the coefficient of the original (unaugmented) surface. A parallel objective, however, is to register this improvement without causing a damaging increase in the pumping power demanded by the forced convection arrangement. These two objectives reveal the conflict that accompanies the application of any augmentation technique: A design modification that improves the thermal contact (for example, roughening the heat transfer surface) is likely to also augment the mechanical pumping power requirement.

The true effect of a proposed augmentation technique on thermodynamic performance may be evaluated by comparing the irreversibility of the heat exchange apparatus before and after the implementation of the augmentation technique (Bejan and Pfister 1980). Consider again the general heat exchanger passage referred to in Equation (7.3), and let $\dot{S}'_{\text{gen},0}$ represent the degree of irreversibility in the reference (unaugmented) passage. Writing $\dot{S}'_{\text{gen},a}$ for the heat transfer–augmented version of the same device, we can define the augmentation entropy generation number $N_{S,a}$ as

$$N_{S,a} = \frac{\dot{S}'_{\text{gen},a}}{\dot{S}'_{\text{gen},0}}. \tag{7.18}$$

Augmentation techniques that are characterized by $N_{S,a}$ values less than one are thermodynamically advantageous.

If the function of the heat exchanger passage is fixed, that is, if $\dot{m}$ and q' are given, the augmentation entropy generation number can be put in the more explicit form

$$N_{S,a} = \frac{1}{1+\phi_0} N_{S,\Delta T} + \frac{\phi_0}{1+\phi_0} N_{S,\Delta P}, \tag{7.19}$$

where ϕ_0 is the irreversibility distribution ratio of the reference design, and $N_{S,\Delta T}$ and $N_{S,\Delta P}$ represent the values of $N_{S,a}$ in the limits of pure heat transfer irreversibility and pure fluid flow irreversibility, respectively. It is not difficult to show that these limiting values are

$$N_{S,\Delta T} = \frac{St_0 D_{h,a}}{St_a D_{h,0}} \tag{7.20}$$

$$N_{S,\Delta P} = \frac{f_a D_{h,0} A_0^2}{f_0 D_{h,a} A_a^2}. \tag{7.21}$$

The geometric parameters (A, D_H) before and after augmentation are linked through the $\dot{m}$ = constant constraint, which reads

$$Re_a \frac{A_a}{D_{h,a}} = Re_0 \frac{A_0}{D_{h,0}}. \tag{7.22}$$

Equations (7.19)–(7.22) show that $N_{S,a}$ is, in general, a function of both the heat transfer coefficient ratio (St_a/St_0) and the friction factor ratio (f_a/f_0). The relative importance of the friction factor ratio is dictated by the numerical value of ϕ_0: This value is known because the reference design is known.

As an example, consider the effect of sand-grain roughness on irreversibility in forced convection heat transfer in a pipe. Because this augmentation technique does not effect the hydraulic diameter and flow cross section appreciably $(D_a \cong D_0,\ A_a \cong A_0)$, the augmentation entropy generation number assumes a particularly simple form

$$N_{S,a} = \frac{St_0}{St_a} + \frac{\phi_0}{1+\phi_0}\left(\frac{f_a}{f_0} - \frac{St_0}{St_a}\right). \tag{7.23}$$

The relationship is displayed graphically in Fig. 7.3 as $N_{S,A}$ versus Re_0 for $\phi_0 = 0$, 1, ∞, and four values of roughness height e/D. To construct this figure, we used the Nikuradse correlation for sand-grain roughness friction factor f_a, the Dippery and Sabersky correlation for heat transfer coefficient St_a, the Karman–Nikuradse relation for friction factor in a smooth tube f_0 and, finally, Petukhov's prediction of heat transfer in a smooth tube (Bejan and Pfister 1980).

Figure 7.3 illustrates the critical position occupied by the irreversibility distribution ratio ϕ_0 in deciding whether the use of rough walls will yield savings in available work. For a fixed Reynolds number and grain size e/D, a critical ratio ϕ_0 exists where the augmentation technique has no effect on irreversibility

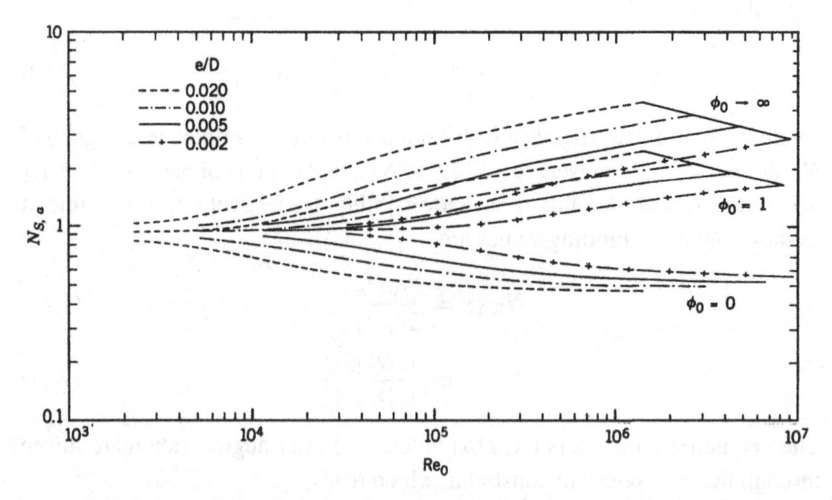

Fig. 7.3. The augmentation entropy generation number for sand-grain roughness in a straight pipe (Bejan and Pfister 1980).

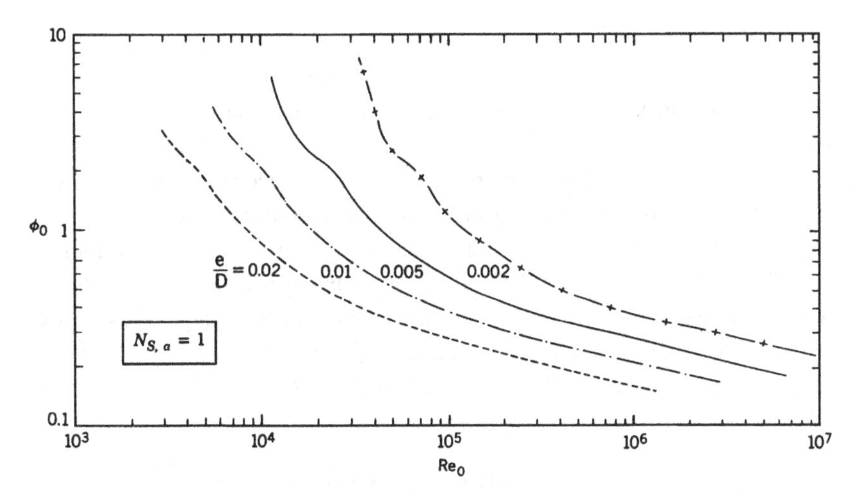

Fig. 7.4. The critical irreversibility distribution ratio for sand-grain roughness in a straight pipe (Bejan and Pfister 1980).

($N_{S,a} = 1$). This critical value is plotted in Fig. 7.4. If in a certain heat exchanger design the irreversibility distribution ratio exceeds the critical ϕ_0, the sand-grain roughening of the wall will not reduce the rate of entropy generation in the tube. This will always be the case despite an automatic increase in heat transfer coefficient. Figure 7.4 also supports the observation that the irreversibility distribution ratio can be considerably smaller than unity at that critical point where the fluid friction irreversiblity begins to be felt in the overall irreversibility minimization effort.

Other heat transfer augmentation techniques can be investigated on the basis of the $N_{S,a}$ criterion described so far. For a more extensive survey of the irreversibility-reduction potential of heat transfer augmentation techniques, the reader is encouraged to consult Ouellette and Bejan (1980).

7.2.3 External flow and heat transfer

The competition between flow and heat transfer irreversibilities also rules the thermodynamic design of external convection heat transfer arrangement, in which the flow engulfs the solid body (walls) with which it exchanges heat. The overall entropy generation rate associated with an external convection configuration is (Poulikakos and Bejan 1982)

$$\dot{S}_{\text{gen}} = \frac{\dot{Q}_B(T_B - T_\infty)}{T_\infty T_B} + \frac{F_D U_\infty}{T_\infty}, \tag{7.24}$$

where $\dot{Q}_B$, T_B, T_∞, F_D, and U_∞ are, respectively, the instantaneous heat transfer rate between the body and the fluid reservoir, the body surface temperature, the fluid reservoir temperature, the drag force, and the relative speed between body and reservoir.

This remarkably simple result proves again that inadequate thermal contact (the first term) and fluid friction (the second term) contribute hand in hand to degrading the thermodynamic performance of the external convection arrangement. One area in which Equation (7.24) has found application is the problem of selecting the size and number (density) of fins for the design of extended surfaces. The thermodynamic optimization of fins and fin arrays of various geometries is described by Poulikakos and Bejan (1982).

7.3 Heat exchangers

7.3.1 The tradeoff between heat transfer and fluid flow irreversibilities

Consider the class of balanced counterflow heat exchangers in the ideal limit of small ΔP's and ΔT. Balance means that the capacity flow rates are the same on the two sides of the heat transfer surface:

$$(\dot{m}c_P)_1 = (\dot{m}c_P)_2 = \dot{m}c_P. \tag{7.25}$$

The two sides are indicated by the subscripts 1 and 2. We write T_1 and T_2 for the fixed (given) inlet temperatures of the two streams and P_1 and P_2 for the respective inlet pressures. The entropy generation rate of the entire heat exchanger is

$$\begin{aligned}\dot{S}_{\text{gen}} &= (\dot{m}c_P)_1 \ln\frac{T_{1,\text{out}}}{T_1} + (\dot{m}c_P)_2 \ln\frac{T_{2,\text{out}}}{T_2}\\ &\quad - (\dot{m}R)_1 \ln\frac{P_{1,\text{out}}}{P_1} - (\dot{m}R)_2 \ln\frac{P_{2,\text{out}}}{P_2},\end{aligned} \tag{7.26}$$

where the working fluid has been modeled as an ideal gas with constant specific heat. The outlet temperature $T_{1,\text{out}}$ and $T_{2,\text{out}}$ can be eliminated by bringing in the concept of heat exchanger effectiveness, ε. If we also assume that $(1-\varepsilon) \ll 1$ and that the pressure drop along each stream is sufficiently small relatively to the absolute pressure levels, the entropy generation rate may be nondimensionalized as (Bejan 1977)

$$N_S = (1-\varepsilon)\frac{(T_2-T_1)^2}{T_1T_2} + \frac{R}{c_P}\left[\left(\frac{\Delta P}{P}\right)_1 + \left(\frac{\Delta P}{P}\right)_2\right], \tag{7.27}$$

where N_S is the entropy generation number:

$$N_S = \frac{1}{\dot{m}c_P}\dot{S}_{\text{gen}}. \tag{7.28}$$

In this form, it is clear that the overall entropy generation rate (N_S) receives contributions from three sources of irreversibility, namely the stream-to-stream heat transfer, regardless of the sign of $(T_2 - T_1)$; the pressure drop along the first stream, ΔP_1; and the pressure drop along the second stream, ΔP_2. The heat transfer irreversibility term can be split into two terms, each describing the contribution made by one side of the heat transfer surface. We are assuming that the stream-to-stream ΔT is due to the heat transfer across the two convective thermal resistances that sandwich the solid wall separating the two streams and that the thermal resistance of the wall itself is negligible. That is, we write

$$\frac{1}{\overline{h}A_1} = \frac{1}{\overline{h}_1 A_1} + \frac{1}{\overline{h}_2 A_2}, \tag{7.29}$$

where A_1 and A_2 are the heat transfer areas swept by each stream and $\overline{h}_1$ and $\overline{h}_2$ are the side-heat transfer coefficients based on these respective areas. The overall heat transfer coefficient $\overline{h}$ is based on A_1. The thermal resistance summation (7.29) also means that

$$\frac{1}{N_{tu}} = \frac{1}{N_{tu,1}} + \frac{1}{N_{tu,2}}, \tag{7.30}$$

where

$$N_{tu} = \frac{\overline{h}A_1}{\dot{m}c_P}, \qquad N_{tu,1} = \frac{\overline{h}_1 A_1}{\dot{m}c_P}, \qquad N_{tu,2} = \frac{\overline{h}_2 A_2}{\dot{m}c_P}. \tag{7.31}$$

In a balanced counterflow heat exchanger, the $\varepsilon(N_{tu})$ relationship is particularly simple (for example, Bejan 1993a):

$$\varepsilon = \frac{N_{tu}}{1 + N_{tu}}. \tag{7.32}$$

Combining Equations (7.27)–(7.32), we find that in the ideal heat exchanger limit (small ΔT and ΔP's), the entropy generation number N_S splits into two groups of terms:

$$N_S = \underbrace{\frac{\tau^2}{N_{tu,1}} + \frac{R}{c_P}\left(\frac{\Delta P}{P}\right)_1}_{N_{S,1}} + \underbrace{\frac{\tau^2}{N_{tu,2}} + \frac{R}{c_P}\left(\frac{\Delta P}{P}\right)_2}_{N_{S,2}} \tag{7.33}$$

The contribution of the ideal-limit analysis is that it separates N_S into all the pieces that contribute to the irreversibility of the apparatus. The first pair of terms on the right-hand side of Equation (7.33) represents the irreversibility contributed solely by Side 1 of heat transfer surface, $N_{S,1}$. The first term in this first pair is the entropy generation number due to heat transfer irreversibility on Side 1, where τ is shorthand for a dimensionless temperature difference parameter fixed by T_1 and T_2,

$$\tau = \frac{|T_2 - T_1|}{(T_1 T_2)^{1/2}}. \tag{7.34}$$

The one-side entropy generation numbers, $N_{S,1}$ and $N_{S,2}$, have the same analytical form; therefore, we can concentrate on the minimization of only one of them (say, $N_{S,1}$) and keep in mind that the minimization analysis can be repeated identically for $N_{S,2}$.

Despite the additive form of $N_{S,1}$ in Equation (7.33), the heat transfer and fluid friction contributions to it are, in fact, coupled through the geometric parameters of the heat exchanger duct (passage) that reside on Side 1 of the heat exchanger surface. This coupling is brought to light by rewriting $N_{S,1}$ in terms of the passage slenderness ratio $(4L/D_H)_1$:

$$N_{S,1} = \frac{\tau^2}{St_1}\left(\frac{D_h}{4L}\right)_1 + \frac{R}{c_P} g_1^2 f_1 \left(\frac{4L}{D_h}\right)_1, \tag{7.35}$$

where f_1 and St_1 are defined according to Equations (7.7) and (7.8). An important ingredient is the relation between N_{tu} and St:

$$N_{tu,1} = \left(\frac{4L}{D_h}\right)_1 St_1, \tag{7.36}$$

which follows from Equations (7.8), (7.11), and (7.31), in combination with $A_1 = L_1 p_1$, where p_1 is the wetted perimeter of Passage 1. Finally, G_1 is a dimensionless mass velocity defined as

$$g_1 = \frac{G_1}{(2\rho p_1)^{1/2}}. \tag{7.37}$$

As summary to the ideal limit, we note that the one-side irreversibility depends on two types of parameters:

$$\tau, \qquad \frac{R}{c_P}, \qquad Pr$$

$$\left(\frac{4L}{D_h}\right)_1, \qquad Re_1, \qquad g_1.$$

The first row contains the parameters fixed by the selection of working fluid and inlet conditions. The second row lists the three parameters that depend on the size and geometry of the heat exchanger passage. How many of these three parameters are true degree of freedom depends on the number of design constraints.

One important constraint concerns the heat transfer area A_1. In dimensionless form, the constant-area condition may be expressed as (Bejan 1982)

$$a_1 = \frac{A_1}{\dot{m}}(2\rho p_1)^{1/2}, \qquad \text{constant}, \tag{7.38}$$

where A_1 is the dimensionless area of side 1 of the surface. It is easy to show

$$a_1 g_1 = \left(\frac{4L}{D_h}\right)_1, \tag{7.39}$$

and only two degrees of freedom remain for the minimization of $N_{S,1}$:

$$N_{S,1}(g_1, Re_1) = \frac{\tau^2}{A_1 g_1 St_1} + \frac{R}{c_P} a_1 f_1 g_1^3. \tag{7.40}$$

Minimizing the entropy generation number subject to fixed (known) Reynolds number yields the optimum mass velocity (Bejan 1988a):

$$g_{1,\text{opt}} = \left[\frac{\tau^2}{(3R/c_P) a_1^2 f_1 St_1}\right]^{1/4} \tag{7.41}$$

$$N_{S,1,\min} = \left[\frac{256\tau^6 (R/c_P) f_1}{27 a_1^2 St_1^3}\right]^{1/4}. \tag{7.42}$$

The minimum entropy generation number varies as $a_1^{-1/2}$; therefore, the thermodynamic goodness of the heat exchanger is enhanced by investing more area in the design of each side.

Other constraints can be taken into account while minimizing the entropy generation rate of the ideal-limit balanced counterflow heat exchanger. Examples illustrated in Bejan (1987, 1988a) are the fixed volume, fixed volume and area, and fixed N_{tu} (Sekulic and Herman 1986).

7.3.2 Flow imbalance irreversibility

The study of balanced counterflow heat exchangers led to the conclusion that the overall irreversibility of the device decreases to zero as the design approaches the ideal limit of infinite overall N_{tu} and zero ΔP on both sides of the surface. In this section, we focus strictly on the ideal design

$$N_{tu} = \infty, \qquad \Delta P_1 = \Delta P_2 = 0, \tag{7.43}$$

and our objective is to show that in this limit the heat exchanger configurations that differ from balanced counterflow are characterized by an unavoidable irreversibility solely due to the flow arrangement. For historical reasons (Bejan 1977), we refer to this remanent irreversibility as the irreversibility due to flow imbalance, or remanent irreversibility.

Consider first an imbalanced counterflow heat exchanger, where

$$\omega = \frac{(\dot{m} c_P)_1}{(\dot{m} c_P)_2} > 1 \tag{7.44}$$

and where the perfect design (7.43) means $P_{1,\text{out}} = P_1$, $P_{2,\text{out}} = P_2$, and $\varepsilon = 1$. The effectiveness – N_{tu} relations for imbalanced counterflow heat exchangers are

$$N_{tu} = \frac{\bar{h}A_1}{(\dot{m}c_P)_2}, \quad \varepsilon = \omega \frac{T_1 - T_{1,\text{out}}}{T_1 - T_2} = \frac{T_{2,\text{out}} - T_2}{T_1 - T_2} \tag{7.45}$$

$$\varepsilon = \frac{1 - \exp[-N_{tu}(1 - \omega^{-1})]}{1 - \omega^{-1}\exp[-N_{tu}(1 - \omega^{-1})]}, \tag{7.46}$$

where $(\dot{m}c_P)_2$ is the smaller of the two capacity flow rates. In this case, the overall entropy generation rate (7.26) has a finite value

$$N_{S,\text{imbalance}} = \frac{\dot{S}_{\text{gen}}}{(\dot{m}c_P)_2} = \ln\left\{\left[1 - \frac{1}{\omega}\left(1 - \frac{T_2}{T_1}\right)\right]^{\omega} \frac{T_1}{T_2}\right\}. \tag{7.47}$$

The imbalance irreversibility of a two-stream heat exchanger with phase-change on one side is a special case of Equation (7.47), namely the limit $\omega \to \infty$, where the stream that bathes Side 1 does not experience a temperature variation from inlet to outlet, $T_{1,\text{out}} = T_1$. For this class of heat exchangers, Equation (7.47) reduces to

$$N_{S,\text{imbalance}} = \frac{T_2}{T_1} - 1 - \ln\frac{T_2}{T_1}, \qquad (\omega = \infty). \tag{7.48}$$

The imbalance irreversibility of two-stream parallel flow heat exchangers is obtained similarly, by combining Equation (7.26) with the perfect-design conditions (7.43) and the $\varepsilon(\omega, N_{tu})$ relation for parallel flow (for example, Bejan 1993a, p. 471). The result is

$$N_{S,\text{imbalance}} = \frac{\dot{S}_{\text{gen}}}{(\dot{m}c_P)_2} = \ln\left\{\left(\frac{T_2}{T_1}\right)^{\omega}\left[1 + \left(\frac{T_1}{T_2} - 1\right)\frac{\omega}{1+\omega}\right]^{1+\omega}\right\}. \tag{7.49}$$

An important observation is that when the two streams and their inlet conditions are given, the imbalance irreversibility of the parallel flow arrangement is consistently greater than the imbalance irreversibility of the counterflow scheme (7.47). Figure 7.5 shows the behavior of the respective entropy generation numbers and how they both approach the value indicated by Equation (7.48) as the flow imbalance ratio ω increases. Taking the $\omega = 1$ limit of Equation (7.49), it is easy to see that the remanent irreversibility of the parallel flow arrangement is finite even in the balanced flow case.

7.3.3 Combined heat transfer, fluid flow, and imbalance irreversibilities

An important structure can be recognized in the heat exchanger irreversibility treatment reviewed in the preceding two sections. First is competition that exists

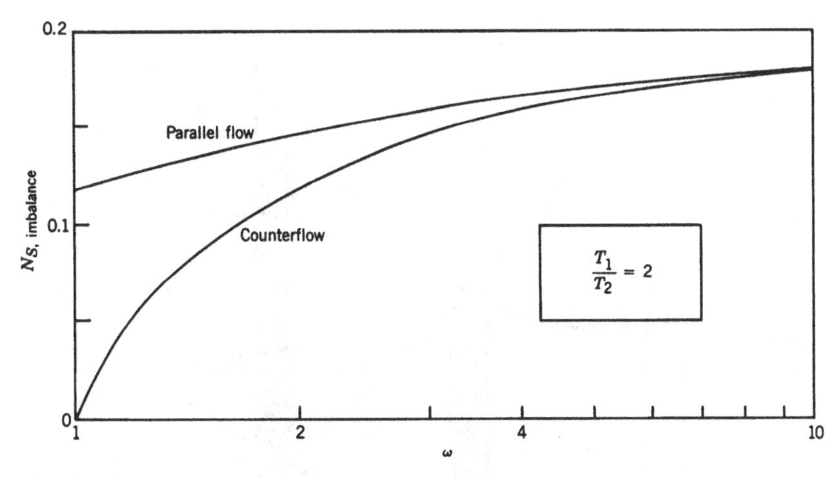

Fig. 7.5. The remanent (flow imbalance) irreversibility in parallel flow is consistently greater than in counterflow (Bejan 1988a).

between heat transfer and fluid flow (pressure drop) irreversibilities, whose various trade-offs were illustrated by considering the analytically simple limit of nearly ideal balanced counterflow heat exchangers. Second is the recognition of remanent or flow imbalance irreversibilities, that is, irreversibilities that persist even in the limit of perfect heat exchangers (7.43).

Figure 7.6 summarizes this structure. The remanent irreversibility deserves to be calculated first in the thermodynamic optimization of any heat exchanger because it establishes the level (order of magnitude) below which the joint minimization of heat transfer (finite ΔT, or N_{tu}) and fluid flow (finite ΔP) irreversibilities falls in the realm of diminishing returns. That is, it would no longer make sense to invest heat exchanger area and engineering into minimizing the sum ($N_{S,\Delta T} + N_{S,\Delta P}$) when this sum is already negligible compared with the remanent irreversibility $N_{S,\mathrm{imbalance}}$.

Only in special cases does the entropy generation rate of a heat exchanger break up explicitly into a sum of three terms so that each term accounts for one of the irreversibilities reviewed above:

$$N_S = N_{S,\mathrm{imbalance}} + N_{S,\Delta T} + N_{S,\Delta P}. \tag{7.50}$$

One such case is the balanced counterflow heat exchanger in the nearly balanced and nearly ideal limit ($\omega \to 1$, $\Delta T \to 0$, ΔP's $\to 0$). In general, these three irreversibilities contribute in a more complicated way to the eventual size of the overall N_S. Deep down, however, the behavior of the three is the same as that of the simple limits singled out for discussion. Figure 7.6 illustrates this behavior qualitatively.

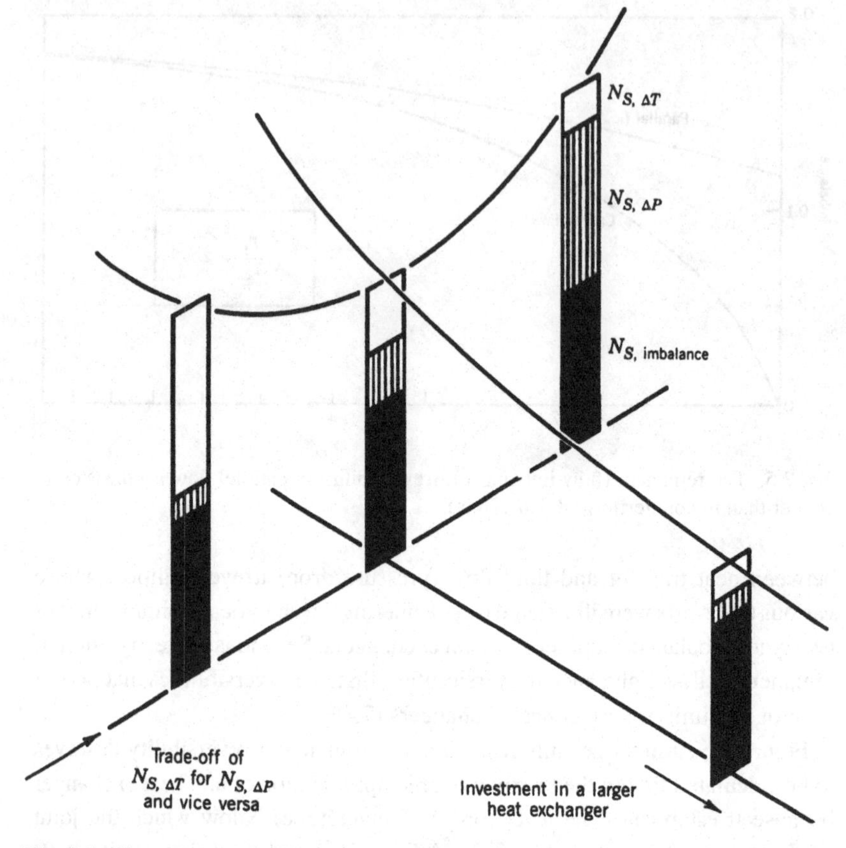

Fig. 7.6. The structure of the total entropy generation rate of a heat exchanger (the book cover of Bejan 1988a).

The entropy generation analysis of other classes of heat exchangers reveals the basic structure outlined in Fig. 7.6. Heat exchangers in which at least one stream is a two-phase mixture are treated in London and Shah (1983), Bejan (1984b, 1987, 1988a), and Zubair, Kadaba, and Evans (1987). The Second-Law optimization of entire heat exchanger networks is presented in Chato and Damianides (1986) and Hesselmann (1984).

7.4 The optimal allocation of a fixed heat exchanger inventory

7.4.1 Power plants

We now turn our attention to the minimization of the entropy generation rate, or the maximization of the power output of a power plant (Novikov 1957)

(El-Wakil 1962) (Curzon and Ahlborn 1975). We will show that when the total heat exchangers inventory is fixed, there is an optimal way of dividing this inventory between the heat exchangers of the power plant (Bejan 1988a,b).

Consider the model outlined in Fig. 7.7 (right). The power plant delineated by the solid boundary operates between the high temperature T_H and low temperature T_L. The modeling challenge consists of providing the minimum construction detail that allows the outright identification of the irreversible and reversible compartments of the power plant.

To see the reasoning behind some of the features included in this drawing, recall that in the traditional treatment of simple power plant models (for example, the Carnot cycle) one tacitly assumes that the heat engine is in perfect thermal equilibrium with each temperature reservoir during the respective heat transfer interactions. In reality, however, such equilibria would require either an infinitely slow cycle or infinitely large contact surfaces between the engine and the reservoirs (T_H and T_L). For this reason the finite temperature differences $(T_H - T_{HC})$ and $(T_{LC} - T_L)$ are recognized as driving forces for the instantaneous heat transfer interactions,

$$\dot{Q}_{HC} = (UA)_H(T_H - T_{HC}), \qquad \dot{Q}_{LC} = (UA)_L(T_{LC} - T_L). \tag{7.51}$$

The proportionality coefficients $(UA)_H$ and $(UA)_L$ are the hot-end and cold-end thermal conductances. In heat exchanger design terms, for example, $(UA)_H$ represents the product of the hot-end heat transfer area A_H times the overall heat transfer coefficient based on that area, U_H.

Because both $(UA)_H$ and $(UA)_L$ are commodities in short supply, recognizing as a constraint the total thermal conductance inventory makes sense (Bejan 1988a,b):

$$UA = (UA)_H + (UA)_L, \tag{7.52}$$

or, in terms of the external conductance allocation ratio x,

$$(UA)_H = xUA, \qquad (UA)_L = (1 - x)UA. \tag{7.53}$$

The heat transfer rates that cross the finite temperature gaps represent two sources of irreversibility. A third source sketched in Fig. 7.7 (right) is the heat transfer rate $\dot{Q}_i$ that leaks directly through the machine and, especially, around the power producing compartment labeled (C). To distinguish this third irreversibility from the external irreversibilities due to $(T_H - T_{HC})$ and $(T_{LC} - T_L)$, we will refer to $\dot{Q}_i$ as the internal heat transfer rate (heat leak) through the plant. The internal heat leak was first identified as a modeling feature of power plant irreversibility by Bejan and Paynter (1976) (see also Bejan 1982, p. 44). The same feature was introduced even earlier in EGM models of refrigeration plants (Bejan and Smith 1974).

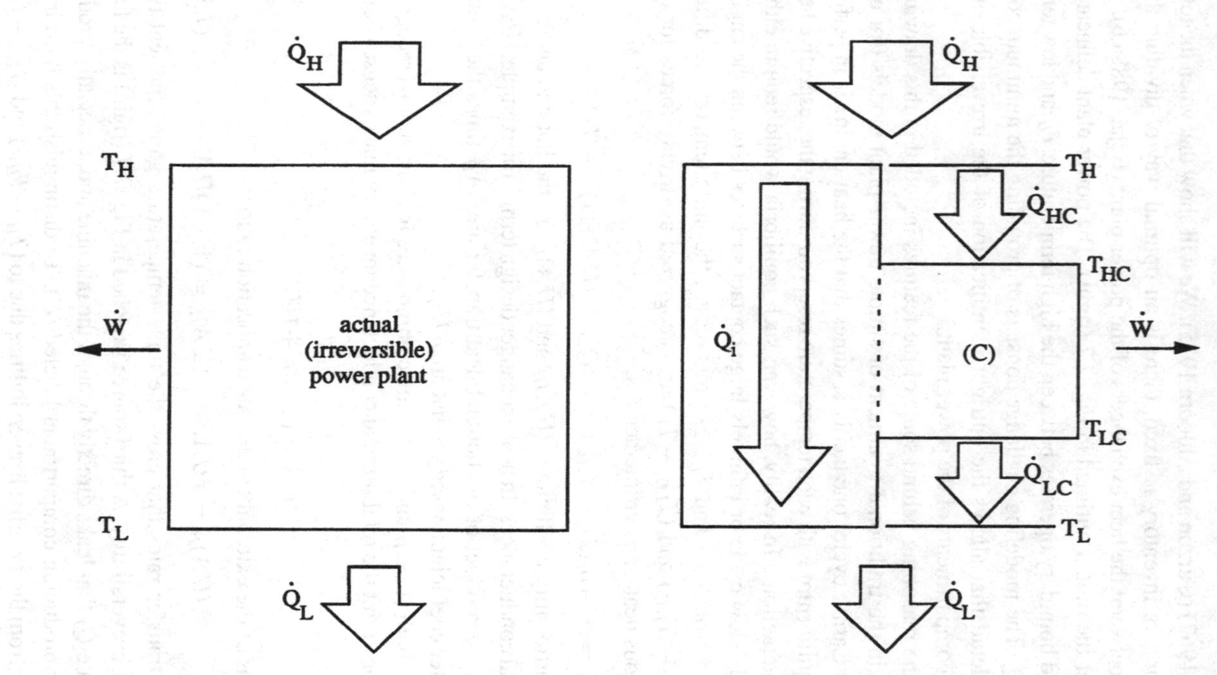

Fig. 7.7. Irreversible power plant (left), and a model with three sources of heat transfer irreversibility (right). (Bejan 1988b).

Many features of an actual power plant fall under the umbrella represented by $\dot{Q}_i$; for example, the heat transfer lost through the wall of a boiler or combustion chamber, the heat transfer removed by the cooling system of an internal combustion engine, and the streamwise convective heat leak channeled toward room temperature by the counterflow heat exchanger of a regenerative Brayton cycle. The simplest internal heat leak model that is consistent with the linear models of Equations (7.51) is

$$\dot{Q}_i = \frac{T_H - T_L}{R_i}, \tag{7.54}$$

where R_i is the internal thermal resistance of the power plant.

To summarize the three irreversibility sources that have been discussed, note that the model of Fig. 7.7 relies analytically on three parameters, namely $[(UA)_H, (UA)_L, R_i]$, or (UA, x, R_i). The remaining power plant compartment, which is labeled C, from Carnot, is assumed irreversibility-free. The Second Law of thermodynamics states that its rate of entropy generation is zero, which means that

$$\frac{\dot{Q}_{LC}}{T_{LC}} - \frac{\dot{Q}_{HC}}{T_{HC}} = 0. \tag{7.55}$$

The analytical representation of Fig. 7.7 is completed by the First-Law statements

$$\dot{W} = \dot{Q}_{HC} - \dot{Q}_{LC} \tag{7.56}$$

$$\dot{Q}_H = \dot{Q}_i + \dot{Q}_{HC} \tag{7.57}$$

$$\dot{Q}_L = \dot{Q}_i + \dot{Q}_{LC}, \tag{7.58}$$

where $\dot{W}$, $\dot{Q}_H$, and $\dot{Q}_L$ are the energy rate interactions of the power plant as a whole, namely the power output, the rate of heat input, and the heat rejection rate.

Consider now under what conditions the instantaneous power output $\dot{W}$ is maximum. Combining Equations (7.53) and (7.55)–(7.58) leads to

$$\dot{W} = (1 - x)UAT_L\left(\frac{T_{LC}}{T_L} - 1\right)\left(\frac{T_{HC}}{T_{LC}} - 1\right), \tag{7.59}$$

in which the unknown is now the ratio T_{LC}/T_L. From Equations (7.55) and (7.53) we find that

$$\frac{T_{LC}}{T_L} = 1 - x + x\frac{\tau}{\tau_C}, \tag{7.60}$$

where

$$\tau = \frac{T_H}{T_L}, \qquad \tau_c = \frac{T_{HC}}{T_{LC}}. \tag{7.61}$$

Taken together, Equations (7.59) and (7.60) yield

$$\dot{W} = x(1-x)UAT_L\left(\frac{\tau}{\tau_c} - 1\right)(\tau_c - 1). \tag{7.62}$$

This expression shows that the instantaneous power output per unit of external conductance inventory UA can be maximized in two ways, with respect to τ_c and x. By solving $\partial\dot{W}/\partial\tau_c = 0$, we obtain the optimal temperature ratio across the Carnot compartment (C),

$$\tau_{c,\mathrm{opt}} = \tau^{1/2} \quad \text{or} \quad \left(\frac{T_{HC}}{T_{LC}}\right)_{\mathrm{opt}} = \left(\frac{T_H}{T_L}\right)^{1/2}, \tag{7.63}$$

and the corresponding maximum power,

$$\dot{W}_{\max} = x(1-x)UAT_L(\tau^{1/2} - 1)^2. \tag{7.64}$$

The optimum represented by Equation (7.63) was first reported by Novikov (1957) and Curzon and Ahlborn (1975), who used models without internal irreversibility. The contribution of the model of Fig. 7.7 is to show that Equation (7.63) holds even when the internal heat leak $\dot{Q}_i$ is taken into account (Bejan 1988b).

An entirely new aspect that is brought to light by the model of Fig. 7.7 is that an optimal way exists to allocate the UA inventory between the hot and cold ends such that the power output is maximized once more. By solving $\partial\dot{W}_{\max}/\partial x = 0$, we obtain the optimum thermal conductance allocation fraction

$$x_{\mathrm{opt}} = \frac{1}{2} \quad \text{or} \quad (UA)_H = (UA)_L \tag{7.65}$$

and the corresponding (twice-maximized) instantaneous power output

$$\dot{W}_{\max,\max} = \frac{1}{4}UAT_L(\tau^{1/2} - 1). \tag{7.66}$$

In conclusion, to operate at maximum power the thermodynamic temperature ratios τ_c and τ and the sizes of the hot-end and cold-end heat exchangers must be balanced. This conclusion was first reported in Bejan (1988a,b). Equation (7.65) also holds when the total thermal conductance inventory UA is minimized subject to fixed power output (Bejan 1993b).

This conclusion changes somewhat if the total thermal conductance constraint of Equation (7.52) is replaced with the new constraint that the physical size (or weight) of the total heat transfer area is fixed,

$$A = A_H + A_L \qquad \text{(constant)}. \tag{7.67}$$

The area constraint can be restated as

$$A_H = yA, \qquad A_L = (1-y)A, \tag{7.68}$$

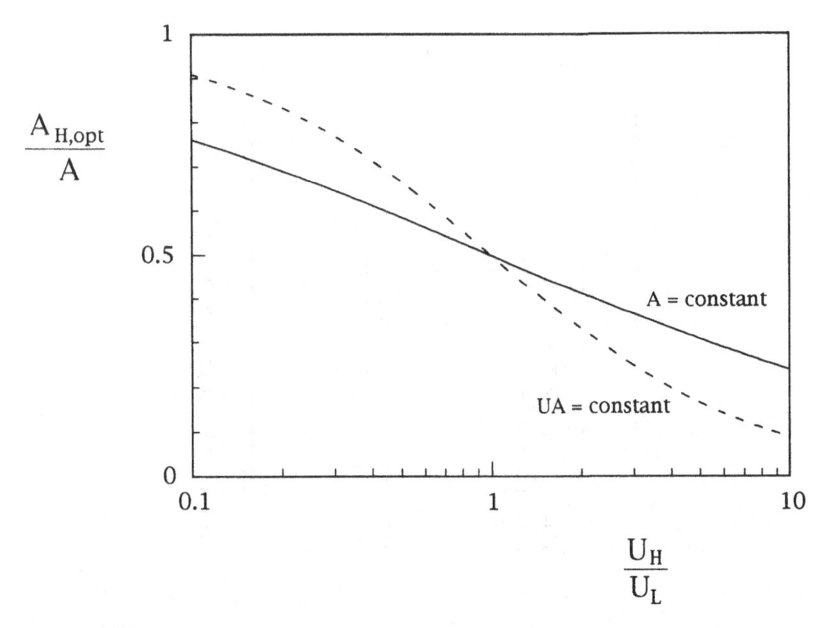

Fig. 7.8. The optimal way to divide the total heat transfer area between the two heat exchangers of a power plant: fixed total area (solid line) vs. fixed total thermal conductance (dashed line) (Bejan 1994).

where y is the area allocation ratio. Substituting Equation (7.68) into the $\dot{W}$ expression derived from Equations (7.55)–(7.58) and solving $\partial \dot{W}/\partial y = 0$ yields the optimal area allocation ratio,

$$y_{\text{opt}} = \frac{1}{1 + \left(U_H/U_L\right)^{1/2}}, \qquad (A = \text{constant}). \tag{7.69}$$

Equation (7.69) is plotted as a solid line in Fig. 7.8. This result shows that a larger fraction of the area supply should be allocated to the heat exchanger whose overall heat transfer coefficient is lower. Only when U_H is equal to U_L should A be divided equally between the two heat exchangers.

7.4.2 Refrigeration plants

In this section we extend to refrigeration plants the optimization method used in the preceding section for power plants. As shown on the left side of Fig. 7.9, the refrigeration plant absorbs the refrigeration load $\dot{Q}_L$ at the temperature level T_L, and rejects $\dot{Q}_H$ to the ambient, T_H. The power required to accomplish this task is $\dot{W}$.

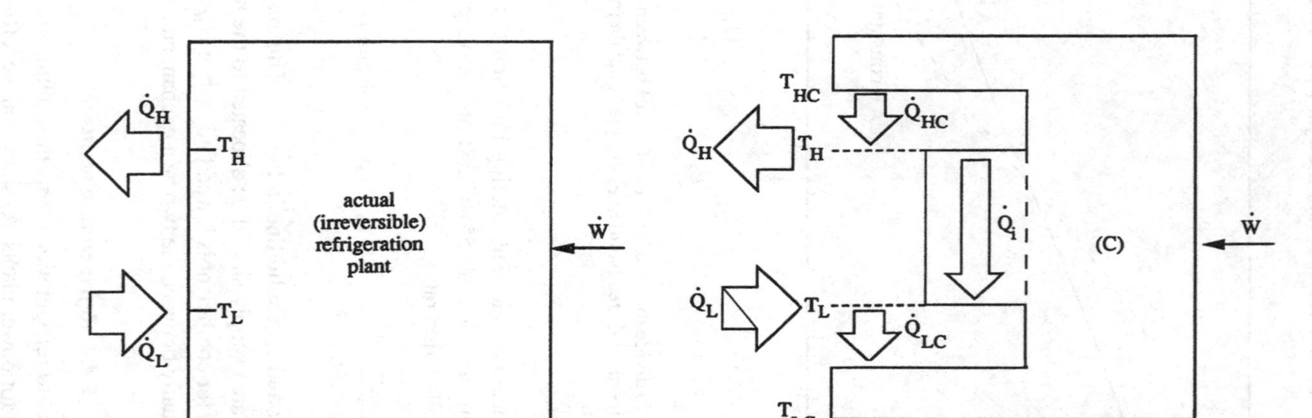

Fig. 7.9. Irreversible refrigeration plant (left), and model with three sources of irreversibility (right) (Bejan 1989).

On the right side of Fig. 7.9 we see a model (Bejan 1989) that accounts for some of the features that make the actual refrigeration plant irreversible. First is the heat leak $\dot{Q}_i$ through the insulation of the machine, from the ambient T_H to the refrigerated space T_L. In a machine of given size, the simplest model for the internal heat leak is given by an expression identical to Equation (7.54), where R_i is the overall thermal resistance between T_H and T_L. This resistance accounts for the actual thermal insulation sandwiched between T_H and T_L, the mechanical supports that connect the cold space to the room-temperature frame, and, in a low-temperature refrigerator, the counterflow heat exchanger (Bejan 1979).

Second, the load $\dot{Q}_L$ can be absorbed from the temperature level T_L only if the working fluid of the refrigeration cycle is at a temperature below T_L. On the right side of Fig. 7.9, that lower temperature is labeled T_{LC}. We assume a linear relation between the heat transfer rate into the cold fluid ($\dot{Q}_{LC}$) and the temperature difference that drives it,

$$\dot{Q}_{LC} = (UA)_L(T_L - T_{LC}). \tag{7.70}$$

Third, the rejection of $\dot{Q}_H$ to the ambient temperature T_H is possible only when the fluid at the hot end of the refrigeration cycle is warmer than the ambient. Assuming that the hot-end temperature of the cycle executed by the working fluid is T_{HC}, we write that

$$\dot{Q}_{HC} = (UA)_H(T_{HC} - T_H). \tag{7.71}$$

The proportionality factors $(UA)_H$ and $(UA)_L$ are the thermal conductances provided by the hot-end and cold-end heat exchangers, respectively. In heat transfer engineering terms, for example, the hot-end conductance $(UA)_H$ scales as the product between the overall heat transfer coefficient and the total area of the hot-end heat exchangers [for example, the battery of aftercoolers (or intercoolers) positioned between the stages of the compressor of a liquid nitrogen or liquid helium refrigerator]. The end conductances are expensive commodities, because they both increase with the size of the respective heat exchangers. For this reason it makes sense to use as a design constraint the total thermal conductance inventory, which is expressed in the same way as in Equation (7.52).

The remaining components in the model of Fig. 7.9 (right) are irreversibility-free. The space occupied by these components is labeled (C), where $\dot{Q}_{HC}/T_{HC} = \dot{Q}_{LC}/T_{LC}$. The analytical description of the refrigeration plant model is completed by the energy conservation statements $\dot{W} = \dot{Q}_H - \dot{Q}_L$, $\dot{Q}_{HC} = \dot{Q}_H + \dot{Q}_i$, and $\dot{Q}_{LC} = \dot{Q}_L + \dot{Q}_i$.

We are interested in how the imperfect features identified in the model influence the overall performance of the refrigeration plant. The latter is described quantitatively by the Second-Law efficiency

$$\eta_{II} = \frac{(COP)_C}{COP}, \tag{7.72}$$

where

$$COP = \frac{\dot{Q}_L}{\dot{W}}, \qquad (COP)_C = \left(\frac{T_H}{T_L} - 1\right)^{-1}. \tag{7.73}$$

It can be shown that

$$\eta_{II} = \frac{\tau - 1}{\tau_C - 1}\left(1 + \frac{\dot{Q}_i}{\dot{Q}_L}\right)^{-1}, \tag{7.74}$$

where

$$\tau = \frac{T_H}{T_L}, \qquad \tau_C = \frac{T_{HC}}{T_{LC}}. \tag{7.75}$$

Note that τ is the actual temperature ratio seen from the outside of the refrigeration plant (Fig. 7.9, left), whereas τ_C is the absolute temperature ratio across the reversible compartment of the model (Fig. 7.9, right). Note further that τ_C is always greater than τ, and that τ_C and τ are both greater than one.

Bejan (1989) showed that when the total thermal conductance is fixed, the instantaneous refrigeration load is maximum when UA is divided equally between the two heat exchangers. More recently, he showed that the same optimization rule applies when the total inventory UA is minimized subject either to a specified refrigeration load $\dot{Q}_L$ or to a specified power input $\dot{W}$ (Bejan 1993b).

Klein (1992) used a refrigerator model similar to that of Fig. 7.9 but without the internal heat leak $\dot{Q}_i$. He invoked a different heat exchanger equipment constraint and showed that under certain conditions the refrigeration load is maximized when $(UA)_H = (UA)_L$. He showed further that the same results emerge out of the numerical optimization of an actual vapor compression cycle.

Finally, it can be shown that if the UA constraint is replaced by the A constraint defined in Equation (7.67), the refrigeration load is maximized by dividing A according to rule (7.69). Another interesting and new result is that Fig. 7.8 applies unchanged to the refrigeration maximization task considered in this section.

7.5 Strategies for optimizing time-dependent processes

7.5.1 Latent heat storage

The simplest way to illustrate the optimization of the latent-heat storage process (Lim, Bejan, and Kim 1992a) is presented in Fig. 7.10. The hot stream $\dot{m}$ of

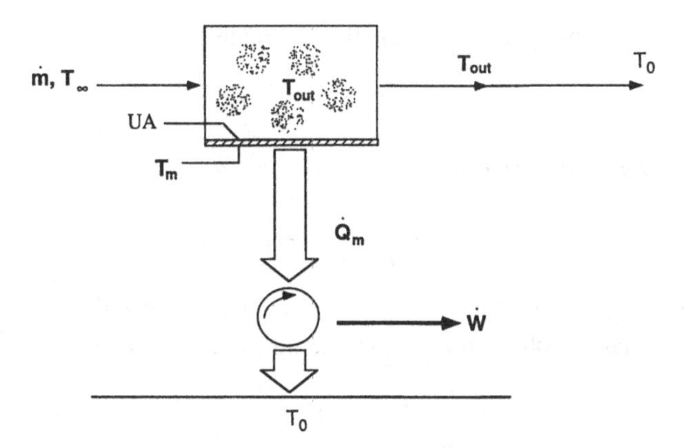

Fig. 7.10. The generation of power using a single melting material and a mixed stream (Lim et al. 1992a).

initial temperature T_∞ comes in thermal contact with the phase-change material through a thermal conductance UA, which is assumed constant. The solid phase-change material is at the melting point, T_m. During the heating and melting of the phase-change material, the stream is well mixed to the temperature T_{out}, which is also the temperature of the stream exhausted into the atmosphere (T_0).

The operation of the installation described in Fig. 7.10 can be visualized as a sequence in which every infinitesimally short storage (melting) stroke is followed by a short energy retrieval (solidification) stroke. During the solidification stroke the flow $\dot{m}$ is stopped, and the recently melted phase-change material is solidified to its original state by the cooling effect provided by the heat engine positioned between T_m and T_0. In this way, the quasi steady model accounts for the complete cycle, that is, storage followed by retrieval.

The cooling effect of the power plant can be expressed in two ways:

$$\dot{Q}_m = UA(T_{\text{out}} - T_m) \tag{7.76}$$

$$\dot{Q}_m = \dot{m}c_P(T_\infty - T_{\text{out}}). \tag{7.77}$$

Eliminating T_{out} between these two equations we obtain

$$\dot{Q}_m = \dot{m}c_P \frac{NTU}{1 + NTU}(T_\infty - T_m), \tag{7.78}$$

in which *NTU* is the number of heat transfer units of the heat exchanger surface:

$$NTU = \frac{UA}{\dot{m}c_P}. \tag{7.79}$$

Of interest here is the maximum rate of exergy, or useful work ($\dot{W}$ in Fig. 7.10), that can be extracted from the phase-change material. To obtain this value, we assume that the power plant is internally reversible:

$$\dot{W} = \dot{Q}_m\left(1 - \frac{T_0}{T_m}\right), \tag{7.80}$$

and after combining with Equation 7.78, we obtain

$$\dot{W} = \dot{m}c_P\frac{NTU}{1+NTU}(T_\infty - T_m)\left(1 - \frac{T_0}{T_m}\right). \tag{7.81}$$

By maximizing $\dot{W}$ with respect to T_m, that is, with respect to the type of phase-change material, we obtain the optimal melting and solidification temperature,

$$T_{m,\text{opt}} = (T_\infty T_0)^{1/2}. \tag{7.82}$$

The maximum power output that corresponds to this optimal choice of phase-change material is

$$\dot{W}_{\max} = \dot{m}c_P T_\infty \frac{NTU}{1+NTU}\left[1 - \left(\frac{T_0}{T_\infty}\right)^{1/2}\right]^2. \tag{7.83}$$

The same results, Equations (7.82) and (7.83), could have been obtained by minimizing the total rate of entropy generation, as pointed out in Section 7.1.

Equation (7.82) was first obtained by Bjurström and Carlsson (1985), who analyzed the heating (melting) portion of the process by using a lumped model. More realistic details of melting and solidification by conduction or natural convection were considered more recently by De Lucia and Bejan (1990, 1991), who showed that $T_{m,\text{opt}}$ is nearly the same as in Equation (7.82).

7.5.2 Sensible heat storage

The corresponding problem of minimizing the entropy generation in a sensible-heat storage process was solved considerably earlier in Bejan (1978b). This instance was the first in which the EGM–FTT method was used to identify an optimal strategy for executing a time-dependent heating process.

The sensible-heat storage process was optimized on the basis of the model shown in Fig. 7.11. The storage system (for example, liquid bath) has the mass M and specific heat c. A hot stream ($\dot{m}c_P$) enters the system through one port, is cooled while flowing through a heat exchanger, and is eventually discharged into the atmosphere. The system temperature T and the fluid temperature T_{out} rise, approaching the hot stream inlet temperature T_∞.

The system M is initially at the ambient temperature T_0. Figure 7.11 shows that the batch heating process is accompanied by two irreversibilities. First, the heat transfer between the hot stream and the cold bath always takes place across a finite temperature difference. Second, the gas stream exhausted into

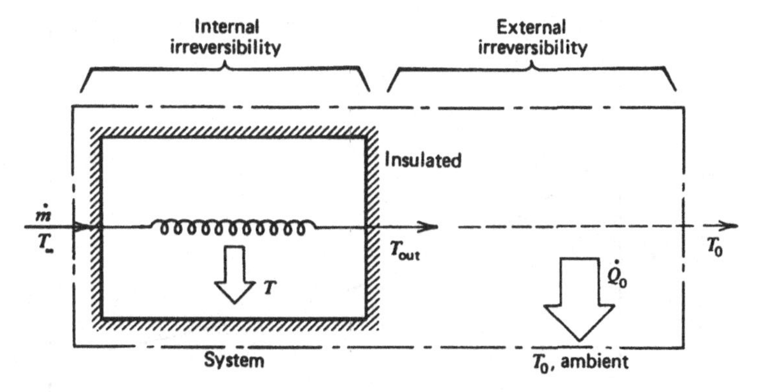

Fig. 7.11. Two sources of irreversibility in the sensible-heat storage process (Bejan 1978b).

the atmosphere is eventually cooled to T_0, again by heat transfer across a finite temperature difference. A third irreversibility source, neglected in the present analysis but considered in Bejan (1978b), is the frictional pressure drop on the gas side of the heat exchanger. The combined effect of these irreversibilities is a basic characteristic of all sensible heat storage systems, namely only a fraction of the exergy brought in by the hot stream is stored in the system M.

It has been shown that the entropy generated from $t = 0$ to t during the storage process reaches a minimum at a certain time (t_{opt}). At times shorter than t_{opt}, the exergy content of the hot stream is destroyed by heat transfer from T_∞ to the initially cold system. In the opposite limit, $t \gg t_{opt}$, the stream leaves M as hot as it enters, and its exergy content is destroyed by direct heat transfer to the atmosphere.

The optimal charging time t_{opt} is reported in Fig. 7.12, where $\theta_{opt} = \dot{m}c_P\, t_{opt}/Mc$, $NTU = UA/\dot{m}c_P$, and UA is the overall thermal conductance between the heating agent and the storage element M. The figure shows that the dimensionless initial temperature difference $(T_\infty - T_0)/T_0$ has a relatively small effect. Furthermore, *NTU* has a negligible effect if it is greater than one, meaning that in a scaling sense $\theta_{opt} \sim 1$, or $t_{opt} \sim Mc/\dot{m}c_P$. In other words, the storage process must be interrupted when the thermal inertia of the hot fluid use ($\dot{m}c_P t$) matches the thermal inertia of the storage material (Mc).

7.5.3 The optimal heating or cooling of a body subject to time constraint

The next time-dependent heat transfer process that was optimized using the EGM–FTT method was the heating or cooling of a body to a prescribed final temperature during a prescribed time interval t_c (Bejan and Schultz 1982). By

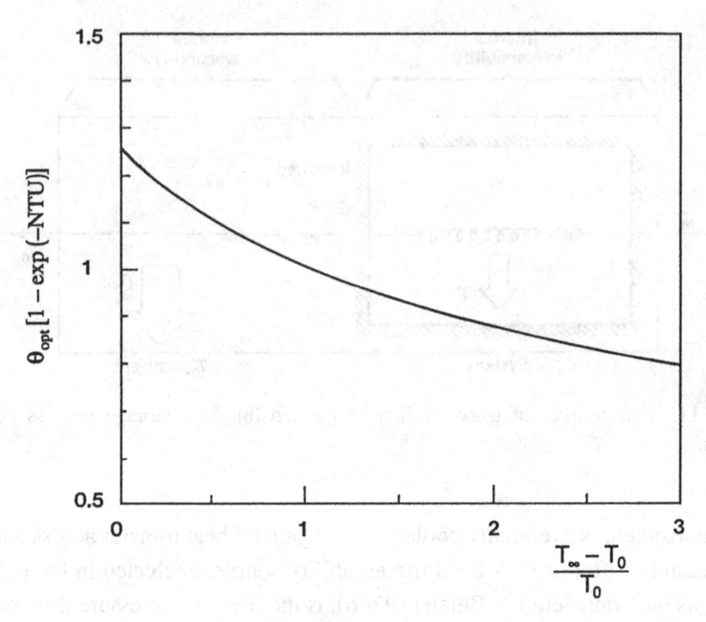

Fig. 7.12. The optimal storage time for sensible heating (Bejan 1978b).

using the cooldown of a body as an example, the total amount of cooling agent (cryogen) used was shown to be minimum when the coolant flow rate $\dot{m}(t)$ varies in a certain optimal way from $t = 0$ to $t = t_c$. Minimizing the total amount of cryogen used is equivalent to minimizing the total refrigerator work input, or the total entropy generated in the cold space during the duration of the process.

Other time-dependent processes optimized by applying the EGM–FTT method are the on-and-off operation of defrosting refrigerators and power plants that experience heat exchanger fouling (Bejan, Vargas, and Lim 1994). Another example is the extraction of exergy from hot dry rock deposits (Lim, Bejan, and Kim 1992b).

7.6 Conclusions

The developments reviewed in this chapter demonstrate that constructing simple but realistic models of processes and devices is possible, by taking into account their inherent irreversibility and combining thermodynamics with heat transfer and fluid mechanics at the earliest stages of modeling. Each individual component or assembly can then be optimized, by minimizing the respective entropy generation. The optima revealed by the entropy generation minimization method mean that similar trade-offs exist (and are worth searching for) in the much more complex systems models that are optimized by design engineers.

References

Andresen, B., Salamon, P. & Berry, R. S. (1984). Thermodynamics in finite time. *Physics Today*, Sept., 62–70.

Bejan, A. (1977). The concept of irreversibility in heat exchanger design: counterflow heat exchangers for gas-to-gas applications. *Journal of Heat Transfer*, **99**, 374–80.

Bejan, A. (1978a). General criterion for rating heat-exchanger performance. *International Journal of Heat and Mass Transfer*, **21**, 655–8.

Bejan, A. (1978b). Two thermodynamic optima in the design of sensible heat units for energy storage. *Journal of Heat Transfer*, **100**, 708–12.

Bejan, A. (1979). A general variational principle for thermal insulation system design. *International Journal of Heat and Mass Transfer*, **22**, 219–28.

Bejan, A. (1980). Second law analysis in heat transfer. *Energy*, **5**, 721–32.

Bejan, A. (1982). *Entropy Generation Through Heat and Fluid Flow*. New York: Wiley.

Bejan, A. (1984a). *Convection Heat Transfer*. New York: Wiley.

Bejan, A. (1984b). Second-law aspects of heat transfer engineering. In *Multi-phase Flow and Heat Transfer III*, ed. T. N. Veziroglu & A. E. Bergles, vol. 1a, pp. 1–22. Amsterdam: Elsevier.

Bejan, A. (1987). The thermodynamic design of heat and mass transfer processes and devices. *International Journal of Heat and Fluid Flow*, **8**, 258–76.

Bejan, A. (1988a). *Advanced Engineering Thermodynamics*. New York: Wiley.

Bejan, A. (1988b). Theory of heat transfer irreversible power plants. *International Journal of Heat and Mass Transfer*, **31**, 1211–9.

Bejan, A. (1989). Theory of heat transfer-irreversible refrigeration plants. *International Journal of Heat and Mass Transfer*, **32**, 1631–9.

Bejan, A. (1993a). *Heat Transfer*. New York: Wiley.

Bejan, A. (1993b). Power and refrigeration plants for minimum heat exchanger size. *Journal Energy Research and Technology*, **115**, 148–50.

Bejan, A. (1994). Power generation and refrigerator models with heat transfer irreversibilities. *Journal of the Heat Transfer Society of Japan*, **33**, 68–75.

Bejan, A. & Paynter, H. M. (1976). *Solved Problems in Thermodynamics*, Department of Mechanical Engineering, Massachusetts Institute of Technology, Cambridge, MA, Problem VII-D.

Bejan, A. & Pfister, P. A., Jr. (1980). Evaluation of heat transfer augmentation techniques based on their impact on entropy generation. *Letters in Heat and Mass Transfer*, **7**, 97–106.

Bejan, A. & Schultz, W. (1982). Optimum flowrate history for cooldown and energy storage processes. *International Journal of Heat and Mass Transfer*, **25**, 1087–92.

Bejan, A. & Smith, J. L., Jr. (1974). Thermodynamic optimization of mechanical supports for cryogenic apparatus. *Cryogenics*, **14**, 158–63.

Bejan, A., Vargas, J. V. C. & Lim, J. S. (1994). When to defrost a refrigerator, and when to remove the scale from the heat exchanger of a power plant. *International Journal of Heat and Mass Transfer*, **37**, 523–32.

Benedetti, P. & Sciubba, E. (1993). Numerical calculation of the local rate of entropy generation in the flow around a heated finned-tube. *ASME HTD*, **266**, 81–91.

Bjurström, H. & Carlsson, B. (1985). An exergy analysis of sensible and latent heat storage. *Heat Recovery Systems*, **5**, 233–50.

Chato, J. C. & Damianides, C. (1986). Second-law–based optimization of heat exchanger networks using load curves. *International Journal of Heat and Mass Transfer*, **29**, 1079–86.

Curzon, F. L. & Ahlborn, B. (1975). Efficiency of a Carnot engine at maximum power output. *American Journal of Physics*, **43**, 22–4.

De Lucia, M. & Bejan, A. (1990). Thermodynamics of energy storage by melting due to conduction or natural convection. *Journal of Solar Energy Engineering*, **112**, 110–6.
De Lucia, M. & Bejan, A. (1991). Thermodynamics of phase-change energy storage: the effects of liquid superheating during melting, and irreversibility during solidification. *Journal of Solar Energy Engineering*, **113**, 3–10.
El-Wakil, M. M. (1962). *Nuclear Power Engineering*, pp. 162–165. New York: McGraw-Hill.
Hesselmann, K. (1984). Optimization of heat exchanger networks. *ASME HTD*. **33**, 95–99.
Klein, S. A. (1992). Design considerations for refrigeration cycles. *International Journal of Refrigeration*. **15**, 181–5.
Kotas, T. J. & Shakir, A. M. (1986). Exergy analysis of a heat transfer process at subenvironmental temperature. *ASME AES*, 2–3, 87–92.
Lim, J. S., Bejan, A. & Kim, J. H. (1992a). Thermodynamic optimization of phase-change energy storage using two or more materials. *Journal of Energy Research and Technology*, **114**, 84–90.
Lim, J. S., Bejan, A. & Kim, J. H. (1992b). Thermodynamics of energy extraction from fractured hot dry rock. *International Journal of Heat and Fluid Flow*, **13**, 71–77.
London, A. L. & Shah, R. K. (1983). Costs of irreversibilities in heat exchanger design. *Heat Transfer Engineering*, **4**, 59–73.
Novikov, I. I. (1958). The efficiency of atomic power stations. *Journal of Nuclear Energy II*, **7**, 125–8 translated from 1957 *Atomnaya Energiya*, **3**, 409.
Ouellette, W. R. & Bejan, A. (1980). Conservation of available work (exergy) by using promoters of swirl flow. *Energy*, **5**, 587–96.
Paoletti, S., Rispoli, F. & Sciubba, E. (1989). Calculation of exergetic losses in compact heat exchanger passages. *ASME AES*, **10**, 21–9.
Poulikakos, D. & Bejan, A. (1982). Fin geometry for minimum entropy generation in forced convection. *Journal of Heat Transfer*, **104**, 616–23.
Sekulic, D. P. & Herman, C. V. (1986). One approach to the irreversibility minimization in compact crossflow heat exchanger design. *International Communications in Heat and Mass Transfer*, **13**, 23–32.
Zubair, S. M., Kadaba, P. V. & Evans, R. B. (1987). Second law based thermoeconomic optimization of two-phase heat exchangers. *Journal of Heat Transfer*, **109**, 287–94.

8
An introduction to thermoeconomics

ANTONIO VALERO AND MIGUEL-ÁNGEL LOZANO
Universidad de Zaragoza

8.1 Why thermoeconomics?

Nicholas Georgescu-Roegen (1971) pointed out in his seminal book, *The Entropy Law and the Economic Process,* that "... the science of thermodynamics began as a physics of economic value and, basically, can still be regarded as such. The Entropy Law itself emerges as the most economic in nature of all natural laws... the economic process and the Entropy Law is only an aspect of a more general fact, namely, that this law is the basis of the economy of life at all levels..."

Might the justification of thermoeconomics be said in better words?

Since Georgescu-Roegen wrote about the entropic nature of the economic process, no significant effort was made until the 1980s to advance and fertilize thermodynamics with ideas taken from economics. At that time most thermodynamicists were polishing theoretical thermodynamics or studying the thermodynamics of irreversible processes.

But the Second Law tells us more than about thermal engines and heat flows at different temperatures. One feels that the most basic questions about life, death, fate, being and nonbeing, and behavior are in some way related to Second Law. Nothing can be done without the irrevocable expenditure of natural resources, and the amount of natural resources needed to produce something is its thermodynamic cost. All the production processes are irreversible, and what we irreversibly do is destroy natural resources. If we can measure this thermodynamic cost by identifying, locating, and quantifying the causes of inefficiencies of real processes, we are giving an objective basis to economics through the concept of cost.

The search for the cost formation process is where physics connects best with economics, and thermoeconomics can be defined as a *general theory of useful energy saving*, where conservation is the cornerstone. Concepts such as thermodynamic cost, purpose, causation, resources, systems, efficiency, structure and cost formation process are the bases of thermoeconomics.

Unlike thermodynamics, thermoeconomics is not closed and finished. It is open for new researchers to improve its bases and extend its applications. As in the way thermodynamics was born, thermoeconomics is now closely related to thermal engineering. Cost accounting, diagnosis, improvement, optimization, and design of energy systems are the main uses for thermoeconomics (Gaggioli and El-Sayed 1987) (Tsatsaronis 1987).

We live in a finite and small world for the people we are and will be, and natural resources are scarce. If we want to survive, we must conserve them. In this endeavor, thermoeconomics plays a key role. We must know the mechanisms by which energy and resources degrade; we must learn to judge which systems work better and systematically improve designs to reduce the consumption of natural resources and we must prevent residues from damaging the environment. These are the reasons for thermoeconomics and its application to engineering energy systems.

8.1.1 The exergetic cost and the process of cost formation

How much exergy is dissipated if we break a glass? Almost none is dissipated, because glass is in a metastable state near thermodynamic equilibrium with the environment. We cannot save useful energy where none exists. However, if a glass is broken, we make useless all the natural resources used for its production. What is important is not the exergy content of the glass but its exergetic cost. Therefore, we will say that the exergetic cost of a functional product is the amount of exergy needed to produce it. And a functional product, according to Le Goff (1979) "*is the product obtained in the energy transformation of its manufacture and defined by the function to which it is destined.*" The set of manufactured objects that allows the manufacturing of other functional products is named a unit or device. And the procedure for fabricating a functional product from a set of functioning units and from other functional products is named a *process or industrial operation*. These processes usually produce residues or by-products.

Knowing the resources sacrificed in making functional products would be a powerful incentive for optimizing processes. First Law analyses discern as losses only the amounts of energy or materials that cross the boundaries of the system. Friction without energy loss, a spontaneous decrease in temperature, or a mixing process are not considered losses. Second Law ascertains losses in energy quality. Combining both laws allows losses in processes to be *quantified* and *localized*. The laws can be combined in many ways. However, production takes materials from the environment and returns products and residues. It is therefore reasonable to analyze exergy, which measures the thermodynamic separation of a product from environmental conditions.

Unfortunately, exergy analysis is necessary but not sufficient to determine the origin of losses. For instance, if the combustion process in a boiler is not well controlled, the volume of air and gases will increase and the fans to disperse them will require additional electricity. The increase in exergy losses from the fans is due to a malfunction of the boiler and not to the fans themselves. Quite commonly, irreversibilities hide costs. Therefore, exergy balances allow localization of losses, but processes and outcomes must also be analyzed. We will term these causality chains *processes of cost formation*, and their study – an additional step to the conventional exergy analysis – we will term *exergetic cost accounting*.

What is important is not the exergy, B in (kilowatts), that the functional products may contain but the exergetic cost, B^*, that is the exergy plus all the accumulated irreversibilities needed to get those products.

8.2 Definitions and concepts

To illustrate the different concepts, we will use a simple example of a thermal system – a cogeneration plant named the **CGAM** Problem. In 1990, a group of concerned specialists in the field of thermoeconomics (**C**. Frangopoulos, **G**. Tsatsaronis, **A**. Valero, **M**. von Spakovsky, and coworkers) decided to compare their methodologies by solving a simple problem of optimization. The objective of the CGAM Problem was to show how the methodologies are applied, what concepts are used, and what numbers are obtained. Details of the results can be found in the literature (Valero et al. 1994) (Tsatsaronis and Pisa 1994) (Frangopoulos 1994) (von Spakovsky 1994).

Industrial installations have a defined aim, to produce one or several *products*. The quantity of resources is identified through mass or energy flows, which are known as *fuel* (Tsatsaronis and Winhold 1985). Each of the components of the plant also has a well-defined objective characterized by its fuel and its product, as Fig. 8.1 shows for the CGAM plant.

To carry out a thermoeconomic analysis of a system, it is necessary to identify its flows with a magnitude sensitive to the changes in quality and quantity of the energy processed. Exergy is an adequate magnitude because it expresses the thermodynamic separation of the intensive properties characterizing the flow (P_j, T_j, μ_j) with respect to those of the environment (P_0, T_0, $\mu_{j,00}$) (Valero and Guallar 1991). The thermoeconomic (exergoeconomic) methodology developed in this book is based on this fact. However, other magnitudes are

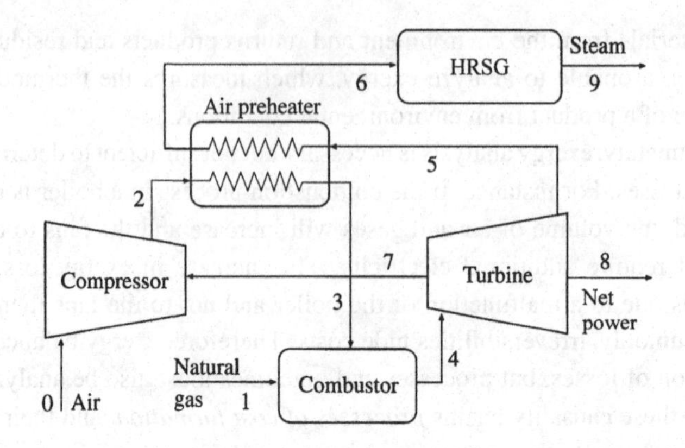

i	Component	Efficiency Fuel (F)	Product (P)
1	Combustor	1	4 – 3
2	Compressor	7	2
3	Turbine	4 – 5	7 + 8
4	Air preheater	5 – 6	3 – 2
5	HRSG	6	9
	CGAM system	1	8 + 9

Fig. 8.1. Physical and productive structures of the CGAM system.

possible. In fact, this problem is currently under study (Valero and Lozano 1992b) (Valero and Royo 1992c) (Royo 1994).

Accordingly, it can be said that *fuel* (F) is the exergy provided for the subsystem through the resources and *product* (P) is the exergy that contains the benefits obtained. Thus, for the CGAM plant, the exergy of process steam and the net power are the products, and the exergy provided by the natural gas is the fuel.

An example of the application of these concepts is the combustion chamber. The aim is to increase the exergy of the air flow that exits from the compressor. The product is, therefore, the difference of exergy between flows 4 and 3 ($P_1 = B_4 - B_3$). The exergy of natural gas ($F_1 = B_1$) is consumed as fuel.

In the case of the turbine, the aim is to obtain mechanical energy; therefore, the product is the exergy employed to drive the compressor and the net power of the plant ($P_3 = B_7 + B_8$). The exergy ($F_3 = B_4 - B_5$) provided by the gas expanded in the turbine is consumed as fuel.

Note that the product of the combustion chamber ($P_1 = B_4 - B_3$), like the fuel of the turbine ($F_3 = B_4 - B_5$), is formed by the flows entering and leaving the subsystems. That is to say, the fuel does not consist exclusively of flows entering the system, nor do the products consist exclusively of flows leaving the system.

We define *losses* (L) as those flows that leave the unit and the plant, are not subsequently used, and do not require a special treatment. When these flows leave the unit, exergy dissipates into the environment. If we suitably enlarge the limits of the unit, these external irreversibilities become internal. We will call the *irreversibility* (I) of the unit "i" the sum of internal exergy destructions plus losses occurring in it, $I_i = L_i + D_i$.

We will call *productive units* those whose objective is to transfer the exergy contained in the fuels to the products. The fuel-product definitions for productive units should be chosen such that the equation $F_i - P_i = I_i$ is an expression of each exergy balance. The *exergy efficiency* of these units is defined as η = exergy in useful products/exergy supplied in fuels = P/F. The inverse is *unit exergy consumption*, $k = F/P$. Using exergy to define F and P guarantees that in any real process, $F - P = I > 0$, $0 < \eta < 1$, and $k > 1$.

Finally, from a formal point of view a *system* can be considered as a complex entity made up of a collection of components and of the relationship existing between them and their environment. Thus, an energy system, such as the plant in the CGAM Problem can be represented as a collection of components interrelated through the mass and energy flows.

The *disaggregation level* is interpreted as the subsystems that compose the total system. Depending on the depth and type of analysis carried out, each subsystem can be a part of a component, the component itself, or a group of components. The same can be said for the interacting energy flows.

The disaggregation level provides a breakdown of the total irreversibility among the plant components. The chosen disaggregation level will affect the conclusions of the analyses. In fact, if we do not have more information about the system than that defined by its disaggregation level, we cannot demand from the obtained set of costs more information than we have introduced. Conversely, the analyst, not the theory, should be required to disaggregate the plant, looking for cause until the information can be used effectively.

The CGAM Problem refers to a cogeneration plant that delivers thirty megawatts of electricity and fourteen kilograms per second of saturated steam at 20 bar. Figure 8.1 shows a convenient disaggregation level just for presenting ideas. The flows have been numbered as follows: (i) The flow of air into the compressor has been eliminated because its energy and exergy are zero. (ii) In the HRSG, the flow corresponding to the outlet gases has been removed because its exergy is not used later and the stream is exhausted into the atmosphere. The same reason applies to combustor heat losses. (iii) We consider a flow of process steam (flow 9) with an exergy value (B_9) equal to the difference of exergy between the flow of steam produced and the flow of feed water entering the HRSG.

Any system, no matter how complex, with n components and m flows, can be represented by an *incidence matrix*, $\mathbf{A}$ $(n \times m)$, relating them. For the CGAM system the incidence matrix takes the form

$$\mathbf{A} = \begin{bmatrix} 1 & 0 & 1 & -1 & 0 & 0 & 0 & 0 & 0 \\ 0 & -1 & 0 & 0 & 0 & 0 & 1 & 0 & 0 \\ 0 & 0 & 0 & 1 & -1 & 0 & -1 & -1 & 0 \\ 0 & 1 & -1 & 0 & 1 & -1 & 0 & 0 & 0 \\ 0 & 0 & 0 & 0 & 0 & 1 & 0 & 0 & -1 \end{bmatrix}. \tag{8.1}$$

Its a_{ij} elements take the value 1 if the flow j enters the component i; -1 if the flow j leaves the component i, and 0 if no direct physical relation exists between them.

In the case of steady-state operation, the incidence matrix allows us to express the mass, energy, and exergy balances as follows: $\mathbf{AM} = \mathbf{0}, \mathbf{AH} = \mathbf{0}$, and $\mathbf{AB} = \mathbf{I}$, where $\mathbf{M}, \mathbf{H}$, and $\mathbf{B}$ are vectors $(m \times 1)$ containing the mass, energy, and exergy of the flows. When the element j of the vectors $\mathbf{M}, \mathbf{H}$, and $\mathbf{B}$ correspond to a mass flow, M_j represents the mass, H_j the enthalpy $m_j(h_j - h_{j,00})$, and B_j the exergy $m_j(h_j - T_0 s_j - \mu_{j,00})$ of this flow. If element j corresponds to a heat or work flow, then M_j is 0, H_j is Q_j or W_j, and B_j is $Q_j(1 - T_0/T_j)$ or W_j. Element I_i of the column vector $\mathbf{I}$ $(n \times 1)$ represents the exergy destroyed in the components because of internal and external irreversibilities.

Using Fig. 8.1, the *fuel* and *product matrices* $\mathbf{A}_F$ $(n \times m)$ and $\mathbf{A}_P$ $(n \times m)$, may be written as

$$\mathbf{A}_F = \begin{bmatrix} 1 & 0 & 0 & 0 & 0 & 0 & 0 & 0 & 0 \\ 0 & 0 & 0 & 0 & 0 & 0 & 1 & 0 & 0 \\ 0 & 0 & 0 & 1 & -1 & 0 & 0 & 0 & 0 \\ 0 & 0 & 0 & 0 & 1 & -1 & 0 & 0 & 0 \\ 0 & 0 & 0 & 0 & 0 & 1 & 0 & 0 & 0 \end{bmatrix},$$

$$\mathbf{A}_P = \begin{bmatrix} 0 & 0 & -1 & 1 & 0 & 0 & 0 & 0 & 0 \\ 0 & 1 & 0 & 0 & 0 & 0 & 0 & 0 & 0 \\ 0 & 0 & 0 & 0 & 0 & 0 & 1 & 1 & 0 \\ 0 & -1 & 1 & 0 & 0 & 0 & 0 & 0 & 0 \\ 0 & 0 & 0 & 0 & 0 & 0 & 0 & 0 & 1 \end{bmatrix}. \tag{8.2}$$

So $\mathbf{A}_F - \mathbf{A}_P = \mathbf{A}$, $\mathbf{F} = \mathbf{A}_F\mathbf{B}$, and $\mathbf{P} = \mathbf{A}_P\mathbf{B}$, where $\mathbf{F}$ and $\mathbf{P}$ are vectors $(n \times 1)$ containing the exergy of the fuel and product of the components of the plant.

8.3 Cost accounting and the theory of exergetic cost

According to the management theory, cost accounting is an economic task for recording, measuring, and reporting how much things cost. Companies and individuals tend to optimize costs because cost is a loss of resources, and problems generally appear when appropriate insight of costs and their causes are lacking. Business managers use cost data for decision making and performance evaluation and control. They have techniques for costing products and services and use differential costs for estimating how costs will differ among the alternatives. Managerial cost accounting became a profession many years ago, and almost every organization uses it.

Energy cost accounting is in addition to a managerial technique for keeping low the use of energy resources, provides a rationale for assessing the cost of products in terms of natural resources and their impact on the environment and helps to optimize and synthesize complex energy systems. Since 1985 we have been developing the theory of exergetic cost (ECT) and its applications to answer these problems. A simplified description of the theory is now presented.

In any energy system the exergy of the resources is greater than or equal to that of the products. For the plant as a whole as well as for any unit, $F - P = I \geq 0$. The amount of exergy needed to obtain the products is equal to the exergy of the resources consumed. This idea permits the introduction of a thermodynamic function called *exergetic cost* that is defined as follows: *given a system whose limits, disaggregation level, and production aim of the subsystems have been defined, we call exergetic cost, B^*, of a physical flow the amount of exergy needed to produce this flow.* We call *unit exergetic cost* of a flow the exergetic cost per unit exergy, $k_i^* = B_i^*/B_i$. B^*, like B, is a thermodynamic function, and its definition is closely related to others that are common in literature, such as materials' energy content, embodied energy, and cumulative exergy consumption (Szargut, Morris, and Stewart 1988).

The cost of a flow is an *emergent* property, that is, it does not exist as a separate thermodynamic property of the flow. Cost is always linked to the production process. And this process links a set of internal and external flows. Therefore, not only the cost of a flow but also a complete set of interrelated costs need to be determined. On the other hand, the classification of flows as internal or external depends on the system limits as for the case of subsystems or the system itself.

8.3.1 Determination of exergetic costs

The fundamental problem of cost allocation (Valero, Lozano, and Muñoz 1986a, b,c), (Valero, Torres, and Serra 1992a), (Valero and Lozano 1992b) (Valero, Serra, and Lozano 1993) can be formulated as follows: *Given a system whose limits have been defined and a disaggregation level that specifies the subsystems that constitute it, how do we obtain the costs of all the flows that become interrelated in this structure*? An initial procedure to solve it can be based on the next four propositions.

P1 The exergetic cost is a conservative property. For each component of a system the sum of the exergetic costs of the inlet flows is equal to the sum of the exergetic costs of the exiting flows. In matrix form, and in the absence of external assessment, the exergetic cost balance for all compounds of the plant is $\mathbf{AB}^* = \mathbf{0}$, where $\mathbf{B}^*$ is a vector $(m \times 1)$ that contains the exergetic cost of the flows. This equation provides as many equations for calculating the exergetic costs as the number of components in the installation.

P2 The exergetic cost is relative to the resource flows. In the absence of external assessment, the exergetic cost of the flows entering the plant equals their exergy. In other words, the unit exergetic cost of resources is one. As many equations can be formulated $(B_j^* = B_j)$ as flows entering the plant.

We now reconsider the problem of cost allocation. If one wants to calculate the cost of each of the m flows relevant to the disaggregation level considered for analysis, it will be necessary to write m independent equations. If all units have only one output flow that is not classified as a loss flow, then the problem is solved by applying the stated propositions. In this case, we say that the system or process analyzed is sequential. In the opposite case, additional equations must be written for each unit equal to the number of output flows that are not loss flows minus one. At this point we need to use exergy to reasonably allocate costs, because this property enables us to compare the equivalence of the flows according to the principles of thermodynamics. Note also that in any structure the number of bifurcations x equals the number of flows minus the number of units and the number of resource flows $(x = m - n - e)$, allowing us to associate the problem of cost allocation to bifurcations. The additional propositions are as follows:

P3 If an output flow of a unit is a part of the fuel of this unit (nonexhausted fuel), the unit exergetic cost is the same as that of the input flow from which the output flow comes.

P4 If a unit has a product composed of several flows with the same thermodynamic quality, then the same unit exergetic cost will be assigned to all of them. Even if two or more products can be identified in the same unit, their formation process is the same, and therefore we assign them a cost proportional to the exergy they have.

Valero, Torres, and Serra (1992a) proved mathematically the validity of these propositions and provided the way for obtaining new ones in complex cases. However, in many cases the application of these propositions is a matter of disaggregation until we can recognize units with products of the same equality and exit flows identified as nonexhausted fuels. Tsatsaronis in Bejan, Tsatsaronis and Moran (1996) provides additional costing considerations for more detailed cases.

We now consider the case of the CGAM system in Fig. 8.1. The plant shows $n = 5$ units and $m = 9$ flows. We assume that the exergy of these flows is known. Propositions **P1–P4** offer a rational procedure for determining the exergetic costs of the m flows of the system. The proposition **P1** says that a balance of exergetic costs can be established for each unit; therefore, n equations are available, as many as components. These will be five for the CGAM system. The proposition **P2** says that the exergetic costs of the e flows that enter the plant coincide with their exergy, thus providing e equations. The fuel flow to the combustion chamber (flow 1) gives the equation $B_1^* = B_1 \equiv \omega_1$ (or $k_1^* = 1$) (ω_1 denotes a given datum). Note that the air entering the compressor should provide another equation. However, the exergy of air is zero, and the air is taken from the environment at no cost.

The propositions **P3** and **P4** yield as many cost equations as the number of bifurcations in the plant ($x = m - n - e$). Three bifurcations are in the CGAM system.

In the case of the turbine, two bifurcations appear, one corresponding to the exiting stream whose exergy has not taken part in the process and corresponds to a nonexhausted fuel. Consequently, proposition **P3** must be applied to this case, resulting in the following equation: $B_5^*/B_5 = B_4^*/B_4$ or $k_5^* = k_4$. Alternatively, if we define $x_{3F} \equiv B_5/B_4$ as the bifurcation exergy ratio corresponding to the fuel (4–5) of the turbine, then $-x_{3F}B_4^* + B_5^* = 0$.

The case of the air preheater is similar. The outlet gas flow 6 can be considered a nonexhausted fuel whose exergy has the same formation process as the input flow 5. Consequently, proposition **P3** yields $B_6^*/B_6 = B_5^*/B_5$ or $k_6^* = k_5^*$. Alternatively, we define $x_4 \equiv B_6/B_5$ as the bifurcation exergy ratio corresponding to the fuel (5–6) of the air preheater, then $-x_4B_5^* + B_6^* = 0$.

The other bifurcation of the turbine corresponds to the output flows 7 and 8 and constitutes its product; therefore, we will apply proposition **P4**, which results in the following equation: $B_8^*/B_8 = B_7^*/B_7$ or $k_8^* = k_7^*$. Alternatively, we define $x_{3P} \equiv B_8/B_7$ as the bifurcation exergy ratio corresponding to the product (7 + 8) of the turbine; then $-x_{3P}B_7^* + B_8^* = 0$.

Now we have the m equations required. A matrix form can be written for the CGAM system as follows:

$$\begin{bmatrix} \\ \\ \mathbf{A}\rightarrow \\ \\ \\ \cdots \\ \alpha_e \rightarrow \\ \cdots \\ \\ \alpha_x \rightarrow \\ \\ \end{bmatrix}\begin{bmatrix} 1 & 0 & 1 & -1 & 0 & 0 & 0 & 0 & 0 \\ 0 & -1 & 0 & 0 & 0 & 0 & 1 & 0 & 0 \\ 0 & 0 & 0 & 1 & -1 & 0 & -1 & -1 & 0 \\ 0 & 1 & -1 & 0 & 1 & -1 & 0 & 0 & 0 \\ 0 & 0 & 0 & 0 & 0 & 1 & 0 & 0 & -1 \\ \cdots & \cdots & \cdots & \cdots & \cdots & \cdots & \cdots & \cdots & \cdots \\ 1 & 0 & 0 & 0 & 0 & 0 & 0 & 0 & 0 \\ \cdots & \cdots & \cdots & \cdots & \cdots & \cdots & \cdots & \cdots & \cdots \\ 0 & 0 & 0 & -x_{3F} & 1 & 0 & 0 & 0 & 0 \\ 0 & 0 & 0 & 0 & -x_4 & 1 & 0 & 0 & 0 \\ 0 & 0 & 0 & 0 & 0 & 0 & -x_{3P} & 1 & 0 \end{bmatrix}\begin{bmatrix} B_1^* \\ B_2^* \\ B_3^* \\ B_4^* \\ B_5^* \\ \\ B_6^* \\ \\ B_7^* \\ B_8^* \\ B_9^* \end{bmatrix} = \begin{bmatrix} 0 \\ 0 \\ 0 \\ 0 \\ 0 \\ \\ \omega_1 \\ \\ 0 \\ 0 \\ 0 \end{bmatrix} \tag{8.3}$$

or in general form:

$$\mathbb{A}\mathbf{B}^* = \mathbb{Y}_e^*$$

where

$$\mathbb{Y}_e^* = \begin{pmatrix} -\mathbf{Y}^* \\ ---- \\ \omega_e \\ ---- \\ \mathbf{0} \end{pmatrix} \quad \text{and} \quad \mathbb{A} = \begin{pmatrix} \mathbf{A} \\ ---- \\ \alpha_e \\ ---- \\ \alpha_x \end{pmatrix}$$

where $\mathbb{A}$ is the *cost matrix* $(m \times m)$ composed for n rows corresponding to the incidence matrix $\mathbf{A}$ $(n \times m)$ of the plant, and $m - n$ rows corresponding to the *production matrix* α $[(m - n \times m]$. $\mathbb{Y}_e^*$ $(m \times 1)$ is the *vector of external assessment* composed of n elements as a result of proposition **P1** ($\mathbf{Y}^* = \mathbf{0}$); e elements with the actual values of the exergies (ω_e) corresponding to the resources of the plant (proposition **P2**); and $m - n - e$ null elements corresponding to the fuel and product bifurcations (propositions **P3** and **P4**).

Inverting matrix $\mathbb{A}$ allows for the exergetic costs to be obtained through $\mathbf{B}^* = \mathbb{A}^{-1}\mathbb{Y}_e^*$, as well as for the exergetic cost of the fuel and the product for each component to be obtained through $\mathbf{F}^* = \mathbf{A}_F\mathbf{B}^*$ and $\mathbf{P}^* = \mathbf{A}_P\mathbf{B}^*$ and the unit exergetic costs to be obtained for all of them.

$$k_i^* = \frac{B_i^*}{B_i}, \text{ with } i = 1 \text{ to } m$$
$$k_{F,j}^* = \frac{F_j^*}{F_j}, \quad k_{P,j}^* = \frac{P_j^*}{P_j}, \text{ with } j = 1 \text{ to } n. \tag{8.4}$$

Note that if $\mathbf{A} = \mathbf{A}_F - \mathbf{A}_P$ and $\mathbf{F} = \mathbf{K}_D\mathbf{P}$ where $\mathbf{K}_D$ is the diagonal matrix $(n \times n)$ whose elements are the unit exergy consumptions of the n components in the plant, the following also holds: $\mathbf{F}^* = \mathbf{P}^*$.

8.3.2 Exergoeconomic cost

Calculating the monetary cost of the internal flows and final products in thermal or chemical plants is a problem of the utmost importance, as the monetary cost is directly linked to the production costs of the different components of the productive process. For these plants the formation of the economic cost of the internal flows and final products is related to both the thermodynamic efficiency of the process and to the depreciation and maintenance cost of the units. Therefore, one can define the exergoeconomic cost of a flow as the combination of two contributions: the first comes from the monetary cost of the exergy entering the plant needed to produce this flow, that is its exergetic cost, and the second covers the rest of the cost generated in the productive process associated with the achievement of the flow (capital, maintenance, etc.).

Accordingly, we call *exergoeconomic cost* of a flow the quantity of resources, assessed in monetary units, needed to obtain this flow, and we denote it by **C**. In the same way we call exergoeconomic cost of the fuel (product) the economic resources necessary to obtain the fuel (product) of the component, and we denote it by $\mathbf{C}_F(\mathbf{C}_P)$. If Z_i is the levelized cost of acquisition, depreciation, maintenance, etc., of the component l in dollars per second the exergoeconomic cost balance for this component can be written as $C_{F,i} + Z_i = C_{P,i}$. In matrix form this equation will be $\mathbf{C}_F + \mathbf{Z} = \mathbf{C}_P$, where vector $\mathbf{Z}$ $(n \times 1)$ contains the levelized acquisition cost of the plant components and $\mathbf{C}_F$ $(n \times 1)$ and $\mathbf{C}_P$ $(n \times 1)$ are the vectors containing the exergoeconomic costs of the fuel and products of the components. Considering the previous equation for all units of the plant we obtain the following system of equations: $\mathbf{AC} = -\mathbf{Z}$, where the unknown quantities C_j are the exergoeconomic costs of the n flows. As for the exergetic costs $(m - n)$, auxiliary equations are required to find the exergoeconomic costs of the flows. The auxiliary equations can be formulated using propositions **P2** through **P4**, namely $\alpha_e\mathbf{C} = \mathbf{C}_e$ (e equations) and $\alpha_x\mathbf{C} = 0$ $(m-n-e$ equations), where the vector $\mathbf{C}_e$ now represents the monetary cost of the flows entering the plant, $C_{e,i} = c_{\omega,i}\,\omega_i$, and $c_{\omega,i}$ is the market price of resource i per unit of exergy.

We conclude that the mathematical problem of calculating exergoeconomic costs of the flows of a plant requires solving the system of equations: $\mathbb{A}\mathbf{C} = \mathbb{Z}$ (m equations with m unknowns), where $\mathbb{Z} = {}^{t}[-\mathbf{Z} \vdots \mathbf{C}_e \vdots \mathbf{0}]$ is the vector that contains the external economic assessments, and $\mathbb{A}$ is the costs matrix used for calculating exergetic costs.

Also note that $\mathbf{C}_F = \mathbf{A}_F\mathbf{C}$ and $\mathbf{C}_P = \mathbf{A}_P\mathbf{C}$ because $\mathbf{AC} = (\mathbf{A}_F - \mathbf{A}_P)\mathbf{C} = \mathbf{C}_F - \mathbf{C}_P = -\mathbf{Z}$. Otherwise, the *unit exergoeconomic cost* of a flow is defined as $c_i = C_i/B_i$, and the *unit exergoeconomic cost* of the *fuel* and *product* is defined as $c_{F,i} = C_{F,i}/F_i$ and $c_{P,i} = C_{P,i}/P_i$.

The uniqueness of the matrix of costs $\mathbb{A}$, when applied to the calculation of the exergetic and exergoeconomic costs, reflects the fact that passing from the former to the latter simply involves modifying the units in which the production factors are expressed (kilojoules or dollars).

8.3.3 Results

Tables 8.1 to 8.3 show the results obtained for the optimal design of the CGAM problem. The description of the physical and economic models and the thermoeconomic methodology applied for its optimization has been presented in a dedicated monograph of *Energy* (Tsatsaronis 1994). Tables 8.1 and 8.2 show that the exergetic and exergoeconomic costs of flows and components can be easily obtained by applying the procedures described in the previous section. Observe that the most irreversible components (combustor and HRSG) cause the higher increase of unit cost between fuel and product. One can easily prove

Table 8.1. *Thermodynamic and thermoeconomic variables of the flows*

Flow	T (K)	P (bar)	B (kW)	k^*(GJ/GJ)	c ($/GJ)
1 Natural gas	298.15	1.013	84 383	1.000	3.857
2 Air	595.51	8.634	27 322	1.844	8.161
3	914.28	8.202	45 954	1.836	8.017
4	1492.63	7.792	103 729	1.627	6.699
5 Gases	987.90	1.099	41 454	1.627	6.699
6	718.76	1.066	20 572	1.627	6.699
7 Power	400.26	1.013	29 692	1.697	7.205
8 Net power			30 000	1.697	7.205
9 Steam		20.	12 748	2.625	11.442

Table 8.2. *Thermodynamic variables and cost of the components*

Component	F (kW)	P (kW)	I (kW)	η	k	r	$Z(10^{-3}\$/s)$	ω (kW)
Combustor	84 383	57 775	26 608	0.685	1.461	0.443	0.98	
Compressor	29 692	27 323	2 369	0.920	1.087		9.03	
Turbine	62 275	59 692	2 583	0.958	1.043		12.91	30 000
Air preheater	20 882	18 631	2 251	0.892	1.121	0.595	5.55	
HRSG	20 572	12 748	7 824	0.620	1.614		8.05	12 748
CGAM system	84 383	42 748	41 635	0.507	1.974		36.52	

Table 8.3. *Unit costs*

	GJ/GJ		$/GJ	
Component	k_F^*	k_P^*	c_F	c_P
Combustor	1.000	1.460	3.857	5.651
Compressor	1.697	1.844	7.205	8.161
Turbine	1.627	1.697	6.699	7.205
Air preheater	1.627	1.824	6.699	7.806
HRSG	1.627	2.625	6.699	11.442

that this difference is given by

$$k_P^* - k_P^* = k_F^* \frac{I}{P}; \qquad c_P - c_F = c_F \frac{I}{P} + \frac{Z}{P} \tag{8.5}$$

These equations reveal that the unit cost of the product is always greater than or equal to that of the fuel. As a consequence the unit costs steadily increase as the production process goes on.

8.3.4 External assessment and additional concepts

Up to now, we have considered the system or plant thermodynamically, without allowing for the physical or economic relationships with other systems or plants. The effects of these relations on costs can be introduced into the analysis by suitably modifying the external assessment vector. In any case, the matrix of costs will remain unaltered. Some important cases are the following:

8.3.4.1 Exergetic amortization

In the balance of exergetic costs, we have not considered the fact that the units that form an installation are functional products and therefore have their own exergetic cost. To keep them in good operation, additional exergy will be required. After determining the exergetic costs of the units, it will be necessary to distribute these costs over the total working lives. In this way, it is possible to obtain with conventional methods a vector of dimension n that corresponds to the exergetic amortization of the units. In a parallel manner, the vector of exergetic maintenance will be obtained. By defining the vector sum of both as $\mathbf{Y}^* = \mathbf{Y}_A^* + \mathbf{Y}_M^*$, it is possible to reformulate the balances of the exergetic cost of the installation (**P1** proposition) in general as $\mathbf{AB}^* = -\mathbf{Y}^*$.

8.3.4.2 Residues

Residual flows require an additional expense of resources for disposal. An example is the flows of slag and fly ash in a coal boiler, which require power-operated units without which the plant could not work. Thus, removing these flows from a coal boiler entails an exergetic cost that equals R^*. Parallel to the process of product formation, there exists a process of residue formation. Residues obviously are not formed in the last device they pass through. For instance, the cost of stack gases in a boiler should not be allocated to the stack but to the combustion chamber. For a proper allocation of the costs of residues we must follow their formation process and assign the cost of their disposal to the unit in which they are formed. The residual structure has its own units-and-flows representation, but the connecting arrows go in the direction opposite to the physical flows to make explicit their negative cost character.

8.3.4.3 Assessment of the plant fuels

The fuel flows consumed by an industrial installation are rarely composed of nontransformed primary resources (fuels, metals, geothermal deposits, etc.) whose values are represented by their thermodynamic disequilibrium with the reference environment, that is to say, from their exergy. Thus, the coal processed by a boiler has an exergetic cost of primary resources V^*, which are higher than the coal's exergy because of different processes: extraction, storage, transport, etc. If we want to incorporate their contribution to the exergetic costs of the flows and products of the plant into our analysis, we must apply the proposition **P2** to coal flow in the following form: $\omega = V^*$.

8.3.4.4 Cumulative exergetic cost or ecological cost (Szargut 1988)

The vector $\mathbb{Y}^*$ incorporates the external information that finally determines the exergetic costs of an installation. In a conventional assessment, we distinguish between the thermodynamic system that constitutes the installation and its thermodynamic environment, ignoring completely the irreversibilities that take place there and that form part of the process using the primary resources to generate the final products. As these assessments are being incorporated, the exergetic costs will include a greater part of the external irreversibilities. The latter have their origin in the manufacture, installation, repair, and maintenance of the units Y^*, in the elimination of residues R^*, and in the previous production of the flows entering the plant V^*. Logically, the most appropriate external assessment will depend on the aim of the analysis that is carried out. The natural assessment of vector $\mathbb{Y}^*$ consists of considering each and every one of the external irreversibilities, and, therefore, Y^*, R^*, and V^* reflect their costs of primary resources. This type of analysis is of foremost importance when considering the ecological cost of products used by our society, and a systematic accounting of each and every ecological cost of products would lead an answer to the viability of our technology and the sustainability of our society (Valero 1993) (Valero 1994a).

8.4 Symbolic exergoeconomics

Suppose we have a conventional Rankine cycle as in Fig. 8.2. What can we do if we would have its global efficiency as a function of the efficiencies of its components? This question poses some previous questions: Does such a formula exist? and if it does exist, how do we assess it in a general way? What conditions must our methods fulfil to assess it? If such a formula exists, we could compute structural influences. We would be able to see how a variation in the efficiency of a unit affects the whole behavior of the plant. Perhaps, new optimization methodologies would arise based on having explicit dependencies. In other words, this type of analysis would open new ways for the study of energy systems such as cost analyses and perturbations, optimization, simulation, and synthesis.

Then, the formula of global efficiency of the plant in Fig. 8.2 is

$$\begin{aligned}\eta_T = \{&(1-y_1)(\eta_3 y_3 + \eta_4 y_4 + \eta_5 y_5)\eta_1 \\ &+ y_1 y_2[(RG - \eta_{10} y_{10})\eta_3 y_3 + (1-\eta_{10} y_{10})(\eta_4 y_4 + \eta_5 y_5)]\eta_2\}/(1-RG),\end{aligned} \tag{8.6}$$

where terms y_i and RG are defined in Table 8.4 and are a function of the efficiencies and bifurcations $\{\eta_i, x_i\}$ of the plant (Torres, Valero, and Cortés 1989). The formula is easy to obtain from the concepts of symbolic exergoeconomics, as we will see in this chapter.

8.4.1 Symbolic computation and exergoeconomics

Analytical or symbolic computation is nowadays a well-known technique. Programs such as Mathematica, Macsyma, Reduce, and Maple can be used to

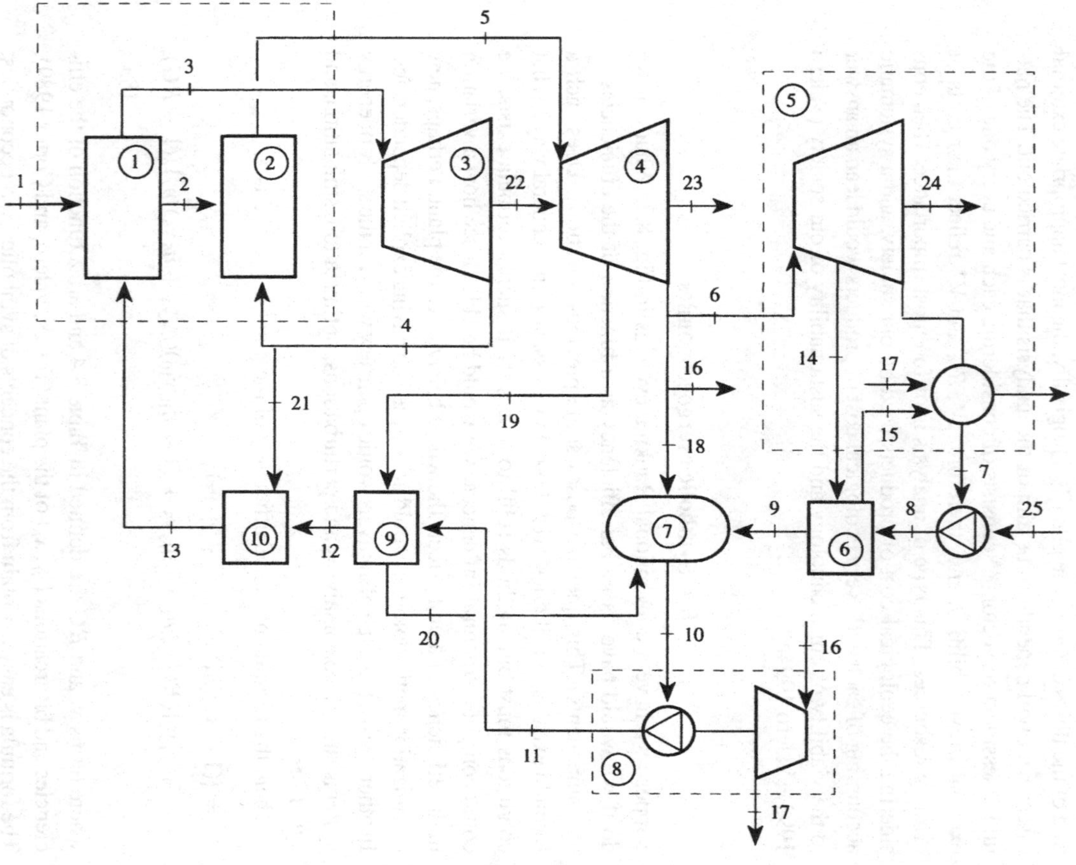

Fig. 8.2. Rankine cycle scheme.

Table 8.4. *Definition of subsystem efficiency and parameters of bifurcation*

Subsystem	Subsystem Efficiency η_i	Parameters of Bifurcation	⟨FP⟩ Coefficients y_i
1 Boiler	$\dfrac{B_3 - B_{13}}{B_1 - B_2}$	$x_1 = \dfrac{B_2}{B_1}$	x_1
2 Reheater	$\dfrac{B_3 - B_6}{B_2}$		$\dfrac{1}{x_2}$
3 HP turbine	$\dfrac{B_{22}}{B_3 - B_4 - B_{21}}$	$x_2 = \dfrac{B_4}{B_3}$ $x_3 = \dfrac{B_{21}}{B_3}$	$1 - x_2 - x_3$
4 MP turbine	$\dfrac{B_{23}}{B_5 - B_6 - B_{18} - B_{19} - B_{16}}$	$x_4 = \dfrac{B_6}{B_5}$ $x_5 = \dfrac{B_6}{B_5}$ $x_6 = \dfrac{B_{18}}{B_5}$ $x_7 = \dfrac{B_{19}}{B_5}$	$x_2(1 - x_4 - x_5 - x_6 - x_7)$
5 LP turbine	$\dfrac{B_{24}}{(B_6 - B_7 - B_{14}) + B_{15} + B_{17}}$	$x_8 = \dfrac{B_7}{B_6}$ $x_9 = \dfrac{B_{14}}{B_6}$	$x_2x_4(1 - x_8 - x_9 - x_9x_{10}) + x_2x_5x_{11}$
6 Heater 1 (LP)	$\dfrac{B_9 - B_8}{B_{14} - B_{15}}$	$x_{10} = \dfrac{B_{15}}{B_{14}}$	$x_2x_4x_9(1 - x_{10})$
7 Deaereator	$\dfrac{B_{10}}{B_9 + B_8 + B_{20}}$		$x_2(x_4x_8 + x_6 + x_7x_{12})$
8 Pump turbine driver	$\dfrac{B_{11} - B_{10}}{B_{16} - B_{17}}$	$x_{11} = \dfrac{B_{17}}{B_{16}}$	$x_2x_5(1 - x_{11})$
9 Heater 2 (MP)	$\dfrac{B_{12} - B_{11}}{B_{19} - B_{20}}$	$x_{12} = \dfrac{B_{20}}{B_{19}}$	$x_2x_7(1 - x_{12})$
10 Heater 3 (LP)	$\dfrac{B_{13} - B_{12}}{B_{21}}$		x_3

$RG = y_6\eta_6\eta_7 + y_7\eta_7 + y_8\eta_8 + y_9\eta_9 + y_{10}\eta_{10}.$

tackle complex manipulations. However, a warning needs to be pointed out when applying such programs to symbolic exergoeconomics: They solve problems because of their great power of manipulating symbols and not because of an intelligent analysis of the physical meaning of intermediate expressions. Analyzing the meaning is our task. Therefore, when solving complex structures we must be aware of unimportant recyclings because they complicate unnecessarily the symbolic solution. Thus, some recyclings could be discarded in a first approximation analysis to clarify the prominent relationships.

8.4.2 The $\mathcal{FP}$ representation

To follow the arguments easily, we will use the CGAM system. The concepts of symbolic exergoeconomics were first introduced in the literature (Valero and Torres 1988) (Valero, Torres, and Lozano 1989) (Valero et al. 1990a,b,c,d, 1992a,b, 1993).

Now a new matrix, $\mathbf{J}$ $(n \times m)$, is introduced. $\mathbf{J} = \mathbf{A}_F - \mathbf{K}_D\mathbf{A}_P$, where the diagonal matrix $\mathbf{K}_D(n \times n)$ contains the unit exergy consumption, k_i, of each component, and the following holds true: $\mathbf{F} = \mathbf{K}_D\mathbf{P}$; $\mathbf{I} = (\mathbf{K}_D - \mathbf{U}_D)\mathbf{P}$. For the CGAM system $\mathbf{J}$ takes the form of

$$\mathbf{J} = \begin{bmatrix} 1 & 0 & k_1 & -k_1 & 0 & 0 & 0 & 0 & 0 \\ 0 & -k_2 & 0 & 0 & 0 & 0 & 1 & 0 & 0 \\ 0 & 0 & 0 & 1 & -1 & 0 & -k_3 & -k_3 & 0 \\ 0 & k_4 & -k_4 & 0 & 1 & -1 & 0 & 0 & 0 \\ 0 & 0 & 0 & 0 & 0 & 1 & 0 & 0 & -k_5 \end{bmatrix}, \tag{8.7}$$

and this matrix can, with the aid of α_e and α_x matrices, be extended to a matrix $\mathbb{J}_{ex} = {}^t(\mathbf{J} \vdots \alpha_e \vdots \alpha_x)$ $(m \times m)$, which has the important property of

$$\mathbb{J}_{ex}\mathbf{B} = \mathbb{Y}_e^*, \qquad \text{or} \quad \mathbf{B} = \mathbb{J}_{ex}^{-1}\mathbb{Y}_e^*, \tag{8.8}$$

and permits one to analytically express the exergy of the flows of a system in terms of the unit exergy consumption of the components, the bifurcation exergy ratios, and the resources entering the plant.

In all exergy analyses performed on plants, by measuring the mass flow rates, pressures, temperatures, compositions, etc., we can get the exergies of the flows. Therefore, there is no point in using this equation to calculate the values of a system's exergies. This equation takes its full meaning when it is used symbolically, because its analytical or symbolic solution obtained by the use of symbolic manipulators provides as many formulae m as there are flows in the plant of the type $B_i = \check{B}_i(\mathbf{k}, \mathbf{x}, \omega_e)$, which tell how the exergy of any flow is a function of the efficiencies, bifurcation exergy ratios, and the plant's exergy resources. Inverting matrix $\mathbb{J}_{ex}$ for the CGAM problem and multiplying by $\mathbb{Y}_e^*$

we get the formulas

$$
\begin{aligned}
&\hat{B}_1 = \omega_1 && \hat{B}_2 = \frac{\zeta_2\omega_1}{\zeta_1 k_1} && \hat{B}_3 = \frac{(\zeta_3 - \zeta_1)\omega_1}{\zeta_1 k_1} \\
&\hat{B}_4 = \frac{\zeta_3\omega_1}{\zeta_1 k_1} && \hat{B}_5 = \frac{\zeta_3 x_{3F}\omega_1}{\zeta_1 k_1} && \hat{B}_6 = \frac{\zeta_3 x_{3F} x_4\omega_1}{\zeta_1 k_1} \\
&\hat{B}_7 = \frac{\zeta_2 k_2\omega_1}{\zeta_1 k_1} && \hat{B}_8 = \frac{\zeta_2 k_2 x_{3P}\omega_1}{\zeta_1 k_1} && \hat{B}_9 = \frac{\zeta_3 x_{3F} x_4\omega_1}{\zeta_1 k_1 k_5},
\end{aligned} \tag{8.9}
$$

where $\zeta_2 = (1 - x_{3F})k_4$, $\zeta_3 = (x_{3P} + 1)k_2k_3k_4$, and $\zeta_1 = (x_{3P} + 1)(k_4 - 1)x_{3F}k_2k_3 + \zeta_3 - \zeta_2$.

An interesting property of matrix $\mathbb{J}_{ex}$ is that it can be converted into matrix $\mathbb{A}$ when all the elements of matrix $\mathbf{K}_D$ are replaced by ones: $\mathbf{J} \to \mathbf{A} = \mathbf{A}_F - \mathbf{U_D}\mathbf{A}_P = \mathbf{A}_F - \mathbf{A}_P$, and $\mathbf{B} = \mathbb{J}_{ex}^{-1}\mathbb{Y}_e^* \to \mathbf{B}^* = \mathbb{A}^{-1}\mathbb{Y}_e^*$.

In other words, the formulas of exergetic costs $\mathbf{B}^*$ $(m \times 1)$ of the type $B_i^* = \hat{B}_i^*(\mathbf{x}, \omega_e)$ are easily obtained from the formulas of exergies by simply substituting the k_i with ones. Also evident is that we can assess the symbolic values of the exergoeconomic costs, because we have the symbolic value of $\mathbb{A}^{-1}$ and $\mathbf{C} = \mathbb{A}^{-1}\mathbb{Z}$, where $\mathbb{Z} = {}^t[-\mathbf{Z} \vdots \mathbf{C}_e \vdots \mathbf{0}]$.

From these results the symbolic values of exergetic unit costs as well as of exergoeconomic unit costs of the type $k_i^* = \hat{k}_i^*(\mathbf{k}, \mathbf{x}, \omega_e)$, $c = \hat{c}_i(\mathbf{k}, \mathbf{x}, \mathbf{Z}, \mathbf{C}_e)$, can easily be obtained from their definitions: $\hat{k}_i^* = \hat{B}_i^*/\hat{B}_i$, $\hat{c}_i = \hat{C}_i/\hat{B}_i$. The same applies to the formula of the efficiency of the system, that is, the CGAM system which takes the symbolic value

$$
\hat{\eta}_{\text{CGAM}} = \frac{\hat{B}_8 + \hat{B}_9}{\hat{B}_1} = \frac{\zeta_2 k_2 k_5 x_{3P} + \zeta_3 x_{3F} x_4}{\zeta_1 k_1 k_5}. \tag{8.10}
$$

We call this form of characterizing a thermal system the $\mathcal{FP}$ representation. Note that all the formulas we got depend on the system's inputs, the local efficiencies, and the bifurcation exergy ratios. Let us denote $\chi_e = \{\mathbf{k}, \mathbf{x}, \omega_e\}$.

8.4.3 The $\mathcal{PF}$ representation

From a purely structural point of view, the number of junctions is related to the number of bifurcations. Thus, if in a system with e input flows and s output flows, we change the direction of arrows, the outputs become inputs, and the bifurcations become junctions.

Therefore, if r is the number of junctions in a structure, it holds $r = m - n - s$. In a bifurcation, whether of fuel or product, the ratio between exergies of the bifurcated and the main flows is by definition x_{ij}. In the same way, as we distinguish fuel and product bifurcation exergy ratios, we can also distinguish between fuel and product junction exergy ratios, r_{ij}. If for the CGAM example the products leaving the systems are $\omega_3 = B_8$ and $\omega_5 = B_9$, and the junction exergy ratios $r_1 = B_3/B_4$ and $r_4 = B_2/B_3$, we can define a matrix $\mathbb{J}_{sr}$ $(m \times m)$ such that

$$\begin{matrix} \\ \\ \mathbf{J} \rightarrow \\ \\ \\ \\ \alpha_s \rightarrow \\ \\ \\ \alpha_r \rightarrow \\ \\ \end{matrix} \begin{bmatrix} 1 & 0 & k_1 & -k_1 & 0 & 0 & 0 & 0 & 0 \\ 0 & -k_2 & 0 & 0 & 0 & 0 & 1 & 0 & 0 \\ 0 & 0 & 0 & 1 & -1 & 0 & -k_3 & -k_3 & 0 \\ 0 & k_4 & -k_4 & 0 & 1 & -1 & 0 & 0 & 0 \\ 0 & 0 & 0 & 0 & 0 & 1 & 0 & 0 & -k_5 \\ \cdots & \cdots & \cdots & \cdots & \cdots & \cdots & \cdots & \cdots & \cdots \\ 0 & 0 & 0 & 0 & 0 & 0 & 0 & 1 & 0 \\ 0 & 0 & 0 & 0 & 0 & 0 & 0 & 0 & 1 \\ \cdots & \cdots & \cdots & \cdots & \cdots & \cdots & \cdots & \cdots & \cdots \\ 0 & 1 & -r_4 & 0 & 0 & 0 & 0 & 0 & 0 \\ 0 & 0 & 1 & -r_1 & 0 & 0 & 0 & 0 & 0 \end{bmatrix} \begin{bmatrix} B_1 \\ B_2 \\ B_3 \\ B_4 \\ B_5 \\ \cdots \\ B_6 \\ B_7 \\ \cdots \\ B_8 \\ B_9 \end{bmatrix} = \begin{bmatrix} 0 \\ 0 \\ 0 \\ 0 \\ 0 \\ \cdots \\ \omega_3 \\ \omega_5 \\ \cdots \\ 0 \\ 0 \end{bmatrix} \tag{8.11}$$

or in general form: $\mathbb{J}_{sr}\mathbf{B} = \mathbb{Y}_s^*$, where $\mathbb{Y}_s^* = {}^t(0 \vdots \omega_s \vdots 0)$ and $\mathbb{J}_{sr} = {}^t(\mathbf{J} \vdots \alpha_s \vdots \alpha_r)$.

The vector $\mathbb{Y}_s^*$ contains the exergy values of the flows leaving the plant, and matrix $\mathbb{J}_{sr}$ is built in a parallel way, because it was $\mathbb{J}_{ex}$ but now has matrices α_s $(s \times m)$ and α_r $(r \times m)$ indicating the structure of outputs and the junction exergy ratios of the system.

If we solve symbolically the expression $\mathbf{B} = \mathbb{J}_{sr}^{-1}\mathbb{Y}_s^*$, we determine the exergy of the flows of the system in terms of the canonical variables of the $\mathcal{PF}$ representation, $\chi_s = \{\mathbf{k}, \mathbf{r}, \omega_s\}$, that is, we obtain expressions of the type $B_i = \overset{>}{B}_i(\mathbf{k}, \mathbf{r}, \omega_s)$.

Thus, the case of the CGAM system results in

$$\begin{array}{lll} \overset{>}{B}_1 = \xi_2(1 - r_1)k_1/\xi_1 & \overset{>}{B}_2 = \xi_2 r_4 r_1/\xi_1 & \overset{>}{B}_3 = \xi_2 r_1/\xi_1 \\ \overset{>}{B}_4 = \xi_2/\xi_1 & \overset{>}{B}_5 = (\xi_2\xi_3/\xi_1) - k_3\omega_3 & \overset{>}{B}_6 = k_5\omega_5 \\ \overset{>}{B}_7 = \xi_2 k_2 r_4 r_1/\xi_1 & \overset{>}{B}_8 = \omega_3 & \overset{>}{B}_9 = \omega_5, \end{array} \tag{8.12}$$

where $\xi_1 = 1 - k_4 r_1(1 - r_4) - k_2 k_3 r_1 r_4$, $\xi_2 = k_3\omega_3 + k_5\omega_5$, $\xi_3 = 1 - k_2 k_3 r_4 r_1$.

The canonical variables of the $\mathcal{FP}$ and $\mathcal{PF}$ representations are interrelated.

The unit exergy consumption of a component k_i is a parameter common to both representations.

If we put $\mathbf{x}$ and ω_e as a function of the canonical variables in the $\mathcal{PF}$ representation, that is,

$$\overset{>}{x}_{3F} = \frac{\overset{>}{B}_5}{\overset{>}{B}_4} = \frac{\xi_3 - k_3\xi_1\omega_3}{\xi_2} \qquad \overset{>}{x}_4 = \frac{\overset{>}{B}_6}{\overset{>}{B}_5} = \frac{k_5}{(\xi_2\xi_3/\xi_1) - k_3\omega_3}$$
$$\overset{>}{x}_{3P} = \frac{\overset{>}{B}_8}{\overset{>}{B}_7} = \frac{\xi_1\omega_3}{\xi_2 k_2 r_4 r_1} \qquad \overset{>}{\omega}_1 = \overset{>}{B}_1 = \frac{\xi_2(1-r_1)k_1}{\xi_1}, \tag{8.13}$$

we may use all the formulas from the $\mathcal{FP}$ representation for deriving its equivalents in the $\mathcal{PF}$ representation.

In general, for the thermoeconomic variable $\mathbf{y}$,

$$\overset{>}{y} = \overset{>}{y}[\mathbf{k}, \overset{>}{\mathbf{x}}(\mathbf{k}, \mathbf{r}, \omega_s), \overset{>}{\omega}_e(\mathbf{k}, \mathbf{r}, \omega_s)] = \overset{>}{y}(\mathbf{k}, \mathbf{r}, \omega_s). \tag{8.14}$$

In so doing, we obtain as an example the symbolic exergetic unit costs in the $\mathcal{PF}$ representation:

$$\overset{>}{k}{}_1^* = 1 \qquad \overset{>}{k}{}_2^* = (1-r_1)k_1k_2k_3/\xi_1 \qquad \overset{>}{k}{}_3^* = \overset{>}{k}{}_4^*/r_1 - k_1(1-r_1)/r_1$$
$$\overset{>}{k}{}_4^* = (1-r_1)k_1 + r_1r_4(1-r_1)k_1k_2k_3/\xi_1 + r_1(1-r_1)(1-r_4)k_1k_4/\xi_1$$
$$\overset{>}{k}{}_5^* = \overset{>}{k}{}_4^* \qquad \overset{>}{k}{}_6^* = \overset{>}{k}{}_4^* \qquad \overset{>}{k}{}_7^* = (1-r_1)k_1k_2/\xi_1 \qquad \overset{>}{k}{}_8^* = \overset{>}{k}{}_7^* \qquad \overset{>}{k}{}_9^* = \overset{>}{k}{}_4^*. \tag{8.15}$$

8.4.4 Fuel and product symbolic analysis

Matrices $\mathbf{A}_F$ and $\mathbf{A}_P$ cannot be inverted because of their dimension $(n \times m)$. To make them square we need a set of $m - n$ auxiliary equations. Because submatrices α_{ex} and α_{rs} are of dimension $(m - n \times m)$, they can be used to get to the following systems of equations:

$$\begin{pmatrix} \mathbf{A}_F \\ \cdots\cdots \\ \alpha_{ex} \end{pmatrix} \mathbf{B} = \begin{pmatrix} \mathbf{F} \\ \cdots\cdots \\ \omega_{ex} \end{pmatrix}; \quad \begin{pmatrix} \mathbf{A}_P \\ \cdots\cdots \\ \alpha_{ex} \end{pmatrix} \mathbf{B} = \begin{pmatrix} \mathbf{P} \\ \cdots\cdots \\ \omega_{ex} \end{pmatrix};$$
$$\begin{pmatrix} \mathbf{A}_F \\ \cdots\cdots \\ \alpha_{rs} \end{pmatrix} \mathbf{B} = \begin{pmatrix} \mathbf{F} \\ \cdots\cdots \\ \omega_{rs} \end{pmatrix}; \quad \begin{pmatrix} \mathbf{A}_F \\ \cdots\cdots \\ \alpha_{rs} \end{pmatrix} \mathbf{B} = \begin{pmatrix} \mathbf{P} \\ \cdots\cdots \\ \omega_{rs} \end{pmatrix}, \tag{8.16}$$

and therefore we can write

$$\begin{pmatrix} \mathbf{A}_F \\ \cdots\cdots \\ \alpha_{ex} \end{pmatrix} \cdot \begin{pmatrix} \mathbf{A}_P \\ \cdots\cdots \\ \alpha_{ex} \end{pmatrix}^{-1} \cdot \begin{pmatrix} \mathbf{P} \\ \cdots\cdots \\ \omega_{ex} \end{pmatrix} = \begin{pmatrix} \mathbf{F} \\ \cdots\cdots \\ \omega_{ex} \end{pmatrix}$$
$$\begin{pmatrix} \mathbf{A}_P \\ \cdots\cdots \\ \alpha_{rs} \end{pmatrix} \cdot \begin{pmatrix} \mathbf{A}_F \\ \cdots\cdots \\ \alpha_{rs} \end{pmatrix}^{-1} \cdot \begin{pmatrix} \mathbf{F} \\ \cdots\cdots \\ \omega_{rs} \end{pmatrix} = \begin{pmatrix} \mathbf{P} \\ \cdots\cdots \\ \omega_{rs} \end{pmatrix}. \tag{8.17}$$

Matrix ${}^t(\mathbf{A}_P \vdots \alpha_{ex})^{-1}$ can be written as $(\mathbf{A}_P^{(-1)} \vdots \alpha_{ex}^{(-1)})$, where $\mathbf{A}_P^{(-1)}$ and $\alpha_{ex}^{(-1)}$ are generalized inverses of $\mathbf{A}_P$ and α_{ex}. These matrices satisfy (Torres 1991): $\mathbf{A}_F\mathbf{A}_P^{(-1)}\mathbf{P} + \mathbf{A}_F\alpha_{ex}^{(-1)}\omega_{ex} = \mathbf{F}$.

The term $\mathbf{A}_F\alpha_{ex}^{(-1)}\omega_{ex}$ is a vector $(n \times 1)$ that refers to the total system's input flows, and this describes which components receive the system's input fuel. We denote by $\mathbf{F}_e \equiv \mathbf{A}_F\alpha_{ex}^{(-1)}\omega_{ex}$ $(n \times 1)$ and $\langle\mathbf{FP}\rangle \equiv \mathbf{A}_F\mathbf{A}_P^{(-1)}$ $(n \times n)$. Then we get the following important equation: $\mathbf{F} = \langle\mathbf{FP}\rangle\mathbf{P} + \mathbf{F}_e$. In a parallel way, matrix ${}^t(\mathbf{A}_F \vdots \alpha_{rs}^{-1})$ can be written as $(\mathbf{A}_F^{(-1)} \vdots \alpha_{rs}^{(-1)})$, where $\mathbf{A}_F^{(-1)}$ and $\alpha_{rs}^{(-1)}$ are generalized inverses of $\mathbf{A}_F$, α_{rs}. These matrices satisfy (Torres 1991) $\mathbf{A}_P\mathbf{A}_F^{(-1)}\mathbf{F} + \mathbf{A}_P\alpha_{rs}^{(-1)}\omega_{rs} = \mathbf{P}$, where $\mathbf{A}_P\alpha_{rs}^{(-1)}\omega_{rs}$ is a vector $(n \times 1)$ that refers to the total system's flows and describes which subsystems the total system's output product leaves; we denote $\mathbf{P}_s \equiv \mathbf{A}_P\alpha_{rs}^{(-1)}\omega_{rs}$ $(n \times 1)$ and $\langle\mathbf{PF}\rangle \equiv \mathbf{A}_P\mathbf{A}_F^{(-1)}$ $(n \times n)$. Then we also get the following important result: $\mathbf{P} = \langle\mathbf{PF}\rangle\mathbf{F} + \mathbf{P}_s$.

The $\langle\mathbf{FP}\rangle$ and $\langle\mathbf{PF}\rangle$ matrices are characteristic of energy systems with a defined productive purpose. The $\langle\mathbf{FP}\rangle$ matrix shows how the fuel of each subsystem is formed from the products of other subsystems; that is, it tells how the fuel is transformed into products throughout the system. In contrast, the $\langle\mathbf{PF}\rangle$ matrix shows how much of the product of each subsystem goes to form part of the fuels of other subsystems; that is, it tells how the product is transformed into fuels throughout the system.

In the particular case of the CGAM problem this matrix takes the form of

$$\langle\mathbf{PF}\rangle = \begin{bmatrix} 0 & 0 & 1-r_1 & 1-r_1 & 1-r_1 \\ 0 & 0 & r_1r_4 & r_1r_4 & r_1r_4 \\ 0 & 1 & 0 & 0 & 0 \\ 0 & 0 & r_1(1-r_4) & r_1(1-r_4) & r_1(1-r_4) \\ 0 & 0 & 0 & 0 & 0 \end{bmatrix} \tag{8.18}$$

This equation allows us to obtain the following symbolic expressions for the products of the components of the CGAM plant:

$$\overset{>}{P}_1 = (1 - r_1)\frac{\xi_2}{\xi_1}, \qquad \overset{>}{P}_2 = r_1 r_4 \frac{\xi_2}{\xi_1}, \qquad \overset{>}{P}_3 = k_2 r_1 r_4 \frac{\xi_2}{\xi_1} + \omega_3,$$
$$\overset{>}{P}_4 = r_1(1 - r_4)\frac{\xi_2}{\xi_1}, \qquad \overset{>}{P}_5 = \omega_5. \tag{8.19}$$

In these expressions the product of each component is related to the plant product or the fuel of other components, through the exergy recycling ratios. Moreover, these expressions allow us to interpret the fuel and product distribution throughout the system. Thus, the product of each subsystem is used as either the plant's product $\mathbf{P}_s$ or the fuel of other subsystems in a ratio indicated by the matrix $\langle\mathbf{PF}\rangle$ coefficients. Each element q_{ij} of the $\langle\mathbf{PF}\rangle$ matrix can be interpreted as the portion of fuel of component j, which comes from component i. From the previous results, one can deduce that the fuel of the i-th component comes from either a portion $1 - \sum q_{ij}$ of the total fuel, or a portion q_{ij} of the product of component j. See the $\mathcal{PF}$ diagram in Fig. 8.3.

These expressions also imply the concept of unit cost, because the unit cost of the fuel is the sum of the unit costs of the internal products and of the external

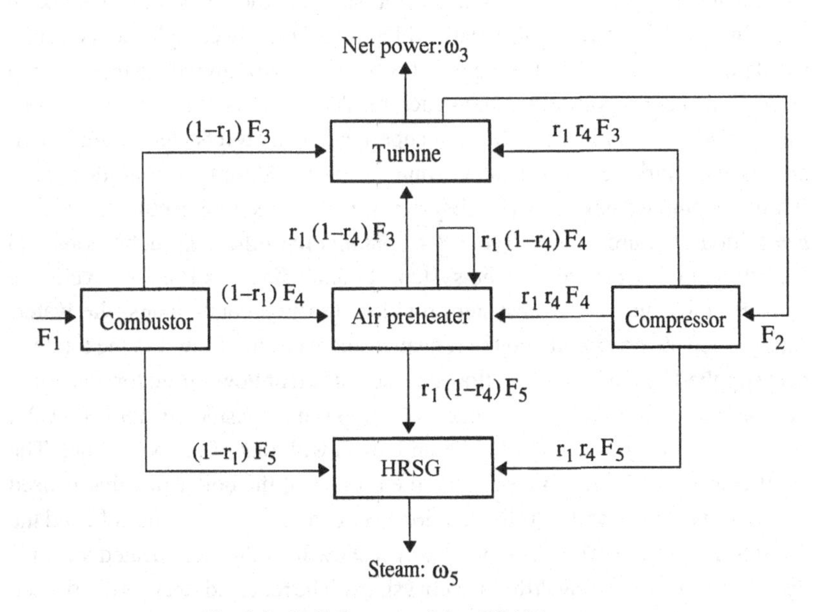

Fig. 8.3. $\mathcal{PF}$ diagram of the CGAM system.

resources of the system that constitute this fuel. In our case,

$$c_{F,1} = c_{Fe,1} = c_{\text{fuel}}, \qquad c_{F,3} = (1 - r_1)c_{P,1} + r_1 r_4 c_{P,2} + r_1(1 - r_4)c_{P,4},$$
$$c_{F,2} = c_{P,3}, \qquad c_{F,4} = (1 - r_1)c_{P,1} + r_1 r_4 c_{P,2} + r_1(1 - r_4)c_{P,4},$$
$$c_{F,5} = (1 - r_1)c_{P,1} + r_1 r_4 c_{P,2} + r_1(1 - r_4)c_{P,4}. \tag{8.20}$$

In general form

$$c_{F,i} = \sum_{j=1}^{n} q_{ij} c_{P,1} + \left(1 - \sum_{j=1}^{n} q_{ij}\right) c_{Fe,i}, \tag{8.21}$$

where $c_{Fe,i}$ is the unit cost of the resources entering the plant in the i-th subsystem. We call $\mathbf{c}_e$ $(n \times 1)$ the vector containing the second term on the right hand side of the previous equation in each component. According to this we can write the unit cost of the fuel in matrix form as $\mathbf{c}_F = \mathbf{c}_e + {}^t\langle\mathbf{PF}\rangle\mathbf{c}_P$. This result is important for the formulation of the optimization problem of a thermal system, which will be described later.

8.4.5 Application to thermal system simulation

The power of the method is shown in the study of the conventional Rankine cycle in Fig. 8.2 (Torres et al. 1989), which is sufficiently complex and close to a real plant, in such a way that the results obtained can be applied in practice and be generalized to a cycle of similar structure. The plant has three turbine sections (high, medium, and low pressure), a superheated and a reheated steam boiler, an auxiliary turbine, three heaters (one for each section), and the deaerator. In this scheme we have $n = 10$ subsystems, $m = 25$ interconnecting flows, and $e = 1$ input resource flows. Table 8.4 contains the efficiency definitions and the bifurcation parameters of the system. In analyzing this Rankine cycle, we consider the boiler as a subsystem divided into two components: the water-superheated steam circuit and the reheated steam circuit. Flow 1 is the total net heat supplied by the boiler, and flow 2 is the portion of flow 1 used for reheating. For the sake of simplicity, the condenser pump is not considered, nor is flow 25, because its exergy is negligible in comparison with that of the boiler fuel. The coefficient y_1 in Table 8.4 represents the portion of the boiler fuel that is used to reheat the steam, and y_2 is the relation between the exergy of the cold and the hot reheated steam. The terms $y_3, \ldots, y_{10}$ show how the superheated steam is distributed as fuel among different subsystems. The reheated steam is distributed among the rest of the subsystems in the ratio $y_2 y_4, \ldots, y_2 y_9$. This steam is not used as fuel in the high-pressure turbine or in the high-pressure heater 3. The products of the regenerative train are distributed among the other subsystems in the same ratio as the superheated steam, except for the low-pressure heater, all of whose product is used as fuel for the deaerator. For instance, factor $(1 - RG)$ indicates the important role played by the regenerative train. We can rearrange the efficiency formula taking into account that the plant's product is the sum of the products of the turbines and that the total efficiency is composed of the contributions of each one associated with a section of the turbine:

$$\eta_T = \frac{P_{\mathrm{HP}}}{\omega_1} y_3 \eta_3 + \frac{P_{\mathrm{MP}}}{\omega_1} y_4 \eta_4 + \frac{P_{\mathrm{LP}}}{\omega_1} y_5 \eta_5. \tag{8.22}$$

With the help of these equations one can distinguish which factors contribute to improving installation efficiency and to reducing production cost, for example, the rate of reheated steam production y_1 or the increase of extraction flow rates RG.

Moreover, this formula can be used for simulation of systems. Torres et al. (1989) studied the connections, analogies, and differences between the information provided by conventional packages such as ASPEN, PROCESS, etc., and the formula for the overall efficiency of thermal systems, expressed as a function of its resources, structure, and the efficiency of the subsystems making it up.

Insofar as the number of parameters or degrees of freedom of the simulator are greater than the number of the parameters of the symbolic formula, the former will be more sensitive to the analysis of variations and to the calculation of marginal costs of the type ($\partial Resources/\partial Parameter$) than the information on average exergy cost provided by the latter. However, when the structure of the systems is complex, structural effects predominate and the modification of characteristic parameters of subsystems have local, not global, effects on all the components, so that the rest of the bifurcation exergy ratios and the efficiency of the subsystems remain constant. Under these conditions we say that a given parameter has stability and independence. The complex case (because of the high degree of branching) of a real power plant was analyzed, and the local effects of modifying parameters were found to spread to the nearest, most directly related subsystem, at the most, while the rest of efficiencies and bifurcation parameters remained unchanged. Thus, for example, the perturbation of one turbine section almost exclusively affects the feed water heater that is fed by it. When this effect occurs, the symbolic formula of the efficiency gives a good picture of the plant behavior. The results from the efficiency formula were good approximations to those obtained with the simulator, followed the same trends, and have relative errors less than 0.01 percent, except where the stability and independence hypotheses did not hold. Using Taylor expansions of the efficiency formula parameters, the errors found were of the order of 10^{-4}

in all the cases studied, and the results were close to those of the simulator, which was carefully fitted to the experimental data of the plant behavior. The power of the formula was thus demonstrated even in a complex case.

Will conventional simulators be replaced with artificial intelligence programs that generate symbolic formulas? This development has no theoretical limitations, and it deserves special attention for the future.

8.5 Thermoeconomic optimization

The problem of optimization of the CGAM system (Fig. 8.1) can be expressed as (Valero et al. 1989) (Lozano and Valero 1993a,b) (Serra 1994) (Valero et al. 1994b)

$$\text{Minimize} \quad C = c_{\text{fuel}}\dot{m}_{\text{fuel}}\text{LHV} + \dot{Z}_{cp} + \dot{Z}_{ap} + \dot{Z}_{cb} + \dot{Z}_{gt} + \dot{Z}_{hrsg} \qquad (8.23)$$

subject to the constraints imposed by the physical and cost models of the installation. Using the structural information provided by the ECT, optimization can be formulated in the following compact form:

$$\text{Minimize} \quad C = {}^t\mathbf{c_e F} + {}^t\mathbf{UZ}$$

subject to the constraints

$$\mathbf{K}_D\mathbf{P} = \mathbf{F} \qquad \mathbf{P}_s + \langle\mathbf{PF}\rangle\mathbf{F} = \mathbf{P} \qquad \mathbf{P}_s = \text{datum} \qquad (8.24)$$

$$\mathbf{k} = \mathbf{f}_k(\tau) \qquad \mathbf{r} = \mathbf{f}_r(\tau) \qquad \omega = \mathbf{f}_\omega(\tau) \qquad (8.25)$$

As can be seen, two types of constraints exist: the structural (8.24) and the local (8.25) ones. The former relate the production of the components with the fuels and the global production objective through the set of parameters $\chi_s = \{\mathbf{k}, \mathbf{r}, \omega_s\}$. The latter relate the parameters χ_s with the physical model of the installation, that is, the set of conventional design variables $\{\tau\}$ such as pressure ratios, isentropic efficiencies, gas turbine inlet temperature, air-preheater exit temperature, and so on.

As is well known, the optimization problem with constraints can be transformed into one without constraints by formulating the Lagrangian function, which in our formulation becomes

$$\begin{aligned} L(\mathbf{F}, \mathbf{P}, \mathbf{k}, r, \omega_s, \tau) = {}&{}^t\mathbf{c}_e\mathbf{F} + {}^t\mathbf{uZ}(\tau) + {}^t\Lambda_F(\mathbf{K}_D\mathbf{P} - \mathbf{F}) \\ &+ {}^t\Lambda_P(\mathbf{P}_s + \langle\mathbf{PF}\rangle\mathbf{F} - \mathbf{P}) + {}^t\Lambda_\omega(\text{datum} - \mathbf{P}_s) \\ &+ {}^t\mathbf{M}_k(\mathbf{f}_k(\tau) - \mathbf{k}) + {}^t\mathbf{M}_r(\mathbf{f}_r(\tau) - \mathbf{r}) + {}^t\mathbf{M}_\omega(\mathbf{f}_\omega(\tau) - \omega_s), \end{aligned} \qquad (8.26)$$

where ${}^t\Lambda_F$ $(1 \times n)$ and ${}^t\Lambda_P$ $(1 \times n)$ are vectors containing the Lagrange multipliers or marginal costs associated with the structural constraints, ${}^t\Lambda_\omega$ $(1 \times s)$ is the vector containing the marginal costs of the products leaving the plant, and ${}^t\mathbf{M}_k$ $(1 \times n)$, ${}^t\mathbf{M}_r$ $(1 \times r)$ and ${}^t\mathbf{M}_\omega$ $(1 \times s)$ are vectors that contain the multipliers associated with the local constraints. Thus, this function becomes for the CGAM problem

$$\begin{aligned}
L(\mathbf{F}, \mathbf{P}, \mathbf{k}, \mathbf{r}, \omega_s, \tau) = {} & c_{\text{fuel}} F_1 + \dot{Z}_{cp}(\tau, P_2) + \dot{Z}_{ap}(\tau, P_4) + \dot{Z}_{cb}(\tau, P_1) \\
& + \dot{Z}_{gt}(\tau, P_3) + \dot{Z}_{hrsg}(\tau, P_5) \\
& + \lambda_{F1}(k_1 P_1 - F_1) + \lambda_{F2}(k_2 P_2 - F_2) + \lambda_{F3}(k_3 P_3 - F_3) \\
& + \lambda_{F4}(k_4 P_4 - F_4) + \lambda_{F5}(k_5 P_5 - F_5) \\
& + \lambda_{P1}[(1 - r_1)(F_3 + F_4 + F_5) - P_1] \\
& + \lambda_{P2}[(r_1 r_4)(F_3 + F_4 + F_5) - P_2] + \lambda_{P3}[(F_2 + \omega_3 - P_3] \\
& + \lambda_{P4}[r_1(1 - r_4)(F_3 + F_4 + F_5) - P_4] + \lambda_{P5}[\omega_5 - P_5] \\
& + \lambda_{\omega 3}[30000 - \omega_3] \\
& + \lambda_{\omega 5}[12748 - \omega_5] + \mu_{k1}[k_{k1}(\tau) - k_1] + \mu_{k2}[k_{k2}(\tau) - k_2] \\
& + \mu_{k3}[k_{k3}(\tau) - k_3] + \mu_{k4}[k_{k4}(\tau) - k_4] \\
& + \mu_{k5}[k_{k5}(\tau) - k_5] \\
& + \mu_{r1}[f_{r1}(\tau) - r_1] + \mu_{r4}[f_{r4}(\tau) - r_4] + \mu_{\omega 3}[f_{\omega 3}(\tau) - \omega_3] \\
& + \mu_{\omega 5}[f_{\omega 5}(\tau) - \omega_5].
\end{aligned} \tag{8.27}$$

The optimum of this function must satisfy

$$\begin{aligned}
&\frac{\partial L}{\partial \mathbf{F}} = \mathbf{0}; \quad \frac{\partial L}{\partial \mathbf{P}} = \mathbf{0}; \quad \frac{\partial L}{\partial \mathbf{k}} = \mathbf{0}; \\
&\frac{\partial L}{\partial \mathbf{r}} = \mathbf{0}; \quad \frac{\partial L}{\partial \omega} = \mathbf{0}; \quad \frac{\partial L}{\partial \tau} = \mathbf{0}.
\end{aligned} \tag{8.28}$$

Applying these conditions we get,

$$\begin{aligned}
&\frac{\partial L}{\partial \mathbf{F}} = \mathbf{c}_e - \Lambda_F + {}^t\langle \mathbf{PF} \rangle \Lambda_P = 0 \quad \text{or} \quad \mathbf{c}_e = \Lambda_F - {}^t\langle \mathbf{PF} \rangle \Lambda_P \quad n \text{ equations} \\
&\frac{\partial L}{\partial \mathbf{P}} = {}^t\mathbf{u} \frac{\partial \mathbf{Z}}{\partial \mathbf{P}} - \Lambda_P + \mathbf{K}_D \Lambda_F = \mathbf{0}, \quad n \text{ equations}
\end{aligned} \tag{8.29}$$

and when in a general way we put $Z_i = P_i^{-1} f_{Z,i}(\tau)$ where a common value of ε_i can be 0.6 (the 0.6 rule), then $\partial Z_i / \partial P_i = \varepsilon_i f_{z,i}(\tau) = \varepsilon_i Z_i / P_i$. Therefore, $\partial \mathbf{Z} / \partial \mathbf{P} = \varepsilon \mathbf{Z} / \mathbf{P}$ and $\Lambda_P = \mathbf{k}_D \Lambda_F + \varepsilon \mathbf{Z} / \mathbf{P}$. Now we can identify the different sets of costs provided by the ECT.

If $\varepsilon = 1$, or Z_i is proportional to P_i, then $\mathbf{c}_e = \Lambda_F - {}^t\langle \mathbf{PF} \rangle \Lambda_P$ and $\Lambda_P = \mathbf{K}_D \Lambda_F + \mathbf{Z}/\mathbf{P}$, which are similar to those obtained for the unit exergoeconomic costs of the $\mathcal{PF}$ representation, that is, $\mathbf{c}_e = \mathbf{c}_F - {}^t\langle \mathbf{PF} \rangle \mathbf{c}_P$ and $\mathbf{c}_P = \mathbf{K}_D \mathbf{c}_F + \mathbf{Z}/\mathbf{P}$. Thus, the marginal exergoeconomics costs are shown to be equal to the average exergoeconomic costs obtained with the ECT when Z_i are proportional to P_i.

If $\varepsilon = \mathbf{0}$ and $\mathbf{c}_e = \mathbf{k}_e^*$, then $\mathbf{k}_e = \Lambda_F - {}^t\langle\mathbf{PF}\rangle\Lambda_P$ and $\Lambda_P = \mathbf{K}_D\Lambda_F$, which are similar to those obtained for the unit exergetic costs of the $\mathcal{PF}$ representation, that is,

$$\mathbf{k}_e^* = \mathbf{k}_F^* - {}^t\langle\mathbf{PF}\rangle\mathbf{k}_P^* \quad \text{and} \quad \mathbf{k}_P^* = \mathbf{K}_D\mathbf{k}_F^*. \tag{8.30}$$

Thus, the marginal exergoeconomic costs are shown to be equal to the average exergetic costs when amortizations of components are not considered.

When solving the optimization problem we also need to apply the following conditions

$$\begin{array}{llll}
\partial L/\partial k_i &= \lambda_{F,i}P_i - \mu_{k,i} = 0 & \text{or} \quad \mu_{k,i} = c_{F,i}P_i & n \text{ equations} \\
\partial L/\partial r_i &= \Lambda_P(\partial\langle\mathbf{PF}\rangle/\partial r_i)\mathbf{F} - \mu_{r,i} = 0 & \text{or} \quad \mu_{r,i} = c_P(\partial\langle\mathbf{PF}\rangle/\partial r_i)\mathbf{F} & r \text{ equations} \\
\partial L/\partial \omega_i &= \lambda_{P,i} - \lambda_{\omega,i} - \mu_{\omega,i} = 0 & \text{or} \quad \mu_{\omega,i} = \lambda_{P,i} - \mu_{\omega i} & s \text{ equations} \\
\partial L/\partial \tau &= 0 & & \tau \text{ equations}
\end{array} \tag{8.31}$$

The simultaneous solution of all these equations provides the optimum variables as well as their costs. Additionally, each cost associated with a variable x_i is interpreted as the derivative $\partial C/\partial x_i$ or the variation of the optimum production when x_i is varied. Valero et al. (1994b) provide details of the solution for the case of the CGAM problem, and Tables 8.1 to 8.3 show the results in the optimum.

The structural information obtained from the thermoeconomic analysis is important for the design of complex thermal systems because it allows us to implement the optimization at a local level (Lozano and Valero 1993b) (Lozano, Valero, and Serra 1993c).

These results were important in the brief history of thermoeconomics, because before this the two branches of study were cost accounting and optimization. Cost accounting provided average exergetic costs by applying a set of reasonable propositions (namely **P1** to **P4**). Because they were not proved, thermoeconomic analysis remained close to a curiosity of thermodynamics rather than solid science. On the other hand, the numerical methods and powerful computers for solving complex optimization energy problems secluded the thermoeconomic optimization practitioners.

Now we can understand the meaning and the role of thermoeconomic costs when analyzing a system. These are some outcomes:

(1) The propositions set by the cost accounting methodology are in fact mathematical consequences of the chain rule for calculating derivatives, that is,

Lagrange multipliers. We are not free to propose *any* accounting rule, and we have the way to propose new ones.

(2) The way in which we conceptually disaggregate systems conditions the values of the costs obtained and therefore the cost accounting rules to be applied. Suppose that in the CGAM problem, we decide to use a different fuel-product definition than the one proposed in this book. Analytically, we would obtain the same theoretical outcomes but different numerical results. This may seem strange, but we can look for a reasonable disaggregation level in the cost accounting case, taking into account the available instrumentation data and our knowledge of the behavior of components. (Valero, Lazano, and Bartolomé 1996)

(3) The marginal exergetic costs obtained from the optimization problem, that is, the Lagrange multipliers, coincide with the average exergetic costs from the cost accounting when the relationships among variables are linear. In other words, a properly chosen disaggregation (causal or productive structure) and a function to *apportion* the energy interactions among components (exergy) play the key role in thermoeconomics and justify the value of thermoeconomics when associating to each cause of loss a cost, this loss being measured in irreversibility. To better understand these ideas, consider the opposite case, that is, the optimization of a system by using the brute force of computers with no tools for understanding the meaning of the Lagrange multipliers.

(4) These results also express the idea of unification of the cost accounting methodologies and the thermoeconomic optimization into a single body of doctrine that connects economics with thermodynamics through the Second Law, namely the science of thermoeconomics.

References

Bejan, A., Tsatsaronis, G. & Moran, M. (1996). *Thermal Design and Optimization.* New York: Wiley Interscience.

Frangopoulos, C. A. (1994). Application of the thermoeconomic functional approach to the CGAM problem. *Energy*, **19**, 323–42.

Gaggioli, R. A. & El-Sayed, Y. M. (1987). A critical review of second law costing methods: Second law analysis of thermal systems. In *International Symposium on Second Law Analysis of Thermal Systems*, ed. M. J. Moran & E. Sciubba. New York: ASME.

Georgescu–Roegen, N. (1971). *The Entropy Law and the Economic Process.* Cambridge, MA: Harvard University Press.

Le Goff, P. (coord.). (1979–1980–1982). Energétique Industrielle: Tome 1. *Analyse thermodynamique et mécanique des économies d'énergie.* Tome 2. *Analyse économique et optimisation des procédes*. Tome 3. *Applications en génie chimique*. Paris: Technique et Documentation.

Lozano, M. A. & Valero, A. (1993a). Theory of exergetic cost. *Energy*, **18**, 939–60.

Lozano, M. A. & Valero, A. (1993b). Thermoeconomic analysis of gas turbine cogeneration systems. In *Thermodynamics and the Design Analysis and Improvement of Energy Systems*, ed. H. J. Richter. Book no. H00874, pp. 311–20, New York: ASME.

Lozano, M. A., Valero, A. & Serra, L. (1993c). Theory of exergetic cost and thermo-economic optimization. In *Energy Systems and Ecology*, ed J. Szargut, Z. Kolenda, G. Tsatsaronis & A. Ziebik, pp. 339–50, Cracow, Poland: University of Cracow.

Royo, F. J. (1994). *Las ecuaciones características de los sistemas térmicos. La energía libre relativa*. Ph.D. dissertation. Department of Mechanical Engineering, University of Zaragoza, Spain.

Serra, L. (1994). *Optimización exergoeconómica de sistemas témicos*. Ph.D. dissertation. Department of Mechanical Engineering, University of Zaragoza, Spain.

Szargut, J. & Morris, D. (1988) *Exergy Analysis of Thermal Chemical and Metallurgical Processes*. New York: Hemisphere Pub. Co.

Torres, C., Valero, A. & Cortés, C. (1989). Application of symbolic exergoeconomics to thermal systems simulation. In *Simulation of Thermal Energy Systems*. AES, Vol. 9/HTD, Vol. 124. Book no. H00527, pp. 75–84, New York: ASME.

Torres, C. (1991). *Exergoeconomía simbólica. Metodología para el análisis termoeconómico de los sistemas energéticos*. Ph.D. dissertation. Department of Mechanical Engineering, University of Zaragoza, Spain.

Tsatsaronis, G. & Winhold, M. (1985). Exergoeconomic analysis and evaluation of energy-conversion plants. I. A new methodology. *Energy*, **10**, 69–80.

Tsatsaronis, G. (1987). A review of exergoecomic methodologies. In *Second Law Analysis of Thermal Systems*, ed. M. J. Moran & E. Sciubba, pp. 81–87. New York: ASME.

Tsatsaronis, G. (ed.) (1994). Invited papers on exergoeconomics. *Energy*, **19**, 279–381.

Tsatsaronis, G. & Pisa, J. (1994). Exergoeconomic evaluation and optimization of energy systems: Application to the CGAM problem. *Energy*, **19**, 287–322.

Valero, A., Lozano, M. A., & Muñoz, M. (1986a). A general theory of exergy saving. I. On the exergetic cost. In *Computer–Aided Engineering of Energy Systems*, ed. R. A. Gaggioli. Book no. H0341A, AES, Vol. 2–3, pp. 1–8, New York: ASME.

Valero, A., Lozano, M. A. & Muñoz, M. (1986b). A general theory of exergy saving. II. On the thermoeconomic cost. In *Computer–Aided Engineering of Energy Systems*, ed. R. A. Gaggioli. Book no. H0341A, AES, Vol. 2–3, pp. 9–16, New York: ASME.

Valero, A., Lozano, M. A. & Muñoz, M. (1986c). A general theory of exergy saving. III. Energy saving and thermoeconomics. In *Computer–Aided Engineering of Energy Systems*, ed. R. A. Gaggioli. Book no. H0341A, AES, Vol. 2–3, pp. 17–22, New York: ASME.

Valero, A. & Torres, C. (1988). Algebraic thermodynamics of exergy systems. In *Approaches to the Design and Optimization of Thermal Systems*, ed. E. J. Wepfer & M. J. Moran, AES, Vol. 7. Book no. G00452, pp. 13–24, New York: ASME.

Valero, A., Torres, C., & Lozano, M. A. (1989). On the unification of thermoeconomic theories. In *Simulation of Thermal Energy Systems*, ed. R. F. Boehm & Y. M. El-Sayed. AES Vol. 6/HTD, Vol. 124. Book no. H00527, pp. 63–74, New York: ASME.

Valero, A. & Carreras, A. (1990a). On causality in organized energy systems. I. Purpose, cause, irreversibility and cost. In *A Future for Energy*, ed. S. Stecco &

M. J. Moran, pp. 387–96, Florence, Italy: Pergamon Press.
Valero, A. & Torres, C. (1990b). On causality in organized energy systems. II. Symbolic exergoecomics. In *A Future for Energy*, ed. S. Stecco & M. J. Moran, pp. 397–408, Florence, Italy: Pergamon Press.
Valero, A., Lozano, M. A., & Torres, C. (1990c). On causality in organized energy systems. III. Theory of perturbations. In *A Future for Energy*, ed S. Stecco & M. J. Moran, pp. 409–20, Florence, Italy: Pergamon Press.
Valero, A. & Lozano, M. A. (1990d). *Master Energía y Eficiencia: Módulo de Termoeconomía*. Bilbao, Spain: CADEM.
Valero, A. & Guallar, J. (1991). On the processable exergy: a rigorous approach. In *Analysis of Thermal and Energy Systems*, ed. D. A. Kouremenos, G. Tsatsaronis & C. D. Rakopoulos, Athens, Greece: University of Athens.
Valero, A., Torres, C. & Serra, L. (1992a). A general theory of thermoeconomics. Part I: Structural analysis. In *On Efficiency, Costs, Optimization and Simulation of Energy Systems*, ed. A. Valero & G. Tsatsaronis. Book no. I00331, pp. 137–46, New York: ASME.
Valero, A. & Lozano, M. A. (1992b). A general theory of thermoeconomics. Part II: The relative free energy function. In *On Efficiency, Costs, Optimization and Simulation of Energy Systems*, ed. A. Valero & G. Tsatsaronis. Book no. I00331, pp. 147–54, New York: ASME.
Valero, A. & Royo, F. J. (1992c). Second law efficiency and the relative free energy function. In *Thermodynamics and the Design Analysis and Improvement of Energy Systems*, ed. R. F. Boehm, AES, Vol. 27, Book no. G00717, pp. 271–78, New York: ASME.
Valero, A. (1993). La termoeconomía: ¿una ciencia de los recursos naturales?. In *Hacia una Ciencia de los Recursos Naturales*, ed. J. M. Naredo & J. Parra, ISBN 84-323-9792-0, pp. 57–78, Madrid: Siglo XXI.
Valero, A., Serra, L., & Lozano, M. A. (1993). Structural theory of thermoeconomics. In *Thermodynamics and the Design, Analysis and Improvement of Energy Systems*, ed. H. J. Richter. Book no. H00874, pp. 191–98, New York: ASME.
Valero, A. (1994a). On the energy costs of present day society. Keynote presentation in ASME WAM'94. In *Thermodynamics and the Design, Analysis and Improvement of Energy Systems*, ed. R. J. Krane. Book no. H01045, pp. 1–45, New York: ASME.
Valero, A., Lozano, M. A., Serra, L., & Torres, C. (1994b). Application of the exergetic cost theory to the CGAM problem. *Energy*, **19**, 365–81.
Valero, A., Lozano, M. A., Serra, L., Tsatsaronis, G., Pisa, J., Frangopoulos, C. & von Spakovsky, M. (1994c). CGAM problem: definition and conventional solution. *Energy*, **19**, 279–86.
Valero, A., Lozano, M. A., & Bartolomé, J. L. (1996). On-line monitoring of power plants performance, using exergetic cost techniques. *Applied Thermal Engineering*, **12**, 933–48.
von Spakovsky, M. R. (1994). Application of engineering functional analysis to the analysis and optimization of the CGAM problem. *Energy*, **19**, 343–64.

9

Artificial intelligence in thermal systems design: concepts and applications

BENIAMINO PAOLETTI
Alitalia and the University of Roma 1
ENRICO SCIUBBA
The University of Roma 1

9.1 Artificial intelligence and expert systems

AI (for an exact definition, see Section 9.3.2.1) is in reality a cumulative denomination for a large body of techniques that have two general common traits: they are computer methods and they try to reproduce a nonquantitative human thought process. General AI topics are not addressed here. For information on this topic see the various monographs giving fundamental information, including Charniak and McDemmott (1983), Drescher (1993), Rich (1983), and Widman, Loparo, and Nielsen (1989).

The applications we will deal with in this chapter are related to a smaller subset of general AI techniques: the so-called knowledge-based systems, also called expert systems. Referring the reader to Section 9.3.2.2 for definitions, will say only that an ES is an AI application aimed at the resolution of a specific class of problems. Neither a computer nor an ES can think: the ES is a sort of well-organized and well cross-referenced task list, and the computer is just a work tool. Nevertheless, ES (and in general AI techniques) can result in efficient, reliable, and powerful engineering tools and can help advance qualitative engineering just as much as numerical methods have done for quantitative engineering.

ESs have many benefits. They provide an efficient method for encapsulating and storing knowledge so that it becomes an asset for the ES user. They can make knowledge more widely available and help in overcoming shortages of expertise. Knowledge stored in an ES is not lost when experts are no longer available. An ES stores knowledge it can gather from experts in some field, even if they are not computer specialists (and even if they are not computer users!). An ES can perform tasks that would be difficult with more conventional software. Thus, implementing features that were virtually impossible with earlier systems becomes possible.

Developing and installing a large ES is a high added-value operation and therefore an expensive enterprise. The provisional returns should always be carefully scrutinized before launching an ES project.

9.2 Possible versus existing applications of AI in powerplant design and operation

An AI procedure reproduces a series of actions that a human expert (or team of experts) would take in response to a stimulus.

In the design and operation of thermal plants, virtually all engineering tasks could be performed with the help of (or entirely by) a properly programmed computer. Some of the possible applications of these AI techniques include process control (chemical, thermomechanical), maintenance management (in real time), emergency procedures, production scheduling (in real time), components design, process design and optimization, qualitative process modeling, and thermoeconomic optimization.

The list of AI applications actually implemented for these tasks is not as extensive as one would expect. They include several applications in production scheduling (most of them not in real time); several applications in process control, to assist or replace the operator in normal plant-conducting mode (notably, in emission monitoring); some applications in process monitoring; some on-line maintenance management procedures (notably, in the aircraft maintenance field); very few emergency-handling procedures (mostly in the chemical and nuclear industry); several applications in simulators for training purposes; some applications to prototype testing, including automatic procedures for data reduction and interpretation and for intelligent status checking; very few design procedures, the majority of which perform a sort of database scanning activity to select the most proper component from a fixed, preassigned list of possibilities; very few expert, almost real-time cycle and plant simulators; virtually no qualitative modeling, because the tendency is to have the computer do the number-crunching job and have the expert engineer make decisions based on the results of the simulations; no thermoeconomic optimization. Actually, some AI applications exist in the field of economics, but none of them has been translated into a suitable engineering equivalent.

9.3 Artificial intelligence

9.3.1 Artificial intelligence is no intelligence!

What is intelligence? Or rephrasing the question in a form more suitable to the paradigms of propositional calculus, what are the attributes a (not necessarily human) being must possess to qualify as intelligent? Specialists debate about an exact definition, but one that would be accepted by most modern cognitive psychologists is the following:

To be intelligent, the being must

- react in a flexible way to different situations;
- be able to take advantage of fortuitous circumstances;
- correctly assess the objective relative importance of the different elements of a new situation being faced;
- solve with original (that is, hitherto unknown to the being) means a problem not previously encountered;
- correctly understand the thinking patterns of others;
- be open-minded, that is, ready to accept the possibility of having incorrect thinking;
- produce new and innovative ideas.

One does not need to be a computer expert, logician, or AI practitioner to see that at least in the way we conceive computers and computer languages today, there cannot be an intelligent android.

9.3.2 Definitions of terms and concepts

9.3.2.1 Artificial intelligence

No definition is universally accepted for AI. One that is sufficient for the purpose of this chapter is the following (Widman et al. 1989): "AI is that part of computer science that investigates symbolic, nonalgorithmic reasoning processes, and the representation of symbolic knowledge for use in machine inference."

Historically, the first reference to the concept of AI is generally credited to McCarthy (Rich 1983), but his original definition was different from the one we have given and is no longer used.

9.3.2.2 Expert systems

An ES is an application (that is, a series of procedures aimed at the solution of a specific class of problems) built around a direct representation of the general expert knowledge that can be collected (and not necessarily systematized in any way) about a problem. In general, an ES associates a rule (see 9.3.2.5) with every suggestion or piece of advice a specialist expert would give on how to solve the problem.

The term *expert system* is inaccurate and misleading. It carries a sort of mystical flavor. The systems we are going to deal with could be better described as knowledge-based systems (KBSs).

KBSs (or ESs) do not use the same thought process as humans. They do not think, nor do they mimic human expertise. They simply store, in a special database, limited knowledge about some features of the real world that are put into them by human experts. The knowledge is stored in such a way that it can be manipulated formally, used to infer results, and in general, represented and applied in an orderly fashion as needed.

A KBS has a software architecture in which both data and knowledge are separated from the computer program that manipulates them (the latter is called the inference engine, and all knowledge is stored in the knowledge base [KB]).

9.3.2.3 Knowledge base

A KB is the model of the portion of the universe that is addressed by an ES and may contain a description of all elements of the classes under consideration, a list of their mutual relations, or a list of rules (including mathematical formulas) according to which to operate on these elements. A KB can contain procedures that point outside itself and create dynamic links to other procedures or databases.

9.3.2.4 Inference engine

An inference engine is the software that uses the data represented in the KB to reach a conclusion in particular cases. In general, it describes the strategy to be used in solving a problem, and it acts as a sort of shell that guides the ES from query to solution.

9.3.2.5 Rules

A rule is a way of formalizing declarative knowledge. Example: IF a machine is rotating AND IF it has blades AND IF a fluid expands in it, THEN the machine is a turbine. Notice that a rule is a statement of relationships and not a series of instructions. The IF...THEN construct shown here is entirely different from the FORTRAN IF...THEN...ELSE construct. The latter is a procedural conditional instruction: if A is true, then do B, or else do C; whereas a rule is a logical relational statement between propositions: if $\mathbf{p}$ is true and $\mathbf{r}$ is true...and $\mathbf{s}$ is false then $\mathbf{z}$ is true.

9.3.2.6 Facts

Facts are specific expressions each describing a particular situation. These expressions can be numerical, logical, symbolic, or probabilistic. Examples: height = 130 m; IF (outlet pressure) LESS THAN (inlet pressure) THEN (process is an expansion) IS TRUE; $\mathbf{p}$ TRUE $\forall x \in A$; IF (pressure high-frequency oscillations) is TRUE AND (shaft high-frequency vibrations) is TRUE, THEN probability (compressor stall) = 75 percent.

9.3.2.7 Objects

An object is a computer structure that can contain data, data structures, and related procedures. It is best described as "a package of information and the description of its manipulation" (Rich 1983). Example: Francis hydraulic turbine with $H = 300$ meters, $Q = 20$ cubic meters per second whose performance is described in subroutine FRANCIS. Notice that the first portion of this object is a declarative description, whereas the second portion is a procedure (which could be, for instance, written in a conventional language such as FORTRAN or *C*).

The traditional view of software systems is that they are composed of a collection of data (which represent some information) and a set of procedures (which manipulate the data). The code invokes a procedure and gives it some data to operate on. Thus, data and procedures are treated as completely distinct entities, although they are not. Every procedure must make some assumption about the form of the data it manipulates.

9.3.2.8 Classes

A class is a description of one or more similar objects. Example: Hydraulic turbines. Notice that in this case, the object described in 9.3.2.7 would be an instance of the class introduced here. Because Pelton and Kaplan turbines would, of course, also belong to this class, the corresponding objects will bear a significant degree of similarity with the Francis object but will perforce have some different characteristic in the description (for example, Pelton hydraulic turbine), in the data attributes (for example, $H = 1000$ meters and $Q = 5$ cubic meters per second), or in the procedural part (for example, whose performance is described in subroutine PELTON).

9.3.2.9 Induction

Sometimes erroneously called abduction, induction is a logic procedure that proceeds from effects backward to their causes. In spite of some literary examples to the contrary (noticeably, Sherlock Holmes, who used almost exclusively inductive reasoning with an amazing degree of success), induction is far from being infallible and has to be used with care, for it needs a metamethod for scanning all the actual causes an event may have and for properly choosing among them. Example: high-frequency vibrations detected on a gas turbine shaft (effect) may be caused by compressor stall, rotor unbalance in the compressor or turbine, fluid dynamic instabilities in the flow into the turbine, irregular combustion, mechanical misalignment of the shaft, a failure in the lubrication system, or bearings wear (to consider only the most likely causes – a crack in the gas turbine casing could also be one!). Pure induction will not produce a solution in this case, unless other data are available that can be included in the inductive process to eliminate some of the causes from the list.

9.3.2.10 Deduction

A logic procedure that proceeds from causes forward to their effects, deduction is much more reliable than induction (of which it is the logical opposite) but also much more deterministically limited. Example: compressor stall can cause high-frequency vibrations on a gas turbine shaft, but if the only cause we have in our knowledge base is failure of lubrication system, we will deduce (with a degree of certainty of 100 percent) the wrong conclusion. So pure deduction must also be implemented together with some double-checking procedures that leave open the possibility for unexpected different causes.

9.3.2.11 Backward chaining

A procedure that attempts to validate or deny a proposition (goal) by searching through a list of conditional rules and facts to see if they univocally determine the possibility of reaching the goal, backward chaining is also said to be goal-driven. Example:

rule 1: "Francis" IF "turbine" AND "hydraulic" AND "specific speed lower than 2 and higher than 0.3"
rule 2: "turbine" IF "rotating" AND "vaned"
rule 3: "hydraulic" IF "fluid = water"
fact 1: "specific speed = $rpm * \frac{(volume\ flowrate)^{1/2}}{(g * total\ head)^{3/4}}$"
fact 2: "rpm = 300"
fact 3: "volume flowrate = 30 m^3/s"
fact 4: "total head = 500 m"
fact 5: "fluid = water"
fact 6: "number of blades = 13"
goal: "given the knowledge base, determine if the machine is a Francis"
(result: "it is not")

A backward-chaining algorithm methodically applies each rule in turn to scan the knowledge base while attempting to verify the goal.

9.3.2.12 Forward chaining

Forward chaining (FC) is a procedure that attempts to deduce from a given knowledge base some new inferences. It is also said to be data-driven. Example: for the same KB as in 9.3.2.11, as FC algorithm would infer from Facts 1, 2, 3, and 4 that the specific speed is equal to 2.8; from Rule 2, Fact 2, and Fact 6, that the machine is a turbine; from Rule 3 and Fact 5 that it is hydraulic; and from Rule 1 that it is not a Francis. This example clearly shows a fundamental characteristic of AI inference procedures: the fact that the machine is not a Francis is already contained in the KB, but it is not explicit and needs some reasoning to be extracted. The deductions of an FC algorithm are not intuitive leaps. They are rather formally correct (albeit in some cases redundant, trivial, or just plain uninteresting) conclusions, that is, effects compatible with the given KB causes.

9.3.2.13 Decision tree

A decision tree is a graph describing a multiple-choice situation (Fig. 9.1). A given goal (the root of the tree) can be reached through the satisfaction of

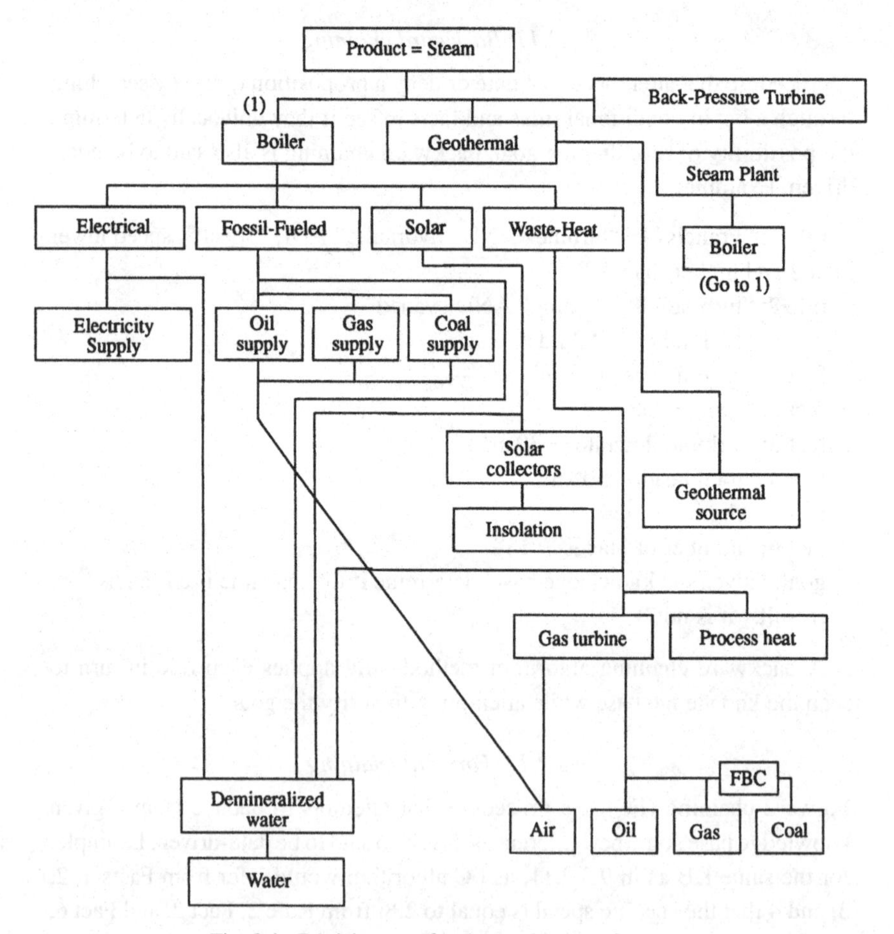

Fig. 9.1. Decision tree for the production of steam.

one or more partial goals, each of which admits its own subgoals, and so on. Because logical or physical constraints can be placed on each one of the nodes (which represent the partial goals) or on each of the branches (which represent the actions that take us from one goal to the next), in general not all paths are acceptable. The problem of finding a path from the root to the leaves (the basic goals, which can be viewed as the original facts representing the knowledge base) constitutes a search problem. A solution is a path connecting the root with one or more leaves. The mere existence of a decision tree does not guarantee that a solution exists or that, if existing, if can be found.

9.3.3 Limitations in logical systems

In spite of the formal completeness and mathematical rigor of logical (symbolic) systems, human thinking has been shown to *exceed* their representation capabilities, that is, anomalous knowledge descriptions exist that, although perfectly acceptable to a human expert, cannot be formalized (and thus cannot be manipulated) by any logical system. The most important examples of such descriptions are vague or approximate descriptions (Example: If the evaporation pressure is too high and there is no reheat, then you might have condensation somewhere in the final stages of the steam turbine) and hidden assumptions (Example: For a given steam throttle temperature, optimize the throttle pressure. Here, the hidden assumption is, "The process is not supercritical").

Vague descriptions can be tackled with a technique based on the so-called fuzzy sets theory, which is a part of logical sets theory and is discussed, for example, in Negoita (1993). The way to formalize approximate reasoning is briefly described in Section 9.3.5.

Hidden assumptions are more difficult to handle, because they are often case-dependent. Simple, general cases can be treated using nonmonotonic logical systems (Brooks 1991) (Shoham and Moses 1989).

9.3.4 Approximate reasoning

If, in a certain application, the formulation of a problem requires vague or approximated rules, one needs an additional set of rules that tell us how to match approximate facts. These rules expand the approximate description to specify all of its implications. For example: "steam near its saturation line" can be made explicit by rephrasing it as "water in a state such that (T) GREATER THAN (T_{sat}) AND ($T - T_{sat}$) LESS THAN ϵ." This is, in essence, a syntactical analysis of the given facts, and, although in principle always successful, it may become cumbersome to implement.

A particular rule that makes coping with approximate knowledge representations easier is the keyword rule. It is again a hierarchical rule, which considers only certain key features of the given facts to be important and tries to match them with the domains of the available inference rules. This approach is powerful, provided the choice of the keywords is proper – a problem in itself (which the domain expert ought to solve).

A common type of knowledge, closely related to approximate reasoning, is that expressed under the form of belief. It is approximate, because it does not result in "IF p THEN q" statements, but it can be treated in an effective way. Without going into details, probably the best way of treating knowledge expressed under the form of a believed fact (for example, "I believe that the number of feed water heaters in a conventional steam plant never exceeds eight.") is to translate into a form of conventional knowledge about the believed fact and some unlikely or unusual counterfact (Shoham and Moses 1989) (for example, "EITHER (number of feed water heaters) $\leq$ eight OR probability (that plant is a conventional steam plant) ≤ 0.01.") Not all such translations have to recur to probabilistic utterances. What is of essence is that the fact that would disprove the belief is a highly unlikely event.

The reason that we should seek ways to handle approximate knowledge is that it is much more powerful than exact knowledge and can allow for inference in cases in which an exact treatment would fail. For example, when designing a Brayton-based gas turbine plant, we may know in advance the maximum allowable cycle temperature, $T_3 = T_{\max}$, but need to know the pressure ratio and the inlet temperature, T_1, to compute the required fuel-to-air mass ratio, α. If the facts we assign are T_1, p_1, p_2, α, the code would be unable to cope with situations in which T_3 is different from the specified $T_{\max}$. But if an approximate fact is substituted for α, such as $0.01 \leq \alpha \leq 0.05$, the domain of application of the code is substantially increased.

Similarly, approximate facts are useful in monitoring and control applications.

9.3.5 Examples and problems

9.3.5.1 Construction of a decision tree

Consider the problem of steam generation. This can be regarded as a basic, apparently elementary, design assignment. The (engineering) reasoning needed to solve the problem is well established, because steam can be generated by only a limited number of processes. Therefore, this example offers a possibility of application for a tree decision diagram, which is represented in Fig. 9.1 and will be developed here.

One can consider the object STEAM to be the fruit of the tree (as in the scheme of Fig. 9.1) or, alternatively, its root (as it would be if the scheme of Fig. 9.1 were read upside-down). For our purposes the way the scheme is read makes no difference, so let us think of STEAM as the only fruit of the process tree.

One natural possibility for steam production is that of using a geothermal source. Correspondingly, we insert a box labeled GEOTHERMAL below the final product, STEAM.

But steam can also be extracted from a backpressure turbine or generated in a boiler. Therefore, two other boxes are parallel with the GEOTHERMAL one, labeled BOILER and BACKPRESSURE TURBINE, respectively.

On closer examination, it appears that the geothermal production is possible only if a geothermal source is available. Therefore, the branch presently ending with GEOTHERMAL will have at its root by necessity GEOTHERMAL SOURCE. Moreover, this branch is clearly terminated here, because the logical path GEOTHERMAL SOURCE → GEOTHERMAL CONVERSION → STEAM is a complete (and logically admissible) one. Whether this choice would be an engineering one depends on factors not considered here.

The branch that starts with BOILER is clearly the most important here. The first choice presented to us is that between electrical, fossil-fueled, solar, and waste-heat boilers. These four boxes will be four branches that collectively bear only one stem (BOILER), which bears the fruit (STEAM). Each of these branches requires some means of energy input and some flow of properly treated WATER, which can be conveniently evaporated into STEAM. By similar reasoning, the meaning and the position of the remaining boxes are readily determined. The roots of the decision tree are all materials found in the environment. Every engineering tree of this kind, which traces back the possible alternatives for the production of some good or service, is bound to have its roots composed by elements of the domain environment. See Sections 9.4.4 and 9.4.5 for some interesting applications of this fact.

The branch that starts with BACKPRESSURE TURBINE must have, as next supporting branch, the STEAM PLANT the turbine belongs to. Because this plant will perforce possess a boiler, a recursive call can be made to the BOILER branch.

The use of such a simple decision tree is trivial. Each complete path from the fruit STEAM to the roots WATER, AIR, FUEL, etc., represents a logically complete production chain. The engineering feasibility of each chain can be assessed by assigning some kind of quantitative information (weight) to each of the boxes (nodes) and of the branches (paths). One speaks of weighted or optimal search (that is, tree-scanning) strategies (Charniak and McDemmott 1985) (Pearl 1984). Additional information can be found in Gelsey (1989) and Papalambros and Wilde (1988).

9.3.5.2 An application example: search methods

Let us consider again the problem of steam generation, to see how the ideal decision tree (which describes a list of the feasible solutions) can be converted into an operative decision tree (which can be scanned to find the optimal solution). We will have to devise a quantitative dependence between each solution path and a cost or weight function whose value will be the discriminating parameter in the decision about the type of solution to choose.

Each one of the black boxes representing a component or a process (BOILER, BACKPRESSURE TURBINE, etc.) can be thought of as an operator, which allows the transition from the state of affairs before its insertion to that after its insertion. From the knowledge point of view, these operators transform the universe included in our database (original or primitive domain) into a different universe (modified or final domain). In the case considered here, these operators act on fluxes, and the set of all possible states that can be generated by the operators is the search space, that is, the decision tree displayed in Fig. 9.1. To search the tree means to scan in a convenient way the graph to find the optimal path between the problem position (generate steam) and the solution (one of the specified processes).

Scanning a tree by considering at each step all of the possible choices is not convenient and would result in an exhaustive search that for large trees, becomes soon intractable. We are looking for a metric (quantitative) description of each move (search step) that identifies locally the most convenient next steps, and that can be used recursively from the design goal to the solution. In every design problem several strategies are possible, and to claim absolute validity for one of them, negating the others, is in general unwise. On the other hand, to examine more than one or two possible search strategies is expensive, in resources.

Let us consider the following heuristic strategy.

A function f_{ijk} exists that quantitatively represents the degree of analogy between the j-th flux of the i-th BOX and the k-th element of the environment. This degree of analogy can be a thermodynamic criterion (the entropy difference or the exergetic destruction, for instance), an economic criterion (the unit cost of a process discharge for instance.). f is normalized so that its values span the interval between 0 and 1. g is the flux that for the i-th component, is nearest in some sense to the environment: $g_{i,k} = \max(j)\ [f_{ijk}]$.

p is the sum of all the weight (or penalty) functions of all the components considered at the present step: $p_k = \sum_i (g_{i,k})$.

The reasoning on which the strategy is based is the following: the more a BOX (i-th component) needs inflows that are nearer (in some measure) to the value they have in nature, the closer this BOX is to the environment, and the fewer are the additional components that will need to be inserted between that BOX and the environment.

This strategy is applied in the following fashion. Starting from the point at which only three components are present (GEOTHERMAL, BOILER, and BACKPRESSURE TURBINE), we recursively follow the path that, step after step, results in the highest value of p. We do this until we reach a BOX for which $p = 1$, that is a BOX that can be fed directly by flows taken as they appear in nature.

Such heuristics (the one given here and others, based for instance on the component cost) can be used not only to establish the most convenient scanning strategy but also to ensure that the resulting process is the one with the fewest components or with the minimal resource consumption.

To assess the relative validity of different heuristic strategies, a proper discriminant is the ratio between the total number of BOXES in the scanned tree, $N_{scanned}$, and the number of components in the final process-path, $N_{process}$. The lower the ratio $N_{scanned}$:$N_{process}$, the better the heuristics.

At the moment in which a particular heuristic strategy is chosen, the graph is not known explicitly, nor is it known in its entirety after the application of the chosen strategy has resulted in an optimal solution. Actually, an efficient search leaves most of the decision tree unscanned.

9.3.6 Techniques for developing expert systems

9.3.6.1 Fundamentals

The problem in building AI software is the need to create a set of tools for the construction, maintenance, manipulation, and use of a possibly large, domain-specific knowledge base.

No universal procedure or standard set of guidelines exists to achieve these goals. To a large extent, the data influence the inference engine, and vice-versa. However, two general principles should be adhered to while building an ES:

- For each possible domain of application (more than one can be foreseen for most practical cases), it should be possible for a domain expert to educate the program interactively, checking and correcting its behavior.
- It should be possible for a non–computer-specialist domain expert to assemble, maintain, and modify the knowledge base.

Traditionally, the task of physically coding the AI constructs is the task of a person who interacts with several domain experts and, being a computer specialist and a general AI practitioner, puts together the building blocks of the code using some of the well-known structured programming techniques. Lately, some attempts to substitute an artificial assistant for the human one have been successful (Drescher 1993) (Varela 1992). So in the following section, we will talk about an assistant without specifying whether human or machine.

The choice of the language to use is naturally fundamental in practice, but, for our point of view here, irrelevant. See, however, Section 9.3.6.3.

Once the language has been chosen, the assistant will have to gather from the expert the fundamental knowledge about the problem to be solved:

- What are the goals?
- What are the premises?
- What are the constraints?
- What are the general laws governing the phenomena?
- Are these laws regular, or do they admit exceptions? Do these exceptions admit a systematization of any sort?
- Can intermediate checks be made during the solution procedure to avoid its running astray?
- Can the basic knowledge about the problem be formulated in chunks? If yes, how well structured are these chunks? Does each subdomain (chunk) admit its own inference engine? Is a single inference engine responsible for the management of the entire body of knowledge?
- Are there implications that explicitly or implicitly require access to a different knowledge base? What kind of expert advice (in a different domain) should be sought?
- Can the expert formulate a macrocode? If the assistant formulates a different one, can the expert critically review it?
- Can the expert bring about one or more test cases for which the solution is known to exist and is – in qualitative or quantitative terms – entirely or partially known?

9.3.6.2 Steps in the development of an ES: a general workplan

Motivation Many reasons may lie behind the decision to commit material and human resources to the development and implementation of an ES. The most common ones are the need for formalizing and clarifying a batch of knowledge that can be gained by having a human expert make his other reasoning explicit, and the need for combining the expertise that can be gathered from several human experts into a body of knowledge that can then be studied for consistency and reliability. Of course, lower-level needs also exist: for instance, operator overloading or shortage of specification engineers. But these can be considered particular instances of the two main motivations stated above.

Stages of ES building It is important to start with small systems (both in the size of their knowledge base and in the complexity of the inferential engine). Empirical validation is mandatory and can be carried out in stages, as the system progresses toward its final form. Various stages in the development of an ES can be recognized:

- Initial knowledge base design phase: This comprises three principal subphases:
  - Problem definition: the specification of goals, constraints, resources, participants, and their roles.
  - Conceptualization: the detailed description of the problem, its possible breakdown into subproblems, the formulation of each of these subproblems (hypotheses, data, intermediate reasoning concepts); study of the possible correlations between the chosen formulation and its feasible computer implementation:
  - Computer representation of the problem: specific choice of representation for the elements identified in the conceptualization phase: possible choice of an existing general-purpose shell; choice of computer language; choice of user interface, choice of the operating system, etc.
- Prototyping: the main purpose in prototyping is to ensure that the project is feasible and to gain some idea of the ultimate costs and of the benefits that can be expected from the code. The prototype should give a feel of how the ultimate system will work, possibly offering also a realistic user interface. Building a prototype will take a nonnegligible portion of the total time allotted for the project, and therefore solving possible fundamental problem-related issues before starting the prototype phase is important.
- Refinement and generalization of the knowledge base: this phase is the one in which, through iterative testing and double-checking, the code is adapted so that it can perform well in *all foreseen circumstances*. Of cource, this result cannot always be ensured, but the number of diverse occurrences is in general finite (and seemingly infinite cases, such as "for continuous variation of the pressure between 5 and 22 MPa," can be reconduced to a finite number of instances by, for example, subdividing the pressure interval in a number n of subintervals), and the general trends of the solution can often be inferred with the help of the ES itself. In any case, this phase is by far the most expensive (in manpower) and can easily occupy a well-trained and experienced design team for months.
- Fielding: In its final version, the code is installed on the site of its final use. As senior and junior computer specialists (and anyone who has ever been on the user's side of a newly developed system) know well, no instance is known of a

code that performed as it should when it was first installed. Aside from system problems, interface incompatibilities with other codes, and obvious bugs, for an ES the possibility exists of a logical fault or incompatibility with the real data of the real problem. The more extensive and careful the prototyping phase has been, the easier will be solving susbtantial installation problems (we consider here trivial any system and compatibility problems, as opposed to substantial problems such as possible faults in the inference engine or in the knowledge base representation). However, it is good planning to consider the possibility of having to assist the user in this installation, not only with system specialists but also with domain experts and knowledge engineers.

A general schedule of activities Let us suppose that we are embarking on a ES design activity. What are the necessary steps that lead us from design inception to the generation of one (or possibly more) final designs? This section offers an answer to this question, with the warning that to be able to encompass a wide range of designs, the procedure becomes of necessity overgeneralized. It is left to the wisdom of the reader to extract from this example all the information that can then be applied to specific cases and to fill in for the obvious gaps that exist between a general method and its particular applications.

The process of generating a design plan has three phases, roughly identifiable as problem specification, analysis of functional relations, and design plan generation (McBride and O'Leary 1993, Reklaitis et al. 1983).

To start a design activity, it is necessary to specify the problem, or determine the parameters and functional relations relevant to the problem. Included are determining the parameters that constitute the design goal, the parameters that constitute the data, and the relations that must be taken into account during design. This phase is divided into three subphases.

First is problem identification and positioning. This task can be performed best by checking a sort of itemized identification list:

- Can the problem be decomposed into a set of smaller and easier (in some sense) independent or loosely dependent subproblems?
- Conversely, can the problem be realized to be a subproblem of a higher-level problem for which the solution is already known?
- If the design problem is composed of an assigned series of design steps, is it possible to ignore or skip steps that are found to be either conflicting with others or superfluous for the general solution strategy?
- Is it possible to foresee the complete problem domain (its universe)?
- Is one solution sought after, or is it necessary to produce and compare several different solutions?
- Is the assigned (or gathered) knowledge base self-consistent?

- Is it known from previous experience that the problem requires a large body of knowledge, or is it sufficient to gather enough knowledge so as to limit somehow the solution space?
- Is the problem such that all of its knowledge base is well defined and unchanging, in which case it can be foreseen that the still-to-be-produced ES will run in batch mode, or is a portion of this knowledge base uncertain or variable, so that an interactive mode should be chosen by necessity?

  The answers to these questions result in a set of guidelines that will indicate which methods are likely to be more useful in solving the problem and will direct the actual implementation of the code.

The second subphase of problem specification is identification of always-relevant parameters and relations. Assuming the configuration of the final product of the design is known, its structure must be analyzed and decomposed to derive a fundamental set of parameters and to define an initial set of relations linking them to each other and to the design data. In many instances, this step is unnecessary, because the related knowledge already exists and has been codified in numerous examples of prior successful designs. In other cases, one can make use of a special class of ES that goes under the name of automated physical modelers (Varela 1992), and that is capable of deriving a fundamental set of parameters and of functional relations among them on the basis of some physical data on the system to be designed.

The third subphase of problem specification is determination of how the given specifications affect the possibly relevant parameters. The specifications (design data) may restrict a possibly relevant parameter, either directly (by setting a range of acceptable values for it) or indirectly (by setting a range of accepted values for another parameter that is correlated to it). In either case, the formerly possibly relevant paramenter becomes relevant, and establishing and investigating the entire chain of functional relations that link all the parameters to one another becomes necessary. These functional relations are the usual design relations: equality, constraint (weak and strong), polynomial dependence, etc.

After the problem has been specified, the next phase is analysis of functional relations and application of heuristics.

The purpose of this phase is to create a skeleton of a feasible (and possibly optimal in some sense) design plan. Both domain-dependent and -independent heuristics are applied to determine the proper grouping of variables and both the order and the method of assigning values to the parameters. Three subphases can be identified.

First is creation of tightly coupled sets of parameters. The task of this subphase is to identify groups of highly interdependent parameters, which should be designed together. The set of relevant parameters is divided into subsets, either with the help of appropriate domain-dependent heuristics or simply by inspection. The goal is to have in one subset all parameters such that, if one of them is varied, then by some simple and predetermined relation all others are varied as well. Generally speaking, the number of subsets will equal the independent degrees of freedom in the design problem.

The second subphase in analysis of functional relations is determination of the order and methods of assignment of values. Starting from the design data, that is from a set of numerical values assigned to some of the problem parameters, it is possible to hierarchically derive an order in which parameters to be derived must be assigned their values and, at the same time, to decide which relations should be used to attribute those values. Usually more than one order is possible, but for each order only one pair exists (order method) and vice-versa, that is, choosing the order and the method is really only one operation.

The third subphase in functional relations analysis is imposition of weak constraints. At this stage, some of the known inequality relations may have been left unused. These can be either internal consistence checks (for example, $p_2 \geq p_1$ in a Brayton cycle) or externally imposed constraints. In either case, they must be logically attached to the parameters they pertain to, so that unfeasible solutions can be formally identified before numerical values are assigned.

In some cases, it is convenient to include in this phase an automatic procedure that varies stepwise one or more of these constraints when assigning numerical values to the parameters. This procedure produces a sensitivity analysis, or forces the generation of several final design configurations, one for each value of the constraint that is being varied.

After the functional relations have been analyzed, the third and final phase is design plan generation. This is a mapping of the skeleton plan to a design plan. The relevant parameters are mapped (that is, set equal to) the variables of a plan. The methods for assigning values are mapped (that is developed into) numerical or functional operators that act on the variables. Then, the goals of the plan are created, and the proper variables are attached to them. Sets of variables corresponding to sets of tightly coupled parameters are attached to individual goals, together with the applicable constraints. At the end of this process, all the elements of the design plan will be connected to one another in a network of goals that possesses the correct semantics of the original design problem and the general rules of logical networks.

A design plan does not necessarily originate one solution. If the plan is correctly laid, it will usually produce more than one feasible design. If needed, optimization procedures can be devised to choose, among different configurations proposed by the ES, the one we consider more desirable. Additional information is needed to perform this extra task. If the configurations proposed by the ES are individually considered, they are all optimal with respect to the original problem formulation and the knowledge imparted to the ES.

9.3.6.3 Artificial intelligence languages and shells

The practice of applying the concepts of artificial intelligence to the solution of practical problems has brought about the necessity of developing programming environments specific to the concepts and procedures of the techniques to be implemented.

In principle, this step is not necessary. Any language can be used to implement any logical procedure or technique. Other considerations make (in a weak sense) necessary the development of new programming instruments. The so-called added value of a new programming language or tool consists in the fact that the new language has been specifically devised to implement a particular set of solution methods with the highest possible efficiency, and therefore its logical patterns are close to the logical patterns of the theoretical inferential approach. Thus, the conceptual abstract level at which the problem is solved in principle is almost the same at which its actual solution is hard-programmed. Conversely, the knowledge and the practice with a particular programming language can induce the user to tackle a problem with a logical approach close to that implemented by the language.

Distinctive characteristics of each programming language are the type and quality of its inferential engine and the way in which the knowledge base can be represented.

An efficient and flexible data representation and retrieval or association is much more important for a language than its inferential engine. For this reason, we must mention an important class of languages not necessarily designed for the procedures of AI but that represent almost-ideal AI implementation environments: the so-called object-oriented languages (see also Section 9.3.2). Object-oriented languages can combine into a single knowledge base both the knowledge data and the dedicated procedures to interrelate or compute these data. Furthermore, they are intrinsically organized into classes and subclasses, so that procedures specific to a class can be passed to all or some of its subclasses by a built-in capability called inheritance. This chapter cannot delve further into these topics, but we must mention that the concept of instance (see Section 9.3.2) is implemented in an object-oriented language by the simple "is-a" operator (**A** is-a **B**). Implementing the same notion in a different language is cumbersome.

As a consequence of these convenient properties, object-oriented languages have become an important part of the AI-language family.

LISP LISP is the AI language that has enjoyed the largest popularity. Its distinctive characteristic is its ability to handle lists of simple or complex objects. It is therefore a flexible language. An unlabelled list can easily be stored, and LISP data manipulation procedures allow for many data-handling operations (label/unlable, split/merge, add/cancel, aggregate/disaggregate), which make passing a property from one object (the list) to another easy. Because homogeneity is not a required attribute for a list, an object can be qualified further by a sublist inserted in the original list that described that object. The method is an easy way to join declarative and procedural knowledge. Also simple is implementing a set of lists of biunivocal correspondences. Property **p** is linked to its value **v** by a list containing just **p** and **v**. This set of property values can be then considered to be a list by itself and attached to the original list describing a particular object. Such a procedure has many possibilities in design applications. An object **P** (a plant) can be described using its general attributes (boiler, turbines, condenser, pumps, feed water heaters), and then a property-value list can be attached to each component (boiler, $T_{max,steam} = 550\,\mathrm{K}$, etc.).

LISP does not have its own inference engine but has a library of procedural operators by which the user can construct the engine of choice.

Several LISP dialects and jargons exist, which are specialized versions of the basic language, developed for particular applications or for commercial reasons (InterLISP, Planner, etc.). LISP machines have been constructed as well. These are computers in which some or all of the list-manipulation operators have been hard-coded.

PROLOG PROLOG is the best known (and used) AI-programming language. It has, like LISP, extensive list-manipulation capabilities. In addition to these, however, PROLOG has an internal inference engine based on two operations: the resolution principle, $\mathbf{p}_1 = \mathbf{fbf}_1$, and the unification principle, $\mathbf{if} < \mathbf{p}_1\ \mathbf{and}\ \mathbf{p}_2\ \mathbf{and} \ldots \mathbf{p}_n >= \mathbf{then}\ \mathbf{p}_{n+1}$.

These are called the Horn rules of first-order logic theory and can be used to establish the truth of single facts (p_1) or to deduce the truth of complex facts through a cause–effect chain (p_{n+1}).

These expressions are all the fundamental logical expressions of PROLOG, and therefore the programmer must base descriptions of the universe of the problem in terms of facts (or truth conditions, or axioms) and then deduce from them a solution to the problem. The inferential engine is of the backward type (see Section 9.3.2), so that every problem must be formulated as a verification problem whose aim is to check whether a given task can be performed, that is whether a chain of atomic facts (axioms) exists that leads from the general premises to that task.

In spite of a certain awkwardness of its grammar (it does not possess any of the WHILE...FOR...DO constructs of the usual programming languages), PROLOG is so widely employed that several versions have been developed for specific platforms or for specific applications (often, with the addition of extensive metastructures). One finds, among others, PROLOG2, Arity-PROLOG, Quintus-PROLOG, VM-PROLOG, and Visual-PROLOG. A remarkable late development is an object-oriented version of Quintus-PROLOG.

SMALLTALK SMALLTALK is an (and some would say "the") object-oriented language. It was conceived to be so and was designed to be used in a visual programming context. The language is procedural, characterised by sophisticated class-handling capabilities, by an extensive and user-friendly interface, and by a superb, proprietary graphic development package. It is useful for rapid prototyping, because (after the necessary initial learning phase) it effectively translates the procedural statement of a problem into a complete working application.

SMALLTALK is not portable, and several versions exist for both UNIX- and DOS (Windows) platforms. Because its best use is in the prototyping phase of a project, SMALLTALK is not so widely popular among design engineers, who prefer to use the same environment to prototype and to implement the final version of a code.

C++ C++ is a special version of the original C language, extended to support object-oriented programming. Therefore, it adds to the power and flexibility of its parent language the advantages of a complete object-oriented environment, such as heredity, embedding, and inhomogeneous predicate procedures. A serious limitation of C++ is the almost total lack of class-handling procedures. Some versions implement this capability through metaprocedures but at the cost of portability. C++ is a complex language that requires a long learning time, but it is bound to constitute a sort of reference basis for all object-oriented AI languages in the foreseeable future because of its direct descendance from C, which is probably the most widely used declarative programming language.

ART ART is a shell for the development of ESs and for knowledge-based models. It was produced to integrate the advantages of both declarative and procedural languages and has a reversible inferential engine that can work in the forward- and backward-chaining mode. ART is, in essence, a system of rules whose activation depends on some events or preconditions. This activation process being dynamic, it can vary not only from application to application but during a single application. Its internal structure makes it ideally suited for the implementation of systems of semantic rules (including the construction of an ES).

9.3.6.4 Examples and problems

Conceptual construction of a thermal-plant–oriented ES This example describes one of the basic problems encountered in the development of an ES: the formal implementation of the model rules. Some of these rules (see Sections 9.3 and 9.3.6.2) represent portions of the knowledge base of the specific tasks to be modeled, whereas others are expressions of a general or specific solution method. In the case discussed here, the objective was that of explicitly formalizing a set of definitions and of operational principles that could lead to the automatic synthesis of a feasible process starting from the specification of its products and from the a priori imposition of some engineering contraints.

No unique way exists to achieve this objective: Section 9.3.6.2 prescribes the general operational steps required to reach a solution, but no universal method has yet been found to formulate models of physical processes. The approach taken here is that of a game. We have tried to define some rules of the game of modeling, starting with initial formulations that were both simple and naive. These sets of rules have subsequently been sharpened and deepened by systematic use of counterexamples, that is, by the mathematical technique that goes under the name of reduction ad absurdum.

The general procedure can be described as follows: first, an initial commonsense definition of a rule is formulated, then it is implemented mechanically and rigidly in all the types of practical engineering situations that can be imagined, trying to discover some absurd, nonphysical, or improper consequences of its application. The original rule is then iteratively amended to avoid its impossible consequences and to make it more robust to slight perturbations in the engineering scenario with reference to which it was employed. This method may seem intrinsically slow, but it converges quickly to feasible, physically sound, and logically robust rules with no implementation problems. At least two types of cultural backgrounds are needed in a team trying to follow this route. The physical–engineering know-how must be merged with its mathematical–philosophical counterpart. What is actually more convenient is that the compentences be totally separated, that is, that the mathematicians do not know anything (or know little) about the physical processes to be synthesized.

The problem is not that of understanding the principles of engineering material science, thermodynamics, and chemistry according to which a process is designed. These principles constitute the domain knowledge that can be easily captured in a set of possibly complex engineering rules. The real problem is that of capturing the formal principles that constitute the method (or the art) of engineering design. This reason is why the adoption of a design game strategy is convenient: because the rules of the game also define its outcome for a given set of premises, it is possible to adjust these rules to obtain the proper results from every other game one can think of playing under different premises.

A fundamental step in the game strategy we are trying to develop is this: In constructing of a process, we have to start from the final product and proceed backward toward the basic fluxes of constituent materials. The most basic materials are, of course, those contained in what we call the environment, so one of the general rules of our game will be the following:

Rule 1: The process is to be thought of as a path connecting the final product $\mathbf{F}$ with the fluxes $\mathbf{f}_{\mathbf{env}}$ available from the environment.

A mathematical proof can be represented by a graph (like a decision tree) in which the nodes are the successive states of the propositions that constitute the proof material and the branches are the mathematical inferential symbols that connect these states (implies, negates, etc.). Similarly, a process can be thought of as a graph whose nodes are the components and whose branches are the fluxes. So the second general rule of our game will be the following:

Rule 2: Process synthesis is topologically similar to the construction of a decision tree whose roots are in the environment and whose fruit is the final product.

This rule constitutes another important step in the search for a solution, but it has an important drawback. If the fluxes components tree is indeed a suitable representation for a process, in the sense that it can in principle be used to proceed backward from the final product to the environment, in practice the extensive or intensive parameters of some of the fluxes required by one component may not match the corresponding parameters of the available fluxes (at that level of the tree). Common engineering sense tells us that this reason is indeed why we insert pumps, compressors, expanders, throttling valves, and heat exchangers in real processes. So a third rule must be introduced:

Rule 3: A generic flux is specified (that is, identified) both by its qualitative and by its quantitative parameters.

This rule assures us that a mismatch will be immediately detected. A first corrective action in such a case would be that of defining some auxiliary components (pumps, compressors, expanders, throttling valves, and heat exchangers) and of inserting them on the skeleton of a process constructed by some rules of inference that lead from the product to the environment. But this sequence is not what happens in reality. In the real engineering game, we are prepared in advance not only to insert, say, a feed water pump upstream of a boiler to match the pressure level of the feed water with that required by the boiler, but also to insert an air-to-water heat exchanger (intercooler) between the first and the second stage of a compressor, to improve the process even if the sensible heat of the water stream is subsequently wasted into the environment.

The problem can be solved by recurring again to the complexity principle. As a mathematical proof is applied whenever it reduces the complexity of the problem, so an auxiliary component should be inserted whenever possible, because its application introduces compatibility where none existed and simplifies the process by avoiding the construction of more complicated alternatives to circumvent the mismatch. So a fourth general rule of our game will be as follows:

Rule 4: Whenever a mismatch is detected in the process parameters, auxiliary components must be introduced to eliminate it.

This rule is practically meaningless unless used in conjunction with yet another general rule:

Rule 5: The fluxes required as inputs by the auxiliary components must be sought first among those available in the environment and then among those made available by the process at its present stage of definition. Only if both of these searches are unsuccessful should new required external inputs be specified.

If we consider the set of game rules assembled so far, we realize that the entire game of process synthesis has been reduced to a top-down inference procedure, which starts from the final product (the final effect, the symptom) and traces its way back to the initial constituent materials (the first cause, the disease). The local subrules that drive the choice of the single components, of the type and structure of the connections, etc., are the usual engineering rules and can be easily implemented in local procedures.

The problem is solved, at least in principle. Or is it? Again, the game approach offers some further insight. This type of procedure can be applied without excessive difficulty to the synthesis of energy conversion processes, as long as they do not require extensive use of heat exchanger chains (also called heat exchanger networks or HEN). Repeated failures in designing a heat exchanger chain led to the discovery that often, the top-down strategy alone cannot handle these cases. When several streams, some hot and some cold, must exchange heat among them in a thermodynamically acceptable fashion, a compromise must be struck between the top-down and the bottom-up approach. Suppose, in fact, that the first of these heat exchangers is introduced. It will fix one or more mismatches and produce less hot and less cold outputs than they were before its insertion in the process. But what if we now introduce a second heat exchanger? It, too, will reduce some mismatches and modify the temperature levels of some of the fluxes. Would the result have been the same if we had inserted the second heat exchanger first? Almost surely not. A heat exchanger reduces a mismatch in a different way depending on the stage of the layout at which its insertion takes place. But this reality is tantamount to negating the top-down approach. It becomes necessary at this point to introduce yet another game rule:

Rule 6: Any time two or more heat exchangers are being inserted in a process diagram, even if the insertion of each additional unit is decided according to a top-down strategy, the procedure to be used is of a bottom-up type. Because every new heat exchanger is subjected to local conditions that depend on the present stage of the process layout, a local iterative procedure must be introduced to fine-tune the heat exchanger chain.

This rule appears once more to cure all problems. Experience has shown, however, that avoiding nonlocal effects of the iterative procedure is difficult. Often, the interconnectivity of a process is so high that small changes propagate throughout the entire structure quickly (that is, within few iterations). This feature is not negative, per se, and a process synthesis procedure to be so sensitive is actually desirable. But the number of slightly different processes that are originated by this strategy grows exponentially with the number of heat exchangers and can become so taxing for the computational resources that the code is prevented from producing any output within reasonable times. Yet another fix is badly needed.

The solution came again from an analysis of the thinking patterns of design engineers. It was noted that when choosing a component, no exact sizing is done by the process engineer. Rather, an approximated or estimated value for the component size is guessed, and some corresponding values for the relevant fluxes connected to that component are derived. It is only at a later stage in the process design that balance equations are used to completely specify all components. The difference between an expert engineer and a novice is mostly that the estimates of the expert are much closer to reality than those of the novice. Then, a similar procedure can be introduced in our game strategy. Each component can be labeled by an approximate indication of its size, and the entire layout can be devised enforcing an approximate match between all components, including heat exchangers. The number of acceptable solutions will increase substantially, but the advantage of reduced iterations will probably more than offset this effect. And once a certain number of complete process structures has been created, it is not difficult to run a mass and energy balance code to size each component exactly. This step brings us to the formulation of another game rule:

Rule 7: When introducing a new component in a process, a fuzzy label must be attached to it, which contains the estimated operational ranges for both its extensive and its intensive parameters. The immediate successor of this component (in the top-down direction that the procedure commonly follows) must have operational ranges compatible with those of its parent component. This rule applies to heat exchangers as well.

The seven game rules defined so far seem to constitute a successful game strategy. Further verification is pending (see Sections 9.4.4 and 9.4.5). We are still playing this game.

9.4 Applications of artificial intelligence methods to the design of thermal systems

9.4.1 Foreword

The four examples included in this section are representative of the applications of AI methods to the design of thermal systems. The last two are especially worth careful consideration. They represent attempts to generate a feasible process or plant configuration with no previous qualitative or quantitative knowledge of what the solution should look like, basing the decision process on qualitative general commonsense engineering rules only.

9.4.2 Choice of a feed water pump

In this example, an automatic procedure must be devised for the choice and specification of a feed water pump. Because the extra work is negligible, we will find it convenient to construct the procedure so that it can handle any kind of pump geometry (axial, mixed, centrifugal) and cover the entire range of possible operation.

The following is a task list that can be constructed by consulting a pump design textbook (or by asking an expert):

- read design data: Q, H
- read constraints (if applicable): NPSH, speed, service, etc.
- is rpm given?

  if Y ⇒ compute n_s

  if N ⇒: is NPSH given?

  if Y ⇒ compute iteratively s and derive $rpm_{\max}$

  if N ⇒ (assume rpm, compute NPSH, check n_s), *iterate*

- check if multistaging is required

- check if multiple pumps in parallel are required
- compute d_s from Cordier line
- compute D_2
- choose $\phi = \frac{V_{2m}}{U_2}$
- compute b_2
- choose $\frac{D_2}{D_{1e}}$ and $\frac{D_1 i}{D_{1e}}$
- compute D_{1e} and D_{1i}
- verify torsional stress on shaft
- compute overall dimensions (approximate)
- check component library for nearest defined pump type
- check characteristics: head at Q_{design}, $Q_{\min}$, and $Q_{\max}$
- produce a technical specification sheet

Solution We will try to reduce the tree of design choices (Fig. 9.2) to a sequel

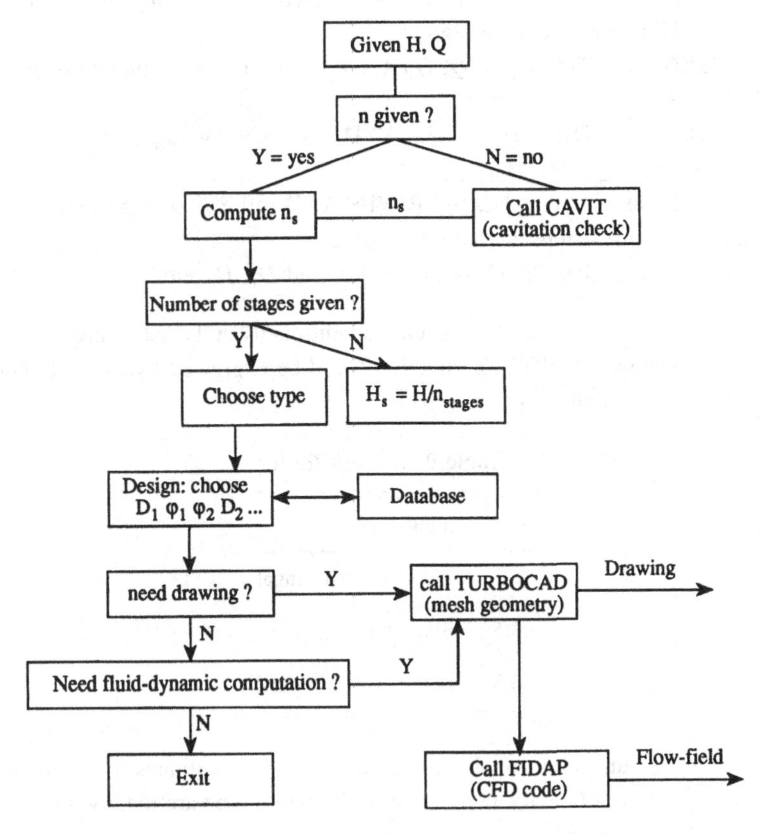

Fig. 9.2. Decision tree for the selection and design of a feedwater pump.

of rules of the type

$$\textbf{IF } p \textbf{ THEN } q,$$

where p is the premise (which can consist of more than one rule) and q is a series of design activities.

In this example, the decision tree (which we will assume to have been drawn following some expert advice) shows that in addition to conditional rules of the type shown above, some other steps imply some quantitative decision, such as *choose* $\mathrm{NPSH}_{\min}$, *choose* D_1, etc. The easiest way to account for these steps is to treat them as functional calls to some external numerical procedure that will return the values of the properly addressed variables. A set of rules that reproduce the decision tree of Fig. 9.2 is the following:

- Rule 1: **IF** *rpm* not given
  THEN call CAVIT (NPSH, s, *rpm*, T) **AND** read resulting *rpm* value.
- Rule 2: **IF** pump type not assigned
  THEN call PTYPE(*rpm*, Q, H) **AND** read n_s and resulting pump type.
- Rule 3: **IF** n_{stages} not assigned
  THEN call PSTAGE(n_s, Q, H) **AND** read resulting n_{stages}.
- Rule 4: **IF** $n_s(Q, rpm, H_{\text{stage}}) \neq n_{s,\text{old}}$
  THEN call CAVIT **AND** call PTYPE **AND** call PSTAGE **AND** iterate.
- Rule 5: **IF** dimensions are not constrained
  THEN call PSHAPE(Q, H_s, n_s) **AND** read D_1, D_2, etc.

The remaining steps can be executed conditionally at the user's prompt.

As an example, the PTYPE function could be expressed by the following truth table ($n_s = rpm \frac{Q^{0.5}}{H^{0.75}}$):

Table 9.1. *Truth Table*

n_s	Pump Type
<0.2	Multistage centrifugal
<2.5	Centrifugal
<3.5	Mixed-flow
<5.5	Axial
>5.5	Multiple axial in parallel

The use of functional maps for the pump greatly facilitates the execution of the procedure. Maps for the choice of the pump type are readily available

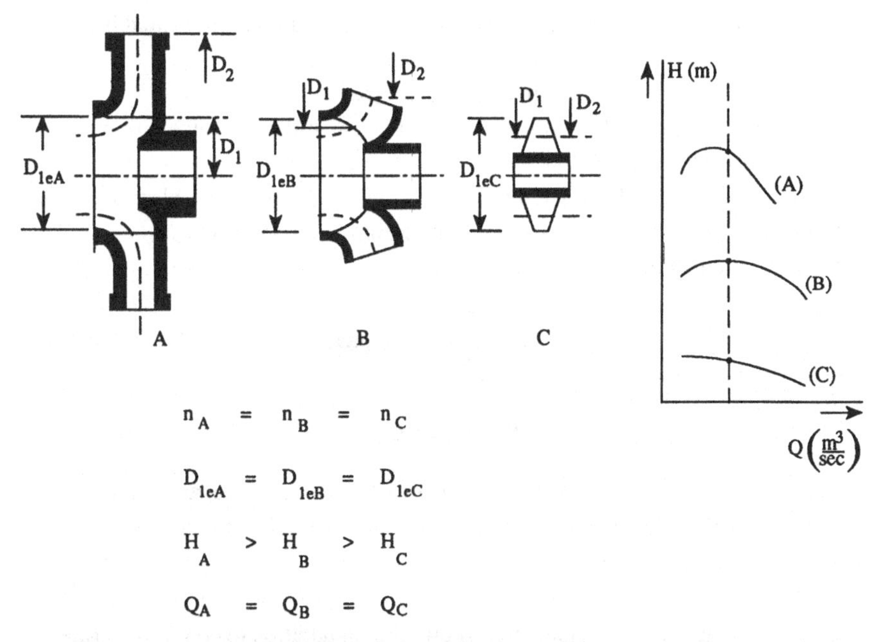

Fig. 9.3. Operational maps of dynamic pumps (schematic): n, rpm; D_1, inlet pump diameter [m]; H, total head [m]; Q, volumetric flow rate [m^3/s].

(see Fig. 9.3), as well as tables from which one can choose D_1, D_2, ϕ_1, etc. as functions of n_s. This point is not marginal, nor is it by chance that such information has been codified for us in this form. These maps and tables represent the combined knowledge of a multitude of experts who have solved the same problems before us and constitute in this case a conveniently structured, ready-to-use knowledge base.

9.4.3 Choice and technical specification of a gas turbine plant

In this example, an automatic procedure must be devised for the choice and specification of an open-cycle gas turbine plant. The procedure will read the technical requirements (design data), choose the proper type of plant from a library of available plants, suggest possible (limited) cycle modifications, and produce a technical specification for the design engineer.

The following is a task list that can be constructed by consulting an energy system design textbook (or by asking an expert):

- read design data: P_{el}, P_{therm}

- read constraints (if applicable): speed, service, fuel types, availability of process water for injection into the combustor, etc.
- check whether $\frac{P_{\text{therm}}}{P_{el}}$ is within a feasible range
- scan library seeking for components (that is, gas-turbine plants) with matching or almost matching P_{el}
- try to match type of service, fuel, rpm range, etc., with the problem data
- match $\frac{P_{\text{therm}}}{P_{el}}$ of the proposed plant with design requirements
- produce a technical specification form

Solution The flowchart of the procedure is depicted in Fig. 9.4. We will try to reduce the tree to a sequel of rules of the type

$$\textbf{IF } p \textbf{ THEN } q,$$

where p is the *premise* (which can consist of more than one rule) and q is a consequence. Notice that premise and consequence are functionally equivalent to cause and effect here.

In this example, the decision tree shows that in addition to conditional rules, the procedure includes other steps that imply some quantitative decision, such as: simulate process, check P_j, etc. These steps will be treated as functional calls to some external numerical procedure that will return the values of the properly addressed variables. A set of rules that reproduce the decision tree of Fig. 9.4 is the following:

- Rule 1: **IF** $\frac{P_{\text{therm}}}{P_{el}} \leq 0.5$
  THEN abort procedure **AND** display warning **ELSE** check library by P_{el}.
- Rule 2: **IF** $0.9P_{el} \leq P_j \leq P_{el}$
  THEN make suggestions for matching P_j and P_{el} **AND** check library by $P_{el} = P_j$ and type of service.
- Rule 3: **IF** available fuels match required fuel
  THEN check whether $\left(\frac{P_{\text{therm}}}{P_{el}}\right)_{\text{library}} = \left(\frac{P_{\text{therm}}}{P_{el}}\right)_{\text{required}}$.
- Rule 4: **IF** check is negative (i.e., no match)
  THEN call GTSIM, change $T_{\max}, \dot{m}_w$ to obtain match (iterate) **ELSE** write technical specification.

The remaining steps can be executed conditionally at the user's prompt.

As an example, the suggestions invoked by Rule 2 to force a match between P_j and P_{el} could be derived as follows.

In a small neighborhood of the design point, a variation in the output power

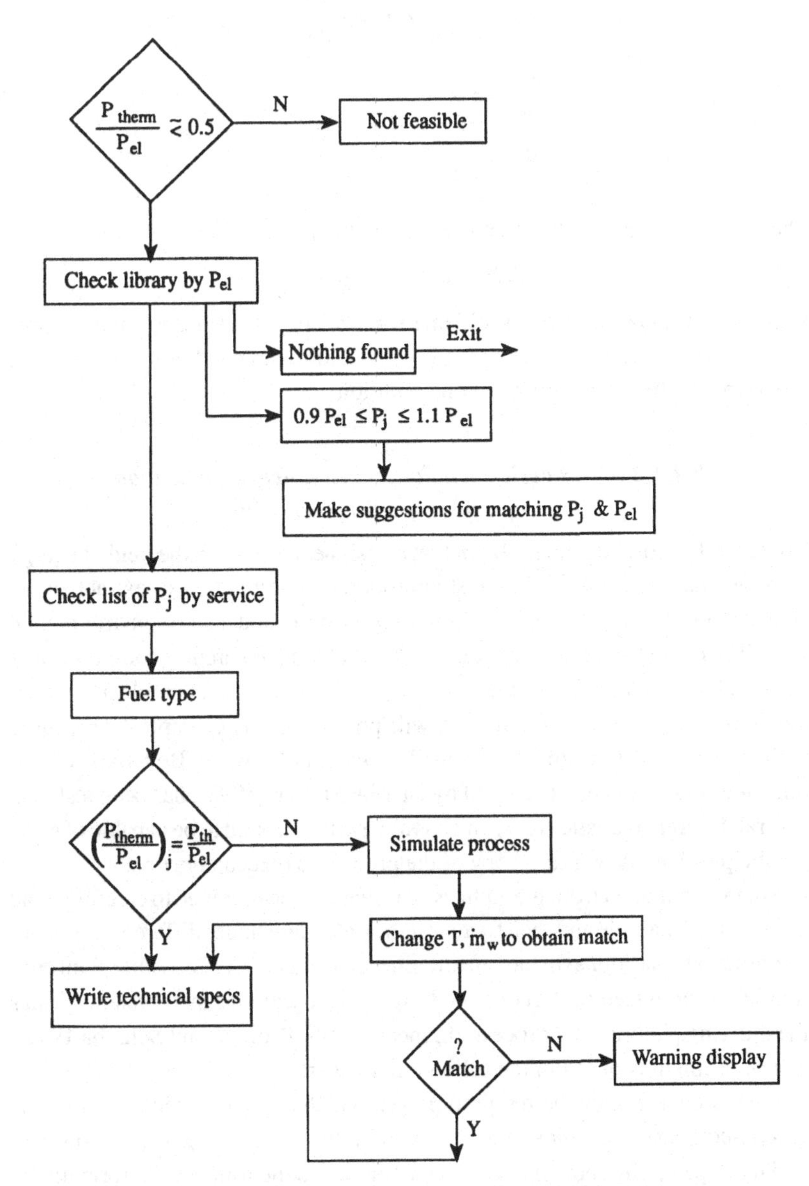

Fig. 9.4. Decision tree for the choice of a gas-turbine plant.

is linearly proportional to a variation of $T_{\max}$ and of $\dot{m}_a$:

$$dP = \frac{\partial P}{\partial \dot{m}_a} d\dot{m}_a + \frac{\partial P}{\partial T} dT$$

$$\frac{\partial P}{\partial \dot{m}_a} = \alpha$$

$$\frac{\partial P}{\partial T} = \beta.$$

The matching requires then that (ΔP being the mismatch):

$$\Delta P = \alpha \Delta \dot{m}_a + \beta \Delta T,$$

which is a linear optimization problem in two independent variables and requires an additional condition to be solvable. This additional condition can be a cost criterion, an efficiency optimization condition, etc.

9.4.4 Process design: choice and technical specification of a fossil-fueled powerplant

Consider the most general problem that a designer can face in the field of energy systems, that of choosing the most appropriate configuration to extract an assigned amount of energy from a given resource base, under some environmental as well as configurational constraints. This problem, apparently complex, turns out to admit a relatively simple qualitative solution, but its complexity makes a quantitative solution difficult. We will present in this example a procedure (called COLOMBO from COnstrained, LOgically Modified Boundaries Optimization) that has been developed by De Marco et al. (1993), has been tested in several different realistic cases, and has always been found to be capable of solving the problem. A working copy of the procedure (executable on a UNIX system) can be obtained from the authors. The reader is encouraged to carefully read through this example and then try to test the procedure under different sets of environmental conditions (in the sense defined here). The list of possible plant configurations produced by the code sheds light on the way an expert system mimics the thinking patterns of a process engineer (Melli, Paoletti, and Sciubba 1992).

The problem is posed in the following manner:

Design data include the required power, which must be specified both in its type (electrical, mechanical, thermal) and in its amount (megawatts installed), and the type of environment. The latter comprises the following information:

- type and amount of available resources (fuel, air, other materials that can take part in the process);
- amount of water available (m^3/sec);

- any environmental characteristic that can be foreseen to influence the choice of the process (for example, "station is mounted on a barge," or similar);
- all legal or technical constraints known to apply at the site under consideration.

Design data also include a set of negative instances, that is, of processes that cannot be accepted as a solution (for example, "nuclear plant not acceptable").

The components library includes back-pressure steam turbine, combustion chamber, compressor, concentrated pressure loss, condenser, condensing steam turbine, contact (mixing) heat exchanger, distributed pressure loss, electric generator, feedwater heater, flow merger, flow splitter, gas turbine, hydrostatic head, mechanical generator, power merger, power splitter, pump, surface heat exchanger, steam generator, and throttling valve.

New components can be added. A component is identified by a set of input and output streams (specifying what kind of fluid it works on, what type of power it adsorbs or delivers, and which output streams are the main streams and which ones are the by-products), by a set of design parameters that uniquely identify it (inlet and outlet pressure, temperature, mass-flow rate, power, etc.), and by a set of mathematical rules that allow one to completely specify both the inlet and the outlet streams given certain proper subsets of the design parameters. A specific component is, in AI terms, an instance of a class; this class can therefore be divided into subclasses according to some peculiar characteristic of each subclass. Here, the power range is chosen, so the class gas turbines can be considered as comprising, say, four subclasses: less than five megawatts, between five and thirty megawatts, between thirty and 100 megawatts, and above 100 megawatts. These sub-classes inherit of course all properties of the class they belong to but can have different values (or ranges) of the design parameters. Each subclass is further divided (where feasible) into three subsubclasses: high tech, state-of-the-art (standard), and low tech. Again, this classification allows for different values of specific design parameters (in the case of a gas turbine, for instance, inlet gas temperature ranges will differ).

The procedure works backward, considering the design goal as an effect, and trying to find its cause. Given the design goal, the inferential engine scans the component library to seek the components that would deliver the goal (which is a given amount of thermal, mechanical, or electrical power). The first component that satisfies the goal is placed in the first slot of the working memory as the last component of a possible process P_1 that is still to be identified. If n components are found, the procedure will have n possible process trees to work on. Then, for each P_i ($i = 1, \dots n$), the inputs required by the component just found are taken to be the design goal, and the library is scanned again seeking for components whose outputs match the goal. The procedure is repeated until all the required inputs are available from the class environment and the configuration developed so far meets all specified constraints (in this case, a process has been found and is displayed to the user in a proper form; if more processes have been found, all of them are displayed) and at a certain level in the process tree, no match can be found (under the specified constraints) for the required inputs. The procedure is aborted.

To match a design goal means that the output of the component under consideration must be of the same type as the design goal (mechanical power, etc.) and that the range attached to that subclass shares at least a subrange with the range required by the design goal. In some cases, only the first of these two conditions is met, that is, the output range of the component under consideration does not match the required input. In these cases, the component is added to the process for future action, but the mismatch is recorded. In fact, the procedure at each step checks on the possibility that one or more by-products or secondary streams can be used to force the match (for instance, increasing the temperature of the water flowing out of the condenser to match the boiler inlet temperature).

A relative cost is attached to each of the constructed processes. This cost is a function of the number of components and of the number and amount of external resources used. So, for instance, a steam power plant with an externally fueled feed water heater train will have a higher cost than the same process with regenerative feed water heating.

COLOMBO displays the solution in the form of an interconnection matrix, **M**, in which the rows and the columns represent, respectively, components and streams. The elements of **M** can be

$$M_{i,j} = 0 \Rightarrow \text{stream } j \text{ does not interact with component } i$$

$$M_{i,j} = 1 \Rightarrow \text{stream } j \text{ is an input of component } i$$

$$M_{i,j} = -1 \Rightarrow \text{stream } j \text{ is an output of component } i$$

The procedure can be directly tested by running it. We give a detailed description of its structure by discussing a specific example of its application.

Consider the following problem:

Find the most appropriate plant of nameplate power $P = 100$ megawatts, if P has to be delivered under the form of electrical energy.

The environment is composed of natural gas of given chemical composition, with LHV = 41000 kJ/kg and maximum, available mass flow rate $\dot{m} = 50$ kg/s; air of standard chemical composition, with $T_o = 32°\text{C}$ and relative humidity $H_\tau = 5\%$; and no water. The plant factor, P_f, is 0.75 (6,600 hours/year equivalent at nameplate load), and the minimum conversion efficiency, $\eta_{I,\text{process}}$, is 0.3.

These data have to be imparted to the code in a proper form. COLOMBO needs a mix of numerical values ($\dot{m}_w = 0$, $T_o = 32°C$. etc.) and propositional strings (plant_output_is_electrical). The specific format of the database can vary from case to case and is more a matter of convenience than an issue of substance.

The logical steps of the inference are then the following:

1. *action*: scan components library (CL) to find suitable components
   result: electrical generator; input: shaft-work, approximately 102 megawatts
2. *action*: scan CL to find suitable components
   result: gas-turbine high-tech; input: combustion-gas at $T_{g,\max} = 1{,}500°C$
   gas-turbine standard; input: combustion-gas at $T_{g,\max} = 1{,}300°C$
   gas-turbine low-tech; input: combustion-gas at $T_{g,\max} = 1{,}100°C$
   steam-turbine high-tech; input: steam at $T_{s,\max} = 650°C$
   steam-turbine standard; input: steam at $T_{s,\max} = 550°C$
   steam-turbine low-tech; input: steam at $T_{s,\max} = 450°C$
   diesel-engine; input: fuel-oil, $\dot{m}_{oil} \sim$ six kilograms per second
   power-splitter: input: shaft work

(We will not follow the power-splitter branch now, but as we will see, this branch is also capable of generating valid configurations.)

3. *action*: check constraints and environment
   result: discard diesel-engine (no fuel oil in environment)
4. *action*: new Goal 1 = hot pressurized gas at $T_{g,\max} = 1{,}500°C$
   new Goal 2 = hot pressurized gas at $T_{g,\max} = 1{,}300°C$
   new Goal 3 = hot pressurized gas at $T_{g,\max} = 1{,}100°C$
   new Goal 4 = hot pressurized steam at $T_{s,\max} = 650°C$
   new Goal 5 = hot pressurized steam at $T_{s,\max} = 550°C$
   new Goal 6 = hot pressurized steam at $T_{s,\max} = 450°C$
   action: scan CL to find suitable components
   result: for new Goal 1: combustion-chamber high-tech; input: fuel and pressurized air, $p_{\max}$ = four megapascals
   result: for new Goal 2: combustion-chamber high-tech; input: fuel and pressurized air, $p_{\max}$ = four megapascals
   result: for new Goal 2: combustion-chamber standard; input: fuel and pressurized air, $p_{\max}$ = three megapascals
   result: for new Goal 3: combustion-chamber high-tech; input: fuel and pressurized air, $p_{\max}$ = four megapascals
   result: for new Goal 3: combustion-chamber standard; input: fuel and pressurized air, $p_{\max}$ = three megapascals
   result: for new Goal 3: combustion-chamber low-tech; input: fuel and

pressurized air, p_{max} = two megapascals
result: for new Goal 4: boiler high-tech; input: fuel, air, and preheated water, $T_{fw,max} = 300°C$
result: for new Goal 5: boiler standard; input: fuel, air, and preheated water, $T_{fw,max} = 275°C$
result: for new Goal 6: boiler low-tech; input: fuel, air, and preheated water, $T_{fw,max} = 200°C$

(At this point, six possible partial processes exist, and no constraint has been violated. For the sake of simplicity, we will follow Goals 1 and 4 and neglect the standard and low-tech possibilities.)

Following the branch "new Goal 1":

5. *action*: new Goal 7 = fuel
   new Goal 8 = pressurized air at p_{max} = four megapascals
   action: scan CL to find suitable components
   result: fuel available in environment: branch search terminated
   result: air compressor: input: air, shaft-work
6. *action*: new Goal 9 = air
   new Goal 10 = shaft-work
   action: scan CL to find suitable components
   result: air available in environment: branch search terminated
   result: shaft-work is a by-product of gas-turbine
   result: shaft-work is a main product of gas-turbine
   result: shaft-work is a main product of power-splitter
   result: shaft-work is a by-product of power-splitter
   Considering the shaft-work as a gas-turbine by-product:
7. *action*: try matching compressor required input with gas-turbine output
   action: check constraints and environment. If no constraint is violated and all inputs come from the environment, the produced configuration is a legal one and is displayed. A new branch is explored.
   Considering the shaft-work as a gas-turbine main product:
8. *action*: go to 4)
   Considering the shaft-work as a power-splitter main product:
9. *action*: go to 2)
   (A built-in rule does not allow a power-splitter to be fed by another power-splitter. This avoids looping.)
   Considering the shaft-work as a power-splitter by-product:
10. *action*: go to 2)
   (... and so on. There can be an infinite series of ever more complicated

plants in which the compressor is powered by a gas-turbine plant whose compressor is in turn powered by yet another gas turbine plant. Some specific rules in the inference engine avoid excessive recursivity.)

Following the branch "new Goal 4":

11. *action*: new Goal 11 = air
    new Goal 12 = fuel
    new Goal 13 = preheated water
    action: scan CL to find suitable components
    result: air available in environment: branch search terminated
    result: fuel available in environment: branch search terminated
    result: deaerator; input: water, medium-pressure steam
12. *action*: new Goal 14 = water
    new Goal 15 = medium-pressure steam
    action: scan CL to find suitable components
    result: water not available in environment: branch search failure
    (Medium-pressure steam is available from back-pressure steam-turbine. Water is also available from condenser.)
    result: condenser; input: low-pressure steam, water
    result: back-pressure steam-turbine; input: high-pressure steam
13. *action*: new Goal 16 = water
    new Goal 17 = low-pressure steam
    action: scan CL to find suitable components
    result: water not available in environment: branch search failure
    (Low-pressure steam is available from condensing-steam turbine; water for the condenser cannot be extracted from another condenser [to avoid looping].)
    result: condensing-steam turbine: input: medium-pressure steam
14. *action*: go to 12), Goal 15
    (Medium-pressure steam can also be obtained as a by-product of steam turbine, but this case [which would not modify the final outcome] is not considered here.)
    (This branch fails, because a recursive call to medium-pressure steam is terminated by an antilooping rule built into the inference engine. It is important to notice that the physical reason of the failure is the absence of water in the environment, which negates the insertion of a condenser into the process and the construction of a steam plant configuration.)
15. *action*: check if there are any unexplored branches. If none, terminate the run.

A list of the first four configurations produced by this run is shown in Fig. 9.5.

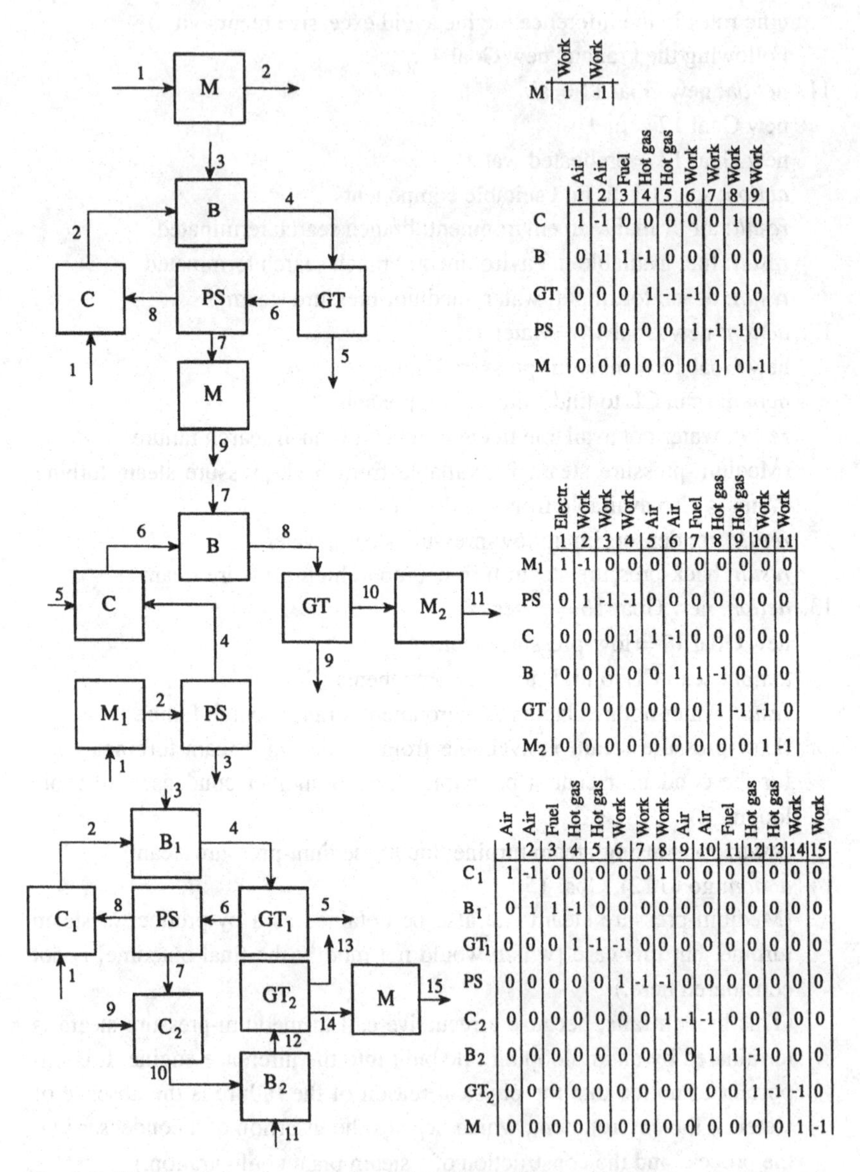

	Work	Work
M	1	-1

	Air 1	Air 2	Fuel 3	Hot gas 4	Hot gas 5	Work 6	Work 7	Work 8	Work 9
C	1	-1	0	0	0	0	0	1	0
B	0	1	1	-1	0	0	0	0	0
GT	0	0	0	1	-1	-1	0	0	0
PS	0	0	0	0	0	1	-1	-1	0
M	0	0	0	0	0	0	1	0	-1

	Electr. 1	Work 2	Work 3	Work 4	Air 5	Air 6	Fuel 7	Hot gas 8	Hot gas 9	Work 10	Work 11
M_1	1	-1	0	0	0	0	0	0	0	0	0
PS	0	1	-1	-1	0	0	0	0	0	0	0
C	0	0	0	1	1	-1	0	0	0	0	0
B	0	0	0	0	0	1	1	-1	0	0	0
GT	0	0	0	0	0	0	0	1	-1	-1	0
M_2	0	0	0	0	0	0	0	0	0	1	-1

	Air 1	Air 2	Fuel 3	Hot gas 4	Hot gas 5	Work 6	Work 7	Work 8	Air 9	Air 10	Fuel 11	Hot gas 12	Hot gas 13	Work 14	Work 15
C_1	1	-1	0	0	0	0	0	1	0	0	0	0	0	0	0
B_1	0	1	1	-1	0	0	0	0	0	0	0	0	0	0	0
GT_1	0	0	0	1	-1	-1	0	0	0	0	0	0	0	0	0
PS	0	0	0	0	0	1	-1	-1	0	0	0	0	0	0	0
C_2	0	0	0	0	0	0	0	1	-1	-1	0	0	0	0	0
B_2	0	0	0	0	0	0	0	0	0	1	1	-1	0	0	0
GT_2	0	0	0	0	0	0	0	0	0	0	0	1	-1	-1	0
M	0	0	0	0	0	0	0	0	0	0	0	0	0	1	-1

Fig. 9.5. The first four processes created by COLOMBO (and their respective interconnection matrices) for the generation of 100 MW of electrical power.

Notice that Configuration 1 has been produced because, since electricity is available from the net, the code has assumed, wrongly, that it can drive a 100-megawatts electrical motor directly from the net.

9.4.5 Choice of optimal steam and power generation for plant expansion

This example requires that an automatic procedure be devised for the choice and specification of a process capable of producing additional power and process steam for a plant already in operation. The procedure will read the technical requirements (design data), build up the proper type of plant from a library of available components, and produce a technical specification for the design engineer.

Referring again to the the example discussed in Chapter 6, we will call here design data and design constraints, respectively, the quantities denoted as musts and wants in that example. Therefore, we will reformulate (and expand a little bit) the problem as follows:

- *Design goal*: devise a process capable of delivering power and process steam
- *Design data*: $P \geq$ thirty megawatts; $\dot{m}_{\text{steam}} \geq$ fourteen kilograms per second
- *Design constraints*: $x_{\text{steam}} > 1$ (superheated steam)

$$p_{\text{steam}} \geq 2 \text{ megapascals}$$
$$\text{NO}_x \leq \text{NO}_{x,r}$$
$$\text{CO} \leq \text{CO}_r$$
$$\text{SO}_2 \leq \text{SO}_{2,r}$$
$$\text{O}_2 \geq \text{O}_{2,r}$$

(where the suffix r denotes the mandatory legal limit)
- *Design environment*: air, standard quality, unlimited quantity water, standard quality, unlimited quantity fuel, natural gas, $\dot{m}_f =$ ten kilograms per second maximum.

Solution The approach we will take in solving the problem is a backward-chaining induction that will start from the design data (symptoms) and will try to reach the design goal (diagnosis) while abiding with all constraints.

The procedure needs some engineering knowledge. Let us assume it possesses a sufficiently complete library **L** of engineering components and one set of property tables for each working fluid. The inductive chain will work as follows (see also Section 9.4.3):

The object POWER can be delivered by the following objects contained in

L:

GAS TURBINE
STEAM TURBINE
DIESEL ENGINE
ELECTRICAL MOTOR

The object STEAM can be delivered by the following objects contained in L:

BOILER
BACK PRESSURE STEAM TURBINE

The procedure now analyzes each single possibility (thus originating a decision tree).

The object GAS TURBINE needs as input the following object:

COMPRESSED HOT GAS,

which can be produced by

PRESSURIZED COMBUSTION CHAMBER
GAS TURBINE

Taking the first path:
PRESSURIZED COMBUSTION CHAMBER needs

FUEL
COMPRESSED AIR

FUEL is available in the environment.
COMPRESSED AIR needs

COMPRESSOR,

which in turn needs

AIR
POWER

AIR is available in the environment.
POWER needs

GAS TURBINE
STEAM TURBINE
DIESEL ENGINE
ELECTRICAL MOTOR

Now, one of these objects has been already selected, and the procedure makes use of a priority rule:

IF last object found belongs to a previously selected list **THEN** terminate search **AND** place object in CONFIGURATION LIST.

This rule is applied every time an object is found after a library scan. Another rule in the inference engine states that

IF all inputs required by CONFIGURATION LIST are available

in ENVIRONMENT

OR are outputs of objects in CONFIGURATION LIST

THEN terminate branch **AND** store CONFIGURATION LIST in POSSIBLE PLANTS LIST

AND start searching next branch

We can see that the inductive chain has produced a path from the design goal to the environment and that this path represents the scheme of a process:

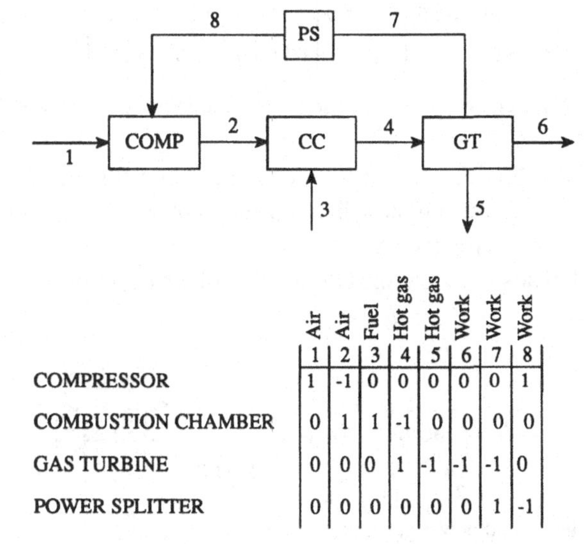

	Air 1	Air 2	Fuel 3	Hot gas 4	Hot gas 5	Work 6	Work 7	Work 8
COMPRESSOR	1	-1	0	0	0	0	0	1
COMBUSTION CHAMBER	0	1	1	-1	0	0	0	0
GAS TURBINE	0	0	0	1	-1	-1	-1	0
POWER SPLITTER	0	0	0	0	0	0	1	-1

Fig. 9.6. Scheme of a gas-turbine cycle proposed by COLOMBO.

Similarly, the search executed with the same methodology on the branch starting with STEAM TURBINE would produce the following two process schemes:

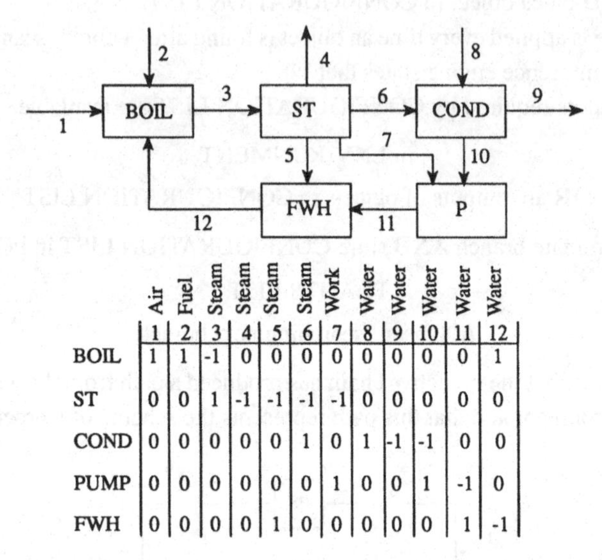

	Air	Fuel	Steam	Steam	Steam	Steam	Work	Water	Water	Water	Water	Water
	1	2	3	4	5	6	7	8	9	10	11	12
BOIL	1	1	-1	0	0	0	0	0	0	0	0	1
ST	0	0	1	-1	-1	-1	-1	0	0	0	0	0
COND	0	0	0	0	0	1	0	1	-1	-1	0	0
PUMP	0	0	0	0	0	0	1	0	0	1	-1	0
FWH	0	0	0	0	1	0	0	0	0	0	1	-1

Fig. 9.7. Scheme of a steam power cycle proposed by COLOMBO.

The other two branch searches would fail, because both DIESEL ENGINE and ELECTRICAL MOTOR need inputs that are not in the environment (diesel oil and electricity, respectively).

The search that starts from the object STEAM results in the following four processes:

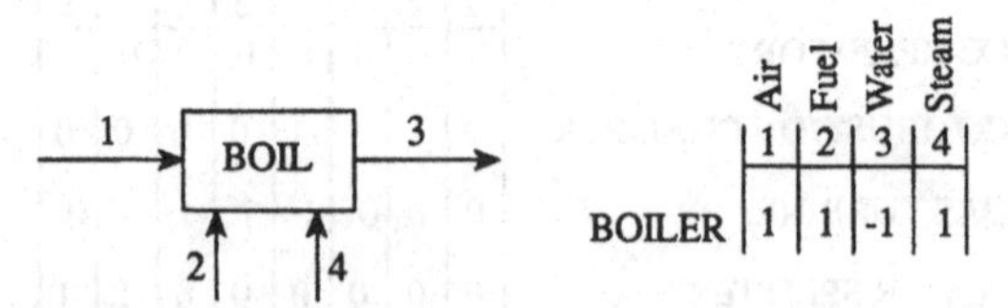

	Air	Fuel	Water	Steam
	1	2	3	4
BOILER	1	1	-1	1

(a) Separate steam- and power generation.

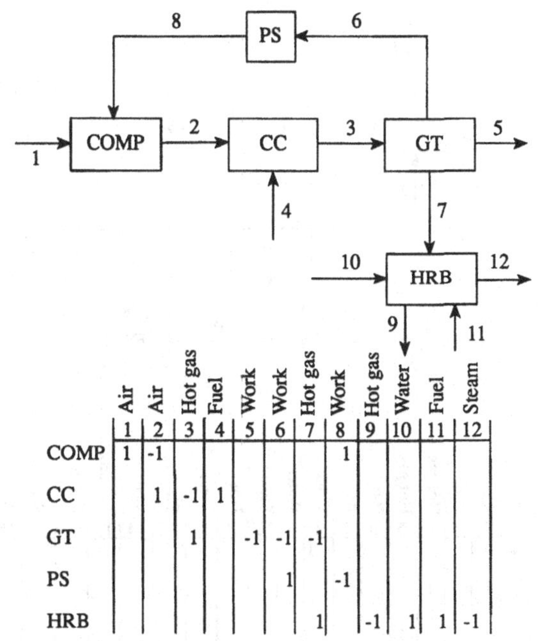

	Air 1	Air 2	Hot gas 3	Fuel 4	Work 5	Work 6	Hot gas 7	Work 8	Hot gas 9	Water 10	Fuel 11	Steam 12
COMP	1	-1						1				
CC		1	-1	1								
GT			1		-1	-1	-1					
PS						1		-1				
HRB							1		-1	1	1	-1

(b) Backpressure steam turbine.

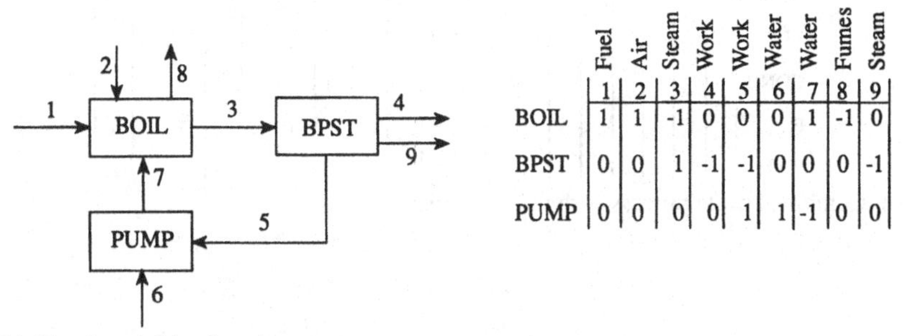

	Fuel 1	Air 2	Steam 3	Work 4	Work 5	Water 6	Water 7	Fumes 8	Steam 9
BOIL	1	1	-1	0	0	0	1	-1	0
BPST	0	0	1	-1	-1	0	0	0	-1
PUMP	0	0	0	0	1	1	-1	0	0

(c) Simple combined cycle process.

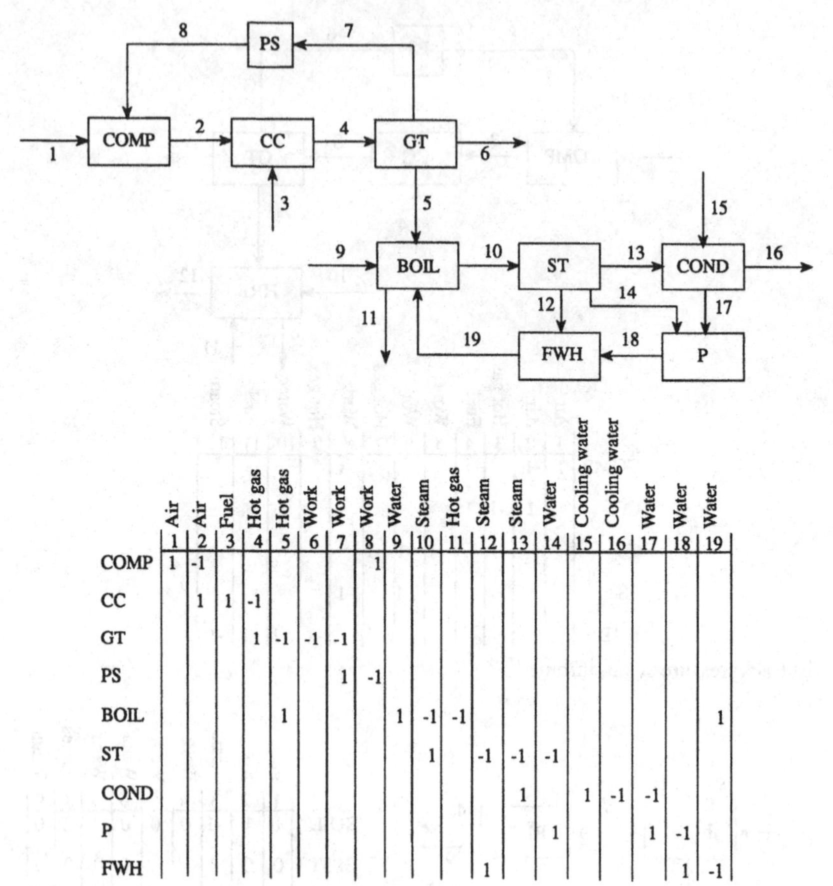

	Air	Air	Fuel	Hot gas	Hot gas	Work	Work	Work	Water	Steam	Hot gas	Steam	Steam	Water	Cooling water	Cooling water	Water	Water	Water
	1	2	3	4	5	6	7	8	9	10	11	12	13	14	15	16	17	18	19
COMP	1	-1						1											
CC		1	1	-1															
GT				1	-1	-1	-1												
PS							1	-1											
BOIL					1				1	-1	-1								1
ST										1		-1	-1	-1					
COND													1		1	-1	-1		
P														1			1	-1	
FWH												1						1	-1

(d) Combined cycle process with re-fired boiler.

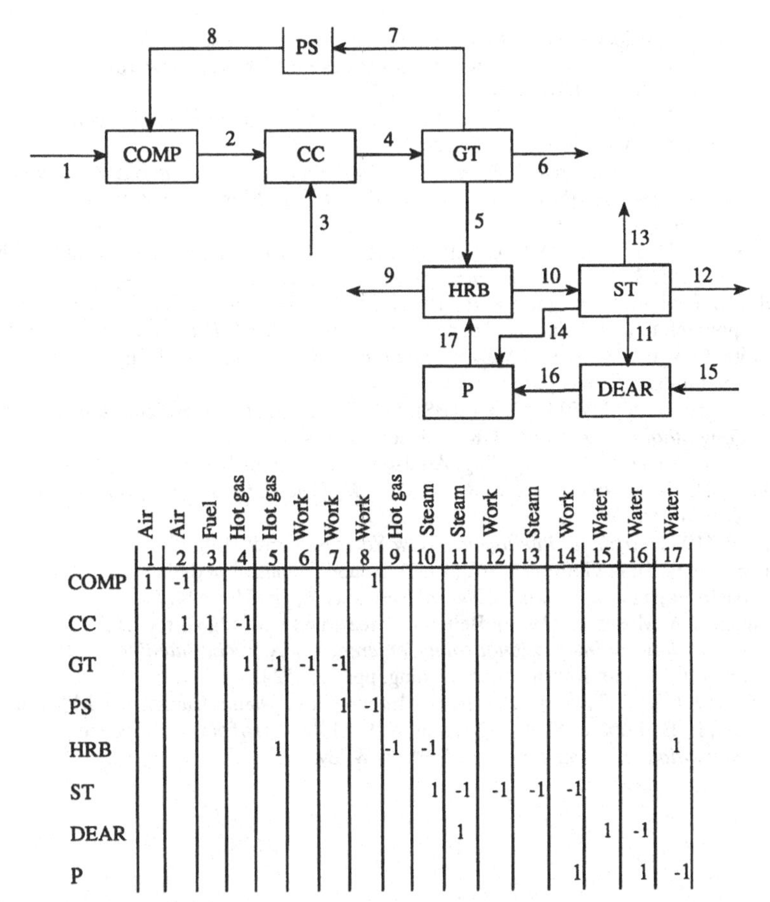

	Air 1	Air 2	Fuel 3	Hot gas 4	Hot gas 5	Work 6	Work 7	Work 8	Hot gas 9	Steam 10	Steam 11	Work 12	Steam 13	Work 14	Water 15	Water 16	Water 17
COMP	1	-1						1									
CC		1	1	-1													
GT				1	-1	-1	-1										
PS							1	-1									
HRB					1				-1	-1							1
ST										1	-1	-1	-1	-1			
DEAR											1				1	-1	
P														1		1	-1

(e) Standard combined-cycle process.

Fig. 9.8. The first four processes devised by COLOMBO for combined steam- and power generation.

To solve the second part of the problem, that is, to choose the optimal process, one must use on each configuration proposed by the code approximate optimization techniques.

References

Brooks, R. A. (1991). Intelligence without representation. *Artificial Intelligence*, **47**, pp. 139–144.

Charniak, E. McDemmott, D. (1985). *Introduction to Artificial Intelligence*. Reading, Addison-Wesley.

De Marco, M., Falcetta, M. F., Melli, R., Paoletti, B. & Sciubba, E. (1993). COLOMBO: an expert system for process design of thermal powerplants. ASME-AES Vol. 1/10, New York: ASME.
Drescher, G. L. (1993). *Made-up minds: a constructivist approach to artificial intelligence*, Cambridge: MIT Press.
Gelsey, A. (1989). Automated physical modeling. *Proceedings of International Joint Conference on Artificial Intelligence*, 1989. Detroit: Morgan Kaufmann Publishing, pp. 1225–1230.
McBride, R. D. & Leary, D. E. O. (1993). The use of mathematical programming with A.I. and E.S., *European Journal of Operations Research* **70**, pp. 1–15.
Melli, R., Paoletti, B. & Sciubba, E. (1992). SYSLAM: an interactive expert system approach to powerplant design and optimization, *IJEEE Journal*, **3**, pp. 165–175.
Negoita, C. V. (1993). *Expert Systems, and Fuzzy Systems*. Redwood City, California: Bejamin/Cummings Pub.
Papalambros, P. Y. & Wilde, D. J. (1988). *Principles of Optimal Design: Modeling and Computation*, Cambridge, UK: University Press.
Pearl J. (1984). *Heuristics*, Reading: Addison-Wesley Publisher.
Reklaitis, G. V., Ravindran. A. & Ragsdell. K. M. (1983). *Engineering Optimization*, New York: J. Wiley.
Rich, E. (1983) *Artificial Intelligence*. New York: McGraw-Hill.
Shiffmann, W. H. & Geffers, H. W. (1993). Adaptive control of dynamic systems by back-propagation networks. *Neural Networks*, **6**, pp. 517–524.
Shoham, Y. & Moses, Y. (1989). Belief as defeasible knowledge, Proc. IJCAI-89. *Proceedings of International Joint Conference on Artificial Intelligence*, 1989. Detroit: Morgan Kaufmann Publishing, pp. 517–524.
Varela, F. J. (1992). *Toward a Practice of Autonomous Systems*. Cambridge: MIT Press.
Widman, L. E., Loparo, K. A. & Nielsen, N. R. (1989). *Artificial Intelligence, Simulation and Modeling*, New York: J. Wiley.

Subject Index

Author Index